THERMODYNAMICS

THERMODYNAMICS

$$S\,dT - V\,dp + \sum_B n_B\,d\mu_B = 0$$

$$S = k \ln \Omega$$

$$J_i = \sum_k L_{ik} X_k, \quad L_{ik} = L_{ki}$$

Patrick Jacobs

University of Western Ontario, Canada

Imperial College Press

ICP

Published by

Imperial College Press
57 Shelton Street
Covent Garden
London WC2H 9HE

Distributed by

World Scientific Publishing Co. Pte. Ltd.
5 Toh Tuck Link, Singapore 596224
USA office: 27 Warren Street, Suite 401-402, Hackensack, NJ 07601
UK office: 57 Shelton Street, Covent Garden, London WC2H 9HE

British Library Cataloguing-in-Publication Data
A catalogue record for this book is available from the British Library.

THERMODYNAMICS
Copyright © 2013 by Imperial College Press

ISBN 978-1-84816-970-8
ISBN 978-1-84816-971-5 (pbk)

Typeset by Stallion Press
Email: enquiries@stallionpress.com

Printed in Singapore.

Contents

Preface   ix

1. Temperature and Energy   1

   1.1   Introduction   1
   1.2   The Zeroth Law of Thermodynamics   4
   1.3   The Ideal Gas   6
   1.4   Work   10
   1.5   The First Law of Thermodynamics   20
   1.6   Heat Capacity and Internal Energy of an Ideal Gas . . .   22
   1.7   Generalized Forces   23

2. Entropy   27

   2.1   The Second Law of Thermodynamics   27
   2.2   The Fundamental Equations for a Closed Phase   29
   2.3   Calculation of Entropy Changes   31
   2.4   Physical Meaning of the Thermodynamic Potentials . .   35
   2.5   Conditions for Equilibrium in a Closed System   39
   2.6   Entropy of Mixing   41
   2.7   Thermodynamic Potentials for an Open Phase   42
   2.8   Other Forms of the Second Law and the Unattainability
       of $T = 0\,\mathrm{K}$ 45

3. An Introduction to Statistical Thermodynamics   53

   3.1   Need for a Microscopic Description   53
   3.2   Postulates of Quantum Mechanics   54
   3.3   A Model for the Ideal Gas   56
   3.4   The Number of Translational States   59
   3.5   The Boltzmann Distribution   62
   3.6   Translational Motion in an Ideal Gas   66
   3.7   Some Applications of the Boltzmann Distribution   68

3.8	The Partition Function	71
3.9	Evaluation of Thermodynamic Functions from Q	74
3.10	Internal Degrees of Freedom	78
3.11	Thermodynamic Functions for a Monatomic Ideal Gas	91

4. The Third Law of Thermodynamics — 95

4.1	Entropy and Probability	95
4.2	The Boltzmann Entropy Equation	96
4.3	The Third Law of Thermodynamics	98
4.4	Experimental Observations Concerning the Third Law	100
4.5	The Thomsen and Berthelot Rule	104
4.6	Unattainability of $T = 0\,\mathrm{K}$	105

5. Systems of One Component — 113

5.1	Real Gases	113
5.2	Intermolecular Potentials	117
5.3	The Joule–Thomson Effect	120
5.4	Thermodynamic Functions for Gases	124
5.5	Gases at High Pressures	127
5.6	Liquids	128
5.7	Solids	136
5.8	Triple Point	139
5.9	Higher-order Phase Transitions	140

6. Systems of More Than One Component — 145

6.1	The Phase Rule	145
6.2	Partial Molar Properties	146
6.3	Mixtures of Gases	149
6.4	Liquid Mixtures	152
6.5	Liquid–Solid Equilibrium	172
6.6	Solutions	179

7. Surfaces and Interfaces — 189

7.1	Plane Interface	189
7.2	Curved Interface	191
7.3	Thermodynamics of Interfaces	193

7.4 Some Typical Results 197
7.5 Thermodynamics of Wetting 200
7.6 The Gas–Solid Interface 201
7.7 Thermodynamics of Adsorption 206
7.8 Nucleation 213

8. Chemical Equilibrium 219

8.1 Stoichiometry 219
8.2 Affinity and Reaction Enthalpy 220
8.3 Temperature Dependence of the Reaction Enthalpy . . . 227
8.4 The Standard Affinity, Reaction Isotherm
    and Equilibrium Constant 232
8.5 Effect of Pressure on Reactions Involving Gases 243
8.6 The Giaque Function 244
8.7 Theoretical Calculation of Equilibrium Constants 246
8.8 Reaction Rate Theory 249

9. Electrolytes 255

9.1 Electrolysis 255
9.2 Conductivity of Electrolyte Solutions 257
9.3 Ionic Mobility and Transport Number 260
9.4 Electrochemical Potential 266
9.5 Electrode Potential 266
9.6 The Cell Potential of an Electrochemical Cell 270
9.7 Activity Coefficients in Electrolyte Solutions 272
9.8 The Standard Cell Potential 282
9.9 pH 284
9.10 Thermodynamic Functions of Ions
     in Aqueous Solution 285
9.11 Electrochemical Cells with Transport 287
9.12 Some Applications 290

10. Fermi–Dirac and Bose–Einstein Statistics 295

10.1 Ensembles 295
10.2 The Canonical Ensemble 296
10.3 The Microcanonical Ensemble 298
10.4 The Grand Canonical Ensemble 299
10.5 Fermi–Dirac and Bose–Einstein Statistics 302

10.6 The Ideal Bose–Einstein Gas 307
10.7 The Ideal Fermi–Dirac Gas 312
10.8 Temperature Dependence of Thermodynamic
       Functions for a FD gas 314

11. Thermodynamics of Solids   319

11.1 Symmetry and the Physical Properties of Crystals . . . 319
11.2 Stress and Strain 321
11.3 Thermodynamics of Stress and Strain 324
11.4 Point Defects in Ionic Crystals 329
11.5 Thermodynamics of Ionic Crystals
       with Point Defects 332
11.6 Semiconductors 337

12. Thermodynamics of Non-equilibrium Systems   343

12.1 Reprise 343
12.2 Entropy Production 346
12.3 Assumptions Upon Which TDNEP is Based 350
12.4 Examples of Non-equilibrium Processes 351
12.5 Thermoelectric Effects 353
12.6 Diffusion and Conduction in Electrolyte Solutions 355

Solutions to the Problems   363

Appendices   417

A1 Partial Differentiation 417
A2 The Classical Electromagnetic Field 423
A3 Sources of Thermodynamic Data 429

References   431

Index   437

Preface

This book contains the fundamental thermodynamics that chemists, as well as many physicists and engineers, need in the second, third and fourth years of a four-year program in North America. This material would normally be covered in the first, second and third years of the chemistry honours degree in the UK. My own preference is for a continuous set of lectures (and/or tutorials and problem classes according to the teaching model employed), constituting a segment of the physical chemistry course. But the book would serve equally well as a text for those courses in which classes on thermodynamics do not form a separate unit but are given as the need arises, rather than in two or three continuous segments. The book is also well suited to the needs of those learning or re-learning thermodynamics on their own.

It is assumed that those specializing in physical chemistry will also be taking a course or a set of classes in statistical mechanics. I do not presume to have met their needs in this book, but for the sake of those who are not taking such a course and because this approach is an essential adjunct to thermodynamics, I have included two chapters (3 and 10) which provide an introduction to statistical thermodynamics and which may be sufficient for the non-specialist.

The book commences with a discussion of thermal equilibrium and the Zeroth Law. This leads on to the definition of the thermodynamic scale of temperature, using only one fixed point, the triple point of water. This method is much shorter and easier to understand than the traditional approach using empirical temperature scales. It also facilitates and shortens considerably the introduction of the Second Law of Thermodynamics. While the first part of the book was being written, I believed this formulation of the Zeroth Law to be original. However, when the first five chapters were completed, my colleague Professor A.R. Allnatt pointed out to me that a very similar approach had been suggested by J.S. Thomsen (1962) though it seems not to have been adopted in a book.

All three laws (Zero, One and Two) are associated with the introduction of a thermodynamic function of state, the temperature T, the internal energy U, and the entropy S. The approach is much shorter and easier to comprehend than the conventional one introducing entropy via heat engines. The mathematical or physical background of some sections of the text has been relegated to appendices. Some flexibility in the use of this text is possible. In some courses it might be wiser to postpone presentation of the material in Chapters 3 to 7 until after Section 8.7.

Fully worked out *examples* are included at appropriate places in the text, which also includes numerous *exercises*. These are questions designed to make the reader stop and think about what he or she has just read. I strongly recommend that exercises not be skipped but be answered before reading further. Most are quite short. They vary in difficulty but are generally quite easy. Answers to exercises are given at the end of each section. There are problems at the end of each chapter and solutions to the problems are given at the end of the book. Again, more benefit will be derived if problems are answered before turning to the solutions. Most of the problems are new but some are taken from my file of problems used when I taught courses in thermodynamics at Imperial College, University of London (as it was then) and at the University of Western Ontario. Some of these problems were devised by colleagues, and I am grateful to Doctors A.R. Allnatt, D.R. Bidinosti, J.W. Lorimer, and M. Spiro for permission to use problems which they devised.

Classical thermodynamics is the study of systems in equilibrium, except for those in which a device due to J.J. Thomson can be applied. This consists in assuming that the non-equilibrium processes are superimposed upon, and independent of, the equilibrium state. Thomson's idea is used only once in the first ten chapters, namely in Chapter 9 to deduce the cell potential of an electrochemical cell with transport. Because I consider the thermodynamics of systems that are not in equilibrium an important, if somewhat neglected, part of the undergraduate thermodynamics course, I have included an introduction to the subject in Chapter 12.

The whole of the manuscript for this book has been read by two of my colleagues, Professor A.R. Allnatt and Professor J.W. Lorimer, and I am extremely grateful to them for their comments and suggestions. Once again, my good friend Professor B. Zapol has drawn all the figures for this book and I am more grateful to him than I can say for his contribution. My wife, Mary Mullin, cheerfully absorbed more than her fair share of domestic duties, making my previous quotation from Pushkin's poem "19 October" (see Jacobs 2005) even more apt than before. The perceptive reader may notice my indebtedness to Professor E.A. Guggenheim, whose writings on thermodynamics are models of rigour and clarity. I have not attempted to emulate him but I hope something of the clarity and rigour has persisted.

In matters of notation, I have been careful to follow, with few exceptions, the recommendations of the International Union of Pure and Applied Chemistry (IUPAC) as set out in the third edition of *Quantities, Units and Symbols in Physical Chemistry* (Cohen *et al.* 2007). In order to economize on symbols I have used $B_2, B_3 \ldots$ instead of $B, C \ldots$ for the virial coefficients. Quantum numbers are set in Roman type to avoid any possible confusion with other well-established symbols, for example, v for vibrational quantum number instead of italic vee, v, which looks like the nu used for frequency. But l is used for the orbital angular momentum quantum number because the Roman ell (l) looks like one (1). When Greek letters such as α and β are used to label phases, they are set in Roman type, but if they are variables or functions they are italic. The solid, liquid and gas phases are distinguished by superscript capital S, L, G, or by cr, liq, gas (often abbreviated to g), in parentheses. I hasten to add that my two readers are not responsible for these departures from the standard notation, which were my own choices and were made purely in the interests of clarity and simplicity. In the small amount of the text devoted to defects in ionic crystals I have used a simpler notation rather than the recommended one, and I trust that this will not offend too many people. When convenient to do so, left and right are abbreviated to L and R, and left side and right side to LS and RS.

Equations in the text are numbered $(n_1.n_2.n_3)$, where n_1 is the chapter number, n_2 is the section number and n_3 the number of the equation in Section $n_1.n_2$. Within the same section, equations are referred to simply by (n_3) or Eq. (n_3). In a few instances it has been necessary to refer to equations in the Problems or Solutions to the Problems, in which case the reference is $(Pn_1.n_2.n_3)$ for Eq. n_3 of problem n_2 in Chapter n_1, or $(Pn_1.n_2.n_3)$ for Eq. n_3 of the solution to problem n_2 in Chapter n_1. In the three appendices, equations are numbered consecutively and referred to as $(An_1.n_2.n_3)$, or by just (n_3) within the same section, $An_1.n_2$. Figures and tables are numbered consecutively within chapters. Figures and tables in the Problems and Solutions to the Problems continue the sequence for that chapter, but with the additional identifier P or S.

London, 2011   Toronto, 2011

1

Temperature and Energy

1.1. Introduction

We begin with some definitions. The science of *thermodynamics* deals with the connections between energy changes and the physical properties and chemical composition of material systems. That part of the physical world which is the focus of attention is called a *thermodynamic system*. When we say "properties" in thermodynamics, we mean *macroscopic properties*, that is the results of measurements on systems containing large numbers of molecules, made over times that are long compared with the periods of molecular vibrations, and involving energies which are much larger than the differences between the energies of the quantum states of the molecules. The immediate neighbourhood of a system — as much as can affect or be affected by the system — is called its *surroundings*. The boundary surfaces of the system, which separate it from its surroundings, are referred to as the *walls* of the system. An *open* system may exchange matter or energy with its surroundings. In a *closed* system, no exchange of matter with its surroundings can occur, although the absorption or loss of energy is still possible.

An *isolated* system has no interaction with its surroundings (Figure 1.1). For example, a leaf on a tree is an open system: during the hours of sunlight the leaf pigments absorb radiation from the sun and this facilitates the reaction of carbon dioxide from the atmosphere with water in the plant leaf to make sugar. This open system thus absorbs both matter and energy from its surroundings. In contrast, the hot coffee in a sealed vacuum flask approximates a closed system until we remove the insulated lid, when the system loses energy to the surrounding atmosphere, and possibly also mass, if we pour some coffee into a mug.

A system is *homogeneous* if it contains within it no apparent surface of discontinuity. A *heterogeneous* system consists of two or more homogeneous regions, which are called *phases*. For example, a system that comprises ice, liquid water and water vapour consists of three phases and has one chemical

1

surroundings

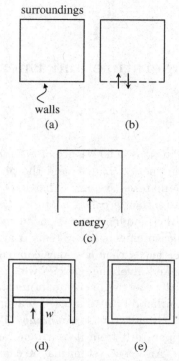

walls
(a)   (b)

energy
(c)

(d)   (e)

Figure 1.1.   Diagrammatic representation of various types of thermodynamic system: (a) system, walls and surroundings; (b) open system; (c) closed system; (d) adiabatically isolated system; (e) isolated system.

component, H_2O. The number of components in a system is the minimum number of independent chemical species necessary to define the composition of all phases in the system. Thus in this example, there is one component because the amounts of H and O may not be varied independently, but must always obey the stoichiometric formula H_2O. Contrast this system with a liquid mixture of chloroform and acetone which is completely miscible at a temperature[a] of 35°C and so the single liquid phase contains two components, $CHCl_3$ and $(CH_3)_2CO$.

---

[a] It is assumed that readers are familiar with the concept of temperature as a measure of hotness or coldness and with the Celsius scale for describing the temperature of a system (a room, or a human body, for example.) The temperature of a system will shortly be defined precisely.

To describe completely a homogeneous chemical system requires, in addition to its mass and chemical composition, the specification of at least two further (macroscopic) properties. This number of properties to be specified is always small and depends on the forces acting on the system. If there are no significant external forces other than those due to the external pressure, then experiment shows that a system of constant mass and composition has only two *degrees of freedom*, that is two is the minimum number of properties which, when held at particular values, ensure constant, reproducible values for all other properties. For example, consider a system of fixed mass that is a mixture of two liquids of constant composition and an inert gas that is not appreciably soluble in the liquid phase. Then at constant temperature and pressure, this system has a constant volume, density, viscosity, surface tension..., which may not be varied independently. The magnitude of an *intensive* property, such as pressure, is independent of the mass of the system, but the magnitude of an *extensive* property, such as volume, is proportional to the mass of the system. (Intensive and extensive properties correspond to the two cases $r = 0$ and $r = 1$ in the application of Euler's theorem in Appendix A1.4. This statement is clarified further in Chapter 2.) If, on examining the state of a thermodynamic system at two different times, a change in any of its properties is observed to have occurred, a *thermodynamic process* or *change in thermodynamic state* is said to have taken place.

Consider now a phase of constant mass and composition and subject to no external forces other than pressure. Such a system has two *degrees of freedom* in that only two macroscopic properties need be specified to describe the system completely. Let these two properties be denoted by X and Y. Then the state of the system can be represented by a point in the XY plane, the coordinates of which are X_1, Y_1. Suppose that the system undergoes a process, as a result of which it adopts a new state X_2, Y_2. During the process it will have passed through a large number of intermediate states X, Y. The locus of the points representing these states is called the *path of the process*. If P denotes any property, then the change in that property when the system undergoes a process by which it changes from a state one to a state two is *independent of the path* of that process. The change in P, written $P_2 - P_1$, is denoted by ΔP.

A system is in *thermodynamic equilibrium* if (a) its properties (e.g. temperature, pressure) are independent of time and (b) no flows of matter or energy exist within the system or at its interface with the surroundings. If only (a) is satisfied, the system is said to be in a *steady state*. For example, a reacting mixture may remain in a steady state for as long as its temperature and pressure are constant and the appropriate mixture of reactants is supplied and products removed so as to maintain constant concentrations of reactants and products. Classical thermodynamics is concerned with equilibrium states and for that reason it is sometimes referred to as *thermostatics* (Callen 1960). If experiment shows that the properties of a system are changing with time then non-equilibrium thermodynamics may be used to study the system. A system not in equilibrium will tend to change spontaneously until it reaches an equilibrium state, unless constraints exist.

1.2. The Zeroth Law of Thermodynamics

A macroscopic system comprising a single homogeneous phase of constant mass and composition and subject to no significant external forces other than pressure can be described by the values of two independent properties (thermodynamic coordinates) X and Y. If the system is in an equilibrium state then X and Y will have constant values independent of time. Consider now two such phases α and β which are separated by a wall which is both rigid and impervious to the transport of matter. If the wall contains a thick layer of insulating material (such as asbestos or vermiculite) the two phases are found to behave quite independently and the coordinates $X^\alpha, Y^\alpha, X^\beta, Y^\beta$ are independent. In this case the two phases are said to be *adiabatically isolated*. But if the wall is a thin metallic sheet, experiment shows that the two phases attain a new equilibrium state in which only three of the variables $X^\alpha, Y^\alpha, X^\beta, Y^\beta$ are independent. In this case the phases are in *thermal contact* and the final state is said to be the state of *thermal equilibrium* between the two phases α and β. If we choose to specify the state of α by the values of X^α and Y^α, then the state of β may be specified by the value of either X or Y and the fact that it is in thermal equilibrium with α. We are now in a position to state the **Zeroth Law of Thermodynamics**: there exists an intensive property called the *temperature* T; two systems, or a system and its surroundings, which are in thermal equilibrium, have the same T. The numbering of the Laws of Thermodynamics is of historical origin. The First and Second Laws of Thermodynamics were formulated in the 19th century before the significance of thermal equilibrium and its relation to the empirical property "temperature" were realized. By then the first and second laws were firmly established so that the new law (which logically must precede them in the development of the subject) was dubbed "the Zeroth Law".[b]

In order to compare the temperatures of two systems, we need to set up a temperature scale. The *thermodynamic temperature* T (from now on "the temperature" will be taken to imply "the thermodynamic temperature") of the triple point of water is defined to be

$$T = 273.16 \, \text{K} \tag{1}$$

K is the symbol for the unit of temperature, namely the kelvin, in SI units. The triple point of water is the system ice + liquid water + water vapour in equilibrium. The Celsius temperature t, measured in °C, is related to the thermodynamic temperature T in kelvin, by the definition

$$t/°\text{C} = (T/K) - 273.15 \tag{2}$$

The difference of 0.01 K between Eqs. (1) and (2) is of historical origin. The Celsius scale of temperature was originally defined so that the temperature of the ice point was 0°C, the ice point being the temperature at which ice is in equilibrium with air-saturated liquid water at a pressure of 1 atm = 101.325 kPa. This is not the same as the triple point of water, because of the presence of air in both the liquid and vapour phases and the difference in pressure. The thermodynamic temperature scale starts at zero, even though a temperature of 0 K is unattainable experimentally. (See Chapters 2 and 4.) The choice of the number 273.16 in (1) makes the thermodynamic scale of temperature numerically identical with the former *gas scale* which was based on two fixed points, the ice point and the steam point.

In addition to providing us with a temperature scale, the Zeroth Law tells us how the temperature of a system may be measured. We require a system I, called a *thermometer*, of much smaller mass than system II, the temperature of which it is required to measure. I must have a property P which is a monotonic (and preferably close to linear) function of T. For example, I could be the familiar Hg in a glass thermometer, a

---

[b]The reason for thus naming the Zeroth Law may be obscure to the uninitiated. When I was a Lecturer in Physical Chemistry at Imperial College London, my teaching duties included a course in chemical thermodynamics for chemical engineers. I included in my lectures a statement of the Zeroth Law. A member of the class told me that, while working in the library, with his lecture notes open at the page describing the Zeroth Law, a friend of his — who was taking a different course in thermodynamics — approached and suggested that they go to lunch. Glancing at the open notebook, his friend remarked "Say, who's this guy Zero?"

Pt resistance thermometer, or a constant volume gas thermometer. The thermometer I is calibrated by measuring the value of P at several fixed points. A "fixed point" is a system which can easily be reproduced and the temperature of which has been measured with great precision in several different laboratories. System I is then brought into thermal contact with II and several measurements of P made and the results averaged. Of course P is useless as a thermometric property unless reproducible results are obtained. (See Zemansky and Dittman (1997) for further details on the measurement of T.) This is a topic to which a great deal of time and effort has been devoted and it may be taken that T is a thermodynamic property which can be measured accurately.

1.3. The Ideal Gas

The *compression factor* Z of a gas is defined by

$$Z = pV_{\mathrm{m}}/RT \tag{1}$$

V_{m} is the *molar volume* of the gas, that is, the volume containing one mole of the gas; the mole being the amount of a substance that contains N_{A} molecules. N_{A} is the symbol for the *Avogadro constant*, that is the number of atoms in 0.012 kg of ^{12}C, which to five significant figures is

$$N_{\mathrm{A}} = 6.0222 \times 10^{23} \, \mathrm{mol}^{-1} \tag{2}$$

In (1) R is the *gas constant* and its value has been fixed by assigning numerical values to T. At limitingly low pressures, $Z \to 1$ independently of the nature of the gas, so that

(1)
$$\lim_{p \to 0} \frac{pV}{nRT} = 1 \tag{3}$$

Exercise 1.3-1

Show that Eq. (3) is the basis of a gas thermometer and that careful measurements of pV at some constant temperature T, provide a numerical value for T.

Equation (3) shows that, at limitingly low pressures, all gases obey the *equation of state*

(3)
$$pV = nRT \tag{4}$$

An *equation of state* is a relation between the variables p, V and T and (4) is referred to as the *ideal gas equation*.

Exercise 1.3-2

What is the amount of oxygen contained in $0.100\,\text{m}^3$ of the gas at a pressure of $1.01 \times 10^5\,\text{Pa}$ and a temperature of $300\,\text{K}$, assuming ideal behaviour?

If we choose to measure the amount of gas in terms of the number of molecules, rather than the number of moles, then

$$(4) \qquad\qquad pV_m = RT = N_A kT \qquad\qquad (5)$$

where $k = R/N_A$ is Boltzmann's constant, which has the value (Mills *et al.* 1988)

$$k = 1.38066 \times 10^{-23}\,\text{J K}^{-1} \qquad\qquad (6)$$

Therefore

$$(2),\ (5),\ (6) \qquad\qquad R = N_A k = 8.3145\,\text{J K}^{-1}\,\text{mol}^{-1} \qquad\qquad (7)$$

Although gases only behave ideally at low pressures, nevertheless ideal behaviour is a convenient reference standard to which we refer the properties of real gases. At higher pressures

$$(1) \qquad Z = pV_m/RT = 1 + B_2(T)/V_m + B_3(T)/V_m^2 + \cdots \qquad (8)$$

which is the *virial equation of state* in terms of the inverse molar volume. $B_2(T)$, $B_3(T)\ldots$ are the second, third, $\ldots$ *virial coefficients*. The virial coefficients depend on T being positive at high temperatures and negative at low temperatures (Figure 1.2). The temperature at which $B_2(T)$ changes sign is the Boyle temperature, T_B. As the third term in (8) is much smaller than the second one, gases behave ideally over a larger range of p at T_B than they do at higher or lower temperatures. For other attempts to describe the p, V, T behaviour of gases, see Ott *et al.* (1971).

Exercise 1.3-3

Obtain a virial equation for the compression factor in terms of the pressure rather than reciprocal volume, as in (1.3.8). Express the coefficients B_2', B_3' in this equation in terms of the coefficients, B_2, B_3 in (8).

An ideal gas obeys Boyle's law, namely that pV is constant at constant temperature and this is approximately true for real gases at moderate

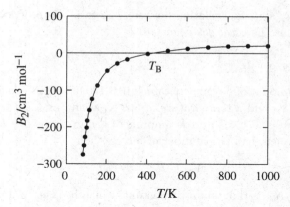

Figure 1.2. Second virial coefficient of argon as a function of T (after Dymond and Smith 1980).

pressures and high temperatures. However, marked deviations from Boyle's law occur if a gas is cooled to temperatures close to that at which condensation to the liquid state occurs. Figure 1.3 shows isotherms for carbon dioxide close to the critical temperature, $T_c = 31.04°C$. The critical temperature is the temperature above which separate gas and liquid phases cannot coexist. Above the critical temperature the distinction between gas and liquid is not meaningful and the substance is called a fluid (when it is not a solid). The shaded area in this figure marks the range of values of p, V and T where liquid and gaseous CO_2 coexist. The isotherms are horizontal in this coexistence region, which diminishes to a point at the critical point, C. Consequently, there is a point of inflexion at C and the $p(V)$ isotherms are parallel to the V axis so that

$$\left(\frac{\partial p}{\partial V}\right)_{T_c} = 0; \quad \left(\frac{\partial^2 p}{\partial V^2}\right)_{T_c} = 0 \quad \text{(critical point)} \qquad (9)$$

Answers to Exercises 1.3

Exercise 1.3-1

Equation (3) tells us that, for the same constant amount of a gas at temperature $T, \lim_{p\to 0} pV \propto T$. Choose a gas, such as H_2 or He, which has a low critical temperature. One needs an apparatus with which both p and V can be measured accurately. If Eq. (4) applied exactly at T, the product pV would be constant when p is varied at constant T, but in

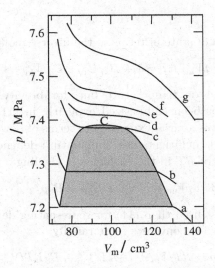

Figure 1.3. Isotherms of carbon dioxide above and below the critical point, C (after Michels, Blaisse and Michels 1937). $t/°C$ for a is 29.929; b, 30.409, c, 31.013; d, 31.185; e, 31.320; f, 31.523; g, 32.054. The lightly-shaded area shows the region of coexistence of liquid in equilibrium with its saturated vapour. In this region, $p(T)$ and the isotherm is a horizontal line, parallel to the V axis. At the critical temperature T_c, this line is reduced to a point of inflexion C, where the gradient $(\partial p/\partial V)_T$ is still parallel to the V axis, and therefore zero.

practice, we would find it to vary slightly, making an extrapolation to $p = 0$ necessary. Therefore, if $\lim_{p \to 0} pV$ is measured at the triple point of water (temperature T_t) and then again for some other system s (at temperature T_s), the ratio

(3) $$[\lim_{p \to 0} pV]_s / [\lim_{p \to 0} pV]_t = T_s/T_t = T_s/273.16\,\text{K} \qquad (10)$$

from which T_s can be evaluated. Clearly, this procedure is far too elaborate to be used on an everyday basis and so various thermometers such as thermocouples, Pt resistance thermometers and Hg in glass thermometers, which are much more convenient to use, have been calibrated against a primary standard gas thermometer.

Exercise 1.3-2

$$n = pV/RT = \frac{1.01 \times 10^5\,\text{Pa} \times 0.0100\,\text{m}^3}{8.3145\,\text{J K}^{-1}\,\text{mol}^{-1} \times 300\,\text{K}} = 0.405\,\text{mol}$$

Exercise 1.3-3

The virial equation of state in terms of the inverse molar volume is

$$(8) \qquad Z = pV_m/RT = 1 + B_2/V_m + B_3/V_m^2 + \cdots \qquad (11)$$

B_2 and B_3 depend on temperature but, for brevity, this temperature dependence has not been emphasized explicitly in (11). Replacing $1/V_m$ by p/RT in (11) gives a virial expansion in terms of powers of the pressure. But since the virial coefficients are temperature dependent, it is usual to incorporate factors of RT in the coefficients, yielding

$$(11) \qquad Z = pV_m/RT = 1 + B_2'p + B_3'p^2 + \cdots \qquad (12)$$

This equation is less used than (11) for representing deviations from ideal behaviour because (11) converges more rapidly

$$(11) \qquad p = RT/V_m + B_2 RT/V_m^2 + B_3 RT/V_m^3 + \cdots \qquad (13)$$

$$(13), (12) \qquad Z = 1 + B_2'RT(1/V_m + B_2/V_m^2 + B_3/V_m^3 + \cdots)$$

$$+ B_3'(RT)^2(1/V_m + B_2/V_m^2 + \cdots)^2 \qquad (14)$$

Equating terms of the same order in (11) and (14) yields

$$B_2' = B_2/RT, \quad B_3' = (B_3 - B_2^2)/(RT)^2 \qquad (15)$$

Deduce the units of the virial coefficients from (11) and (12) and then make units check on (15). There is no difficulty in taking the expansions in (11) and (14) to higher orders of $1/V_m$, if required. For the virial coefficients of many gases, see Dymond and Smith (1980).

1.4. Work

The work done *on a system* is given by

$$dw = X_0 \, dx \qquad (1)$$

where X_0 is the component of the force acting on the system parallel to the displacement dx. When we calculate w by integrating (1) we may find w to be negative, in which case the system is doing work on its surroundings. The force of magnitude X_0 acting on the system is opposed by a force of magnitude X acting on the surroundings (Figure 1.4). If $X = X_0$, the system is in mechanical equilibrium, which is usually just referred to as "*equilibrium*". If $X \neq X_0$, then the system will change spontaneously

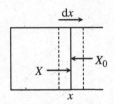

Figure 1.4. The work done on the system as x increases by dx is $dw = -X_0\, dx$. (Note the orientation of the force X_0 and the displacement dx.) Here, $X > X_0$ and dw is negative, because the system is doing work on its surroundings. When $X = X_0$, the system is at equilibrium.

(unless it is prevented from doing so by constraints) until it reaches a new equilibrium state: we call this a *natural* process. If $X = X_0 \pm dX$, then the limit in which $dX \to 0$ is called a *reversible process*. When a system is said to undergo a reversible process, this means that it is arbitrarily close to equilibrium at every instant, so that X_0 may be replaced by X and

$$dw = -X\, dx \tag{2}$$

(1)   (2)

The sign change between (1) and (2) arises because in this process X_0 and dx are oriented in opposite directions (Figure 1.4). X is a *generalized force* and may have several different manifestations, some of which may be acting concurrently. We now consider some examples of the application of Eq. (2).

Example 1.4-1 Change in volume of a phase

If the external pressure is p_0, $X_0 = p_0 A$ where A is the area of the piston in Figure 1.4. Therefore

$$dw = -p_0 A\, dx = -p_0\, dV \quad \text{(natural process)} \tag{3}$$

Notice the operation of the sign convention: if dV is negative, w is positive because work is being done on the system in order to compress it. But if dV is positive, the system expands and does work on its surroundings. For an infinitesimal reversible process

$$dw = -p\, dV \quad \text{(reversible process)} \tag{4}$$

(3)   (4)

For a finite process, the work done on the system (see Figure 1.5) is

$$w = -\int_{V_1}^{V_2} p\, dV \tag{5}$$

(4)   (5)

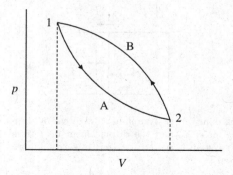

Figure 1.5.   The work done on the system along path A is the negative of the area under curve A. Similarly, the work done along path B is the area under B. Therefore, the net work around the cycle is equal to the area enclosed by the two paths, which is not zero unless the paths are identical.

Now suppose the system returns to its initial state 1 along path B. The work done on the system is

$$w = - \int_{V_2}^{V_1} p\,dV = \int_{V_1}^{V_2} p\,dV \tag{6}$$

If the initial and final states are the same, the system has undergone a *cyclic process*, for which (Figure 1.5)

$$(5),\ (6) \qquad w_{\text{cycle}} = \int_{V_1}^{V_2} p\,dV\,(\text{path B}) - \int_{V_1}^{V_2} p\,dV\ (\text{path A})$$

$$= \text{area enclosed by A and B} \tag{7}$$

If A and B are identical, $w = 0$, but if A and B are different paths

$$(7) \qquad\qquad \oint dw \neq 0 \tag{8}$$

and dw is not a complete differential (Appendix A1). Since w depends on the path, w *is not a thermodynamic property* of a system, but rather a way in which a system interacts with its surroundings.

Exercise 1.4-1

Find an expression for the work done on the gas during the isothermal reversible change in the volume V of an ideal gas from V_1 to V_2. What is the sign of w (a) if $V_2 > V_1$ (expansion); (b) if $V_2 < V_1$ (compression).

Exercise 1.4-2

$10.0\,\text{m}^3$ of hydrogen is produced by dissolving a metal in dilute acid. How much work is done by the gas on pushing back the atmosphere, if the atmospheric pressure in the laboratory is $1.01 \times 10^5\,\text{Pa}$?

Example 1.4-2 Change in area of a surface film

Consider a film on top of a liquid in a trough fitted with a movable bar, so that the area of the film can be varied by moving the bar. There is an inward acting force $X = \gamma L$ on the bar (see Figure 1.6) constraining the film, where γ is the surface tension and L is the width of the trough. This is opposed by the outward acting force X_0 applied to the bar. If the area A of the film is changed reversibly, by ensuring that $X_0 = X + \mathrm{d}X$, then the work done on the film is

$$\mathrm{d}w = \gamma L\,\mathrm{d}x = \gamma\,\mathrm{d}A \qquad (9)$$

$$(9) \qquad w = \int_{A_1}^{A_2} \gamma\,\mathrm{d}A \quad \text{(reversible process)} \qquad (10)$$

If $A_2 > A_1$, w is positive because work must be done against the surface tension to expand the film.

Exercise 1.4-3

The area of a surface film is increased reversibly from $100\,\text{cm}^2$ to $110\,\text{cm}^2$. Calculate the amount of work done on the film if the surface tension γ is $76\,\text{mN}\,\text{m}^{-1}$.

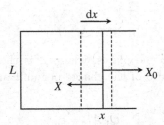

Figure 1.6. Trough of width L containing liquid with a surface film and fitted with a rigid, movable bar so that the area of the film may be varied by changing the force X_0. This force is opposed by the inward acting force on the bar $X = \gamma L$, due to the surface tension γ.

Example 1.4-3 Charging of an electrical capacitor

A capacitor consists of two conducting plates separated by a non-conducting dielectric. A partially-charged capacitor has a potential difference between the plates equal to $\Delta\varphi$, which is caused by a transfer of charge from one plate to the other by the application of a potential difference from an external source. If an extra increment of charge dQ is transferred by increasing the potential difference by an infinitesimal amount, then the work done is

$$dw = \Delta\varphi \, dQ \tag{11}$$

If the external potential difference is increased so that it is maintained infinitesimally greater than the potential difference between the plates and then removed, then the capacitor has been charged reversibly from a charge $+Q_1$ on the one plate and $-Q_1$ on the other, to $+Q_2$ and $-Q_2$, respectively, and the total work done is

$$(11) \qquad w = \int_{Q_1}^{Q_2} \Delta\varphi \, dQ \quad \text{(reversible process)} \tag{12}$$

The ratio of the charge on the plates to the potential difference between them is called the capacity C of the capacitor, so that the work done becomes

$$(12) \qquad w = \frac{Q_2^2 - Q_1^2}{2C} \tag{13}$$

If $Q_2 < Q_1$, then w is negative and the discharging condenser does work on its surroundings. C is a property of the capacitor which does not depend on either Q or $\Delta\varphi$.

The electric field strength $\boldsymbol{E}$ between the plates of the capacitor in vacuum is of magnitude

$$E = \Delta\varphi/L \tag{14}$$

and direction normal to the plates, where L is the distance between the plates of area A. The ratio of the charge per unit area to the electric field strength is

$$Q/AE = \varepsilon_0 \tag{15}$$

where ε_0 is the *electric constant*

$$\varepsilon_0 = 8.85419 \times 10^{-12} \, \mathrm{F\,m^{-1}} \tag{16}$$

If the space between the plates is filled with a non-conducting medium of *permittivity* ε, then

(15) $Q/AE = \varepsilon$ (17)

$\varepsilon/\varepsilon_0$ is the *dielectric coefficient* of the material between the plates of the capacitor

(11), (14), (17) $dw = ALE\,d(\varepsilon E)$ (parallel plate capacitor)   (18)

εE is the *electric displacement D*.

Exercise 1.4-4

Make units checks on Eqs. (15) and (18).

In general, E and D are vectors and (18) must be replaced by

(A2.2.16) $dw = (\boldsymbol{E} \cdot d\boldsymbol{D})\,dV$ (19)

(See Appendix A2 for a general derivation.) The work done per element of volume dV in (19) consists of two terms, namely the work done in creating the field in a vacuum plus the work done in polarizing the material between the plates. The first term is incorporated in dU (see (A2.1.20), (A2.1.22)) so that the term of interest is

$$dw = V\boldsymbol{E} \cdot d\boldsymbol{P} \qquad (20)$$

where $\boldsymbol{P}$ is the *polarization* (electric dipole moment per unit volume) in the material between the plates of the capacitor.

Example 1.4-4 An electrochemical cell

Electrochemical cells are represented by diagrams which show the two electrodes and the solutions in which they are immersed. A vertical bar is used to show a phase boundary. The electric potential difference $\Delta\varphi$ of an electrochemical cell is equal in magnitude and sign to the electric potential of a metallic lead attached to the right-hand electrode minus that of a similar lead attached to the left-hand electrode. The *cell potential* is the limiting value of this electric potential difference when current flow is reduced to zero by opposing $\Delta\varphi$ by an external electric potential difference $(\Delta\varphi)_0$ from a potentiometer. If the cell diagram has been written with the more positive electrode on the right, then the cell chemical reaction is accompanied by the flow of electrons from left to right through the external circuit. If $(\Delta\varphi)_0 = \Delta\varphi - d(\Delta\varphi)$, then the cell does work reversibly on its surroundings. The work done when one mole of elementary charge is transported through the external circuit is

$$w = w_{el} + w_{chem} \tag{21}$$

where the electrical work

$$(11) \qquad\qquad w_{el} = -\int \Delta\varphi\, dQ = -F(\Delta\varphi) \tag{22}$$

The Faraday constant F is the charge carried by one mole of elementary charge, that is

$$F = 6.0222 \times 10^{23} \times 1.6022 \times 10^{-19}\,\mathrm{C\,mol^{-1}}$$
$$= 9.6485 \times 10^4\,\mathrm{C\,mol^{-1}} \tag{23}$$

Note that $-F(\Delta\varphi)$ is negative because the cell is doing work on its surroundings

$$w_{chem} = -\sum_\alpha p^\alpha\, \Delta V^\alpha \tag{24}$$

$$(22),\ (24) \qquad w = -F(\Delta\varphi) - \sum_\alpha p^\alpha\, \Delta V^\alpha \quad \text{(reversible process)} \tag{25}$$

The work calculated from (25) corresponds to the chemical changes that are associated with the transport of one mole of elementary charge from L to R through the cell (or, equivalently, the flow of one mole of electrons from L to R in the external circuit) and the opposite chemical changes would yield a value for w of opposite sign.

Exercise 1.4-5

For the following electrochemical cell, write down the two half-reactions occurring at the two electrodes and then the complete cell reaction. (Further on in this book we shall learn how to write expressions for the cell potential of cells like this one.) (**Hint**: The Ag electrode is the more positive one.)

$$\mathrm{Pt|H_2(g,}p = 10^5\,\mathrm{Pa|HCl(aq)|AgCl(s)|Ag(s)}$$

1.4.1. *Magnetic work*

The magnetization M (magnetic moment divided by volume) is defined by

$$M = (B/\mu_0) - H \tag{26}$$

where B is the *magnetic induction*, H is the *magnetic field strength* and the *magnetic constant*

$$\mu_0 = 4\pi \times 10^{-7}\,\mathrm{H\,m^{-1}} \tag{27}$$

H and M have SI units $\mathrm{A\,m^{-1}}$ and the magnetic constant has SI units $\mathrm{H\,m^{-1}} = \mathrm{V\,s\,A^{-1}}$ so B has SI units $\mathrm{T} = \mathrm{V\,s\,m^{-2}}$. The work done on increasing by dM the magnetization M in a body of volume V, in a uniform external magnetic field H, is

(A2.2.10) $$dw = V H \cdot dM \tag{28}$$

which has SI units of J. The work done actually consists of two terms, one of which is the creation of the magnetic field in vacuum and the second one, the one given in (28), is the magnetization of the material, in close analogy with the electrostatic case (see Eqs. (19), (20)). The total work done on an element of volume dV is

$$dw = (\mu_0 H \cdot dH + H \cdot dM)\,dV \tag{29}$$

Only the second term in (29) is shown in (28), since it is customary to include the work done in creating the field in vacuum (which has nothing to do with the material comprising the system) in dU (see Appendix A2, Eqs. (16)–(19), Pippard 1966, Callen 1960). It is the $H \cdot dM$ term that is of most interest in the study of thermodynamic systems.

Exercise 1.4-6

Why does Eq. (28) not apply to a ferromagnetic material?

Exercise 1.4-7

Check units in Eq. (28).

Exercise 1.4-8

Evaluate $\varepsilon_0 \mu_0$.

Example 1.4-5 A long solenoid

Consider a long, uniform solenoid of length L, with n turns and cross-sectional area A, the length per turn being $l = L/n$, and volume therefore $V = nlA$. The solenoid is filled with an isotropic paramagnetic substance of permeability μ. When the current through the solenoid is I, the magnitudes of $\boldsymbol{B}$ and $\boldsymbol{H}$ inside the solenoid are given by

$$|\boldsymbol{B}| = \mu I/l, \quad |\boldsymbol{H}| = I/l \tag{30}$$

The work done on the system (the material inside the solenoid) is therefore

(29), (30) $$\mathrm{d}w = (A/L)n^2 I \,\mathrm{d}(\mu I) \tag{31}$$

Answers to Exercises 1.4

Exercise 1.4-1

$$w = -\int_{V_1}^{V_2} p \,\mathrm{d}V = -\int_{V_1}^{V_2} \frac{nRT}{V} \,\mathrm{d}V = -nRT\ln(V_2/V_1)$$

If $V_2 > V_1$, the gas expands and does work on its surroundings, so that w is negative. But if $V_2 < V_1$, w is positive, since work must be done on the gas in order to compress it.

Exercise 1.4-2

Since p is constant, the work done *by* the gas in pushing back the atmosphere is $p\Delta V = 1.01 \times 10^5 \,\mathrm{Pa} \times 10.0 \,\mathrm{m}^3 = 1.01 \,\mathrm{MJ}$.

Exercise 1.4-3

$$w = \gamma \Delta A = 0.076 \,\mathrm{N\,m^{-1}} \times 0.0010 \,\mathrm{m}^2 = 0.076 \,\mathrm{mJ}.$$

Exercise 1.4-4

Equation (15) states that the ratio of the charge per unit area to the electric field in a capacitor with vacuum between the plates is $Q/AE = \varepsilon_0$, where ε_0 is the electric constant. In SI units, the charge Q is in C, the area A in m^2 and the electric field strength E in Vm^{-1}. This gives the electric constant ε_0 in $\mathrm{C\,V^{-1}\,m^{-1}}$, or $\mathrm{F\,m^{-1}}$, which is correct. ε in Eq. (17) has the same units as ε_0, so that the relative permittivity (formerly, and somewhat misleadingly, called "dielectric constant") is a dimensionless quantity.

Equation (18) states that

$$dw = ALE \, d(\varepsilon E) \qquad (32)$$

The units of the RS of (18) are: $m^3 \, V \, m^{-1} \, C \, V^{-1} \, m^{-1} \, V \, m^{-1} = V \, C = J$, the SI unit for energy, and therefore, work.

Exercise 1.4-5

The sign of the cell potential indicates that electrons are produced at the LH electrode. Therefore, at the LH (Pt) electrode

$$\frac{1}{2} H_2(g) \rightarrow H^+(aq) + e^-$$

where e^- signifies an electron. At the RH (Ag) electrode, electrons react with silver chloride

$$AgCl(s) + e^- \rightarrow Ag(s) + Cl^-(aq)$$

Therefore the complete cell reaction is

$$\frac{1}{2} H_2(g) + AgCl(s) \rightarrow H^+(aq) + Cl^-(aq) + Ag(s)$$

Exercise 1.4-6

Ferromagnetic materials exhibit hysteresis and therefore a reversible process is not possible with a ferromagnetic material.

Exercise 1.4-7

The RS of Eq. (28) is $V\boldsymbol{H} \cdot d\boldsymbol{B}$, with units of $m^3 \, A \, m^{-1} \, V \, s \, m^{-2} = J$, the units of work.

Exercise 1.4-8

$1/c_0^2 = 1/(2.99792746 \times 10^8 \, m \, s^{-1})^2 = 1.11265 \times 10^{-17} \, s^2 \, m^{-2}$, where c_0 is the speed of light in vacuum. $\varepsilon_0 \mu_0 = 8.85419 \times 10^{-12} \, F \, m^{-1} \times 4\pi \times 10^{-7} \, H \, m^{-1} = 1.11265 \times 10^{-17} \, A \, s \, V^{-1} \, m^{-2} \, V \, s \, A^{-1} = 1.11265 \times 10^{-17} \, s^2 \, m^{-2} = 1/c_0^2$.

1.5. The First Law of Thermodynamics

In general, the work done on a system as it is transformed from one equilibrium state 1 to another equilibrium state 2, depends on the path between one and two. But it is found that if a system is transformed *adiabatically* from state one to state two, then the work done on the system is independent of the adiabatic path. Therefore, there exists a function of state, which is called the *internal energy U*, the change in which between two states is equal to the adiabatic work done on the system. In the form of an equation

$$w = \Delta U \quad \text{(adiabatic process)} \tag{1}$$

Historically, the first convincing illustration of the validity of Eq. (1) was a series of experiments carried out by Joule in 1843–1878 (see, e.g., Denbigh 1955). The adiabatically enclosed system was a calorimeter containing water and by means of a paddle activated by a falling weight, work was expended on the water. Since the system is of constant mass and composition and at constant pressure, $U(T, V)$. However, the $p\Delta V$ term from the work done as the water expands against the constant atmospheric pressure is negligibly small compared to the electrical or mechanical work done on the system and so the temperature of the water serves to specify the state of the system. The rise in temperature observed by Joule was thus a direct measure of the increase in internal energy of the system. Joule found that the expenditure of the same amount of work on the system in various ways resulted in the same rise in temperature and therefore in the same change of state of the system.

Equation (1) holds for adiabatic processes. When the same change in state is brought about under conditions in which the system is not adiabatically isolated, the work required to do this is different from $\Delta U_{\text{adiabatic}}$. This difference is a quantitative measure of the amount of energy exchanged *thermally* with the surroundings by means other than the performance of work on the system. The system is then said to have absorbed heat q from its surroundings (Figure 1.7) where

$$q = \Delta U - w \tag{2}$$

or, for an infinitesimal process

(2) $$dU = dw + dq \tag{3}$$

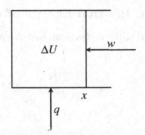

Figure 1.7. w is the work done on the system, q is the heat absorbed by the system and $\Delta U = q + w$ is the change in the internal energy of the system.

which is the **First Law of Thermodynamics.** dw and dq represent the small amounts of work done on the system and heat absorbed by the system during an infinitesimal change in state. They are not differentials of functions of state and the integrals $\int_1^2 dw$ and $\int_1^2 dq$ depend on the path by which the system is brought from state one to state two.

Exercise 1.5-1

What can you assert about the values of ΔU, q and w if a system (a) undergoes a cyclic process; (b) is in a steady state?

If the only work done on the system is that which causes a reversible change in volume, then $dw = -p\,dV$ and

(3) $$dq = dU + p\,dV \tag{4}$$

But $U(T, V)$ and therefore

(4) $$dq = \left(\frac{\partial U}{\partial T}\right)_V dT + \left(\frac{\partial U}{\partial V}\right)_T dV + p\,dV \tag{5}$$

At constant volume

(5) $$\frac{dq}{dT} = \left(\frac{\partial U}{\partial T}\right)_V = C_V = n\,C_{V,m} \tag{6}$$

where C_V is the *heat capacity at constant volume*, with units of $J\,K^{-1}$, and $C_{V,m}$ the *molar heat capacity at constant volume*, with units of $J\,K^{-1}\,mol^{-1}$.

Answer to Exercise 1.5-1

In both (a) and (b), $\Delta U = 0$ and so the heat absorbed by the system is equal to the work done by the system.

1.6. Heat Capacity and Internal Energy of an Ideal Gas

(If not familiar with partial differentiation, see Appendix A1.) For one mole
of a substance

(1.5.6)
$$C_V = \left(\frac{\partial U}{\partial T}\right)_V \tag{1}$$

At constant pressure

(1.5.4)
$$dq = d(U + pV) = \mathrm{d}H \tag{2}$$

where $H = U + pV$ is the *enthalpy H*. Therefore, the *heat capacity at
constant pressure*

$$C_p = \left(\frac{\partial H}{\partial T}\right)_p = \left(\frac{\partial U}{\partial T}\right)_p + p\left(\frac{\partial V}{\partial T}\right)_p \tag{3}$$

But $U(T, V)$ where $V(T, p)$; therefore

(A1.4.3)
$$\left(\frac{\partial U}{\partial T}\right)_p = \left(\frac{\partial U}{\partial T}\right)_V + \left(\frac{\partial U}{\partial V}\right)_T \left(\frac{\partial V}{\partial T}\right)_p \tag{4}$$

(3), (4)
$$C_p - C_V = \left\{ p + \left(\frac{\partial U}{\partial V}\right)_T \right\} \left(\frac{\partial V}{\partial T}\right)_p \tag{5}$$

Equations (1)–(5) are quite general and apply to all material substances.

Early experiments intended to measure the dependence of the internal
energy of a constant mass of gas on its volume failed because the heat
capacity of the measuring system (a large calorimeter filled with water) was
too large to detect the temperature change in the gas. But the dependence
of the internal energy of a gas on pressure can be measured in an experiment
due to Washburn. A steel bomb containing n moles of a gas (such as
nitrogen) at a high pressure p is placed in a thermostat containing water.
The exit tube from the bomb is wound with an electrical heater and it is
then wound around the bomb. When the exit valve on the bomb is opened,
the gas escapes into the atmosphere, and the wattage expended in the
heater measured. This energy required to maintain the bath at a constant
temperature is a direct measure of the heat q absorbed by the gas as it does
work against the atmospheric pressure, p_0. Therefore, from the First Law

$$\Delta U = q - p_0(V_0 - V_b) \tag{6}$$

where V_0 is the volume occupied by n moles of gas at room temperature T and pressure p_0. Results obtained by Rossini and Frandsen (1932) for air at $p \leq 5\,\text{MPa}$ showed that

$$\Delta U = U(p_0, T) - U(p, T) = c(T)[p - p_0] \tag{7}$$

$$(7) \qquad U(p, T) = -c(T)p + U(p_0, T) + c(T)p_0 = -c(T)p + U(T) \tag{8}$$

where $U(T)$ is a function of T only. Consequently, as $p \to 0$

$$(8) \qquad\qquad\qquad U \to U(T) \tag{9}$$

and therefore

$$(9) \qquad\qquad\qquad \left(\frac{\partial U}{\partial p}\right)_T \to 0 \tag{10}$$

It is convenient to describe the properties of real gases in terms of those of a hypothetical *ideal gas* which obeys the equation of state

$$pV = nRT \quad \text{(ideal gas)} \tag{11}$$

at all pressures, and for which U is a function of T only

$$U = U(T) \quad \text{(ideal gas)} \tag{12}$$

$$(5), (11), (12) \qquad C_{p,m} - C_{V,m} = R \quad \text{(one mole of ideal gas)} \tag{13}$$

Exercise 1.6-1

How does the internal energy of an ideal gas depend on its volume?

Answer to Exercise 1.6-1

Since U is a function of T only

$$\left(\frac{\partial U}{\partial V}\right)_T = 0 \quad \text{(ideal gas)} \tag{14}$$

1.7. Generalized Forces

In thermodynamics, we are largely concerned with the quantity dw, which is the work done on a system during an infinitesimal process. This is calculated from the equation

$$dw = \boldsymbol{X}_0 \cdot d\boldsymbol{x} \tag{1}$$

Table 1.1.   Particular examples of a generalized force X_0 in thermodynamics. The conjugate displacement dx in the property x is induced by the force.

Force acting *on* system, X_0	x	work done[a]
surface tension, γL	width of surface film, x	$\gamma\,dA$
due to pressure, $p_0 A$	length (see Figure 1.4) x	$-p\,dV$
electric field strength, $\boldsymbol{E}$	electric displacement, $\boldsymbol{D}$	$V\,\boldsymbol{E}\cdot d\boldsymbol{D}$[b,c]
magnetic field strength, $\boldsymbol{H}$	magnetic induction, $\boldsymbol{B}$	$V\,\boldsymbol{H}\cdot d\boldsymbol{B}$[c]

[a] That is, $dw = X\,dx$, the work done reversibly on the system during an infinitesimal displacement dx. The sign before X depends on the orientation of the vectors X_0 and dx.
[b] In Example 1.4.3., this becomes $dw = \Delta\varphi\,dQ$.
[c] See Appendix A2, especially Eqs. (A2.17) and (A2.22).

where $\boldsymbol{X}_0$ is the *generalized force* acting on a system and dx the *conjugate displacement* of the system due to this force. When the thermodynamic process (the displacement in x) is carried out reversibly, $|\boldsymbol{X}_0|$ may be replaced by $|\boldsymbol{X}|$, the force acting from the system on its surroundings, in opposition to $\boldsymbol{X}_0$. Several examples of this generalized force were given in Section 1.4. In any particular thermodynamic process several different forces may be operating simultaneously, while others are of negligible importance. For this reason, and to maintain generality, it is often useful to employ the idea of a generalized force $\boldsymbol{X}_0$ which may have several different manifestations. Particular examples of a generalized force and its conjugate displacement are given in Table 1.1.

Problems 1

1.1 Find $\frac{\partial f}{\partial x}$, $\frac{\partial f}{\partial y}$ and the four second-order partial derivatives of f if

$$f(x,y) = x^3 - 3xy^2 + 2y^3$$

1.2 For a gas, only two of the variables pressure p, volume V and temperature T are independent. The relation, $f(p, V, T) = 0$ is called an equation of state. Write down the expression for the complete differential df and explain why $df = 0$.
   Hence show that

$$\left(\frac{\partial V}{\partial T}\right)_p \left(\frac{\partial T}{\partial p}\right)_V = -\left(\frac{\partial V}{\partial p}\right)_T \qquad (\textbf{Hint:} \text{ See Section A2.2.})$$

The expansivity α and isothermal compressibility k_T are defined by

$$\alpha = \frac{1}{V}\left(\frac{\partial V}{\partial T}\right)_p, \quad k_T = -\frac{1}{V}\left(\frac{\partial V}{\partial p}\right)_T$$

Express

$$\left(\frac{\partial T}{\partial p}\right)_V$$

in terms of α and k_T

1.3 If $du = y\,dx - x\,dy$, verify that du is not a complete differential and show that $1/y^2$ is an integrating factor for du.

1.4 Find $\int_{x_1,y_1}^{x_2,y_2} du$ along the two paths (a) $x_1, y_1 \to x_2, y_1 \to x_2, y_2$ and (b) $x_1, y_1 \to x_1, y_2 \to x_2, y_2$ for $du = x\,dx + x\,dy$ and confirm that the two results differ by the area enclosed by the two paths of integration.

1.5 Show that $f(x,y,z) = x^2 y - 2xyz - 2xz^2$ is a homogeneous function of degree 3 and verify Euler's theorem for f.

1.6 Obtain an expression of the form $dq = M\,dT + N\,dp$ for the particular case of the change in volume of an ideal gas. Confirm that dq is not a complete differential.

1.7 dw is not a complete differential. Verify this for the particular case of an ideal gas.

1.8 (a) One mole of benzene is vaporized reversibly under a constant pressure, $p_0 = 1.00 \times 10^5$ Pa. Calculate the work done on the benzene during this process, assuming that the molar volume of the liquid is negligibly small in comparison with that of the vapour.

   (b) The heat absorbed during the process described in (a) is 3.08×10^4 J mol^{-1}. Find the change in the internal energy of one mole of benzene that accompanies its vaporization under these conditions.

1.9 An ideal gas is taken through the following cycle:

   A→B Isothermal expansion at T_1 from V_1 to V_2;
   B→C Isochoric change at V_2 from T_1 to $T_2 > T_1$;
   C→D Isothermal compression at T_2 from V_2 to V_1;
   D→A Isochoric change at V_1 from T_2 back to T_1.

   Write down expressions for the work done along AB, BC, CD, DA and around the whole cycle. How does the last result confirm that w is not a thermodynamic property?

1.10 The pressure on one mole of argon that occupies a volume of 2.50 m^3 is suddenly increased to 10000 Pa. The gas attains a new equilibrium state adiabatically, in which its volume is $1.15\,\mathrm{m}^3$. Find (a) the work done; (b) the change in internal energy; and (c) the heat absorbed, during this process.

1.11 van der Waals equation of state for one mole of a gas is

$$\left(p + \frac{a}{V^2}\right)(V - b) = RT \tag{1}$$

where a and b are constants that depend on the gas. This equation was devised in an attempt to correct for deviations from ideal behaviour. Suggest possible reasons for the two correction terms in Eq. (1) that do not appear in the equation of state for an ideal gas

Find

$$\frac{\partial p}{\partial V}, \ \frac{\partial p}{\partial T}, \ \frac{\partial^2 p}{\partial V^2}, \ \frac{\partial^2 p}{\partial T^2}$$

and verify that

$$\frac{\partial}{\partial V}\left(\frac{\partial p}{\partial T}\right) = \frac{\partial}{\partial T}\left(\frac{\partial p}{\partial V}\right)$$

(**Hint:** Solve Eq. (1) for p and then calculate the partial derivatives.)
Find p_C, V_C, T_C and $p_C V_C / RT_C$.

1.12 Prove

$$\left(\frac{\partial \alpha}{\partial p}\right)_T = \left(\frac{\partial \kappa_T}{\partial T}\right)_p \qquad (\textit{\textbf{Hint:}} \ \mathrm{d}\ln V \ \text{is a complete differential.})$$

2

Entropy

2.1. The Second Law of Thermodynamics

Natural processes, which occur spontaneously, are irreversible, that is they cannot be reversed without producing some change in the surroundings or in some other system. Examples of natural adiabatic processes are:

(i) the adiabatic transformation of work into the internal energy of a system (Joule's experiment);
(ii) the transfer of energy in the form of heat from a hotter to a colder body, both systems being enclosed in an adiabatic enclosure;
(iii) the free expansion of a gas from one-half of an adiabatically enclosed system, to fill the entire system;
(iv) the mixing of two dissimilar gases in an adiabatically enclosed system;
(v) a spontaneous phase change, or chemical reaction, occurring in an adiabatically isolated system.

The reverse of any of these processes cannot be made to occur in an adiabatic manner and they are therefore said to be *irreversible*. Yet none of these hypothetical *unnatural processes*, which are the reverse of (i)–(v) and which are never observed to occur in nature, contravene the First Law. We surmise that there must be some function of state, the change in which for any postulated process tells us whether that process is a natural or unnatural one. The introduction of such a function constitutes the **Second Law of Thermodynamics**. There exists an extensive function of state called the *entropy S*; in any thermodynamic process the net entropy change in a system plus its surroundings is

$$dS_{\text{net}} = dS + dS_0 \geq 0 \tag{1}$$

In (1), the inequality applies to natural, irreversible processes and the equality to reversible processes. The entropy change in the system comprises two terms

$$dS = d_iS + d_eS \tag{2}$$

27

where d_iS is due to irreversible processes (e.g. (i)–(v) above) going on within the system and d_eS to the absorption of heat dq from, and to the exchange of matter with, its surroundings. At this stage we consider only closed systems. Provided a system is in thermal equilibrium with its surroundings $(T = T_0)$ the isothermal absorption of heat is reversible and

$$d_eS = \frac{dq}{T} \quad \text{(reversible process, closed system)} \tag{3}$$

Since the amount of heat absorbed depends on the path, dq is an incomplete differential and (3) asserts that T^{-1} is an integrating factor (Appendix A1, Section A1.6) that converts dq into the complete differential dq/T. The justification for this statement is that all results deduced from (3) are found to be in agreement with the experiment. However, the initial realization that T^{-1} was an integrating factor for dq came from 19th century studies of the transformation of internal energy into work, the impetus for such research being provided by the Industrial Revolution. Natural processes that are not adiabatic involve the absorption of heat from, or the rejection of heat to, the surroundings. Experience shows that all such processes cannot be reversed without producing some change in the surroundings or in some other system. Thus although clear examples of irreversibility are provided by adiabatic processes, *all natural processes are irreversible.*

The reader might well question whether the isothermal absorption of heat by a system can ever be reversible, or indeed whether there can ever be such a process as the isothermal absorption of heat. We have to imagine the system in thermal contact with a heat reservoir (its surroundings), that is a very large system, the temperature of which can be varied continuously so that its temperature T_0 can be maintained only infinitesimally different from that of the system, so that

$$T = T_0 + dT \tag{4}$$

If dT is negative, heat will flow into the system until $T = T_0$. By repetition of this process, a finite change in T can be effected. Since dT can be made as small as we please, there are no conceptual difficulties about the isothermal absorption of heat. The other question to be considered is that of reversibility. One could well surmise that the absorption of heat by a system must be accompanied by irreversible processes within the system such as convection or heat conduction. However, if after each step in which sufficient heat is absorbed by the system to effect a rise in its temperature by dT, we pause sufficiently long for these irreversible processes to dissipate, then the final state will be the same as it would have been had the absorption of heat been reversible, because the entropy S (and that of any other function of state) depends on the *state* of the system, not on its previous history.

2.2. The Fundamental Equations for a Closed Phase

In this section we shall be introducing some extensive thermodynamic functions in addition to the internal energy U, the enthalpy H and the entropy S. If the system consists of several phases, then the total value of any extensive property is the sum of its value in each of the phases. Thus considering our closed system to comprise a single phase causes no particular difficulties. It is assumed at this stage that the phase is at rest and that electric and magnetic fields are absent. Provided chemical reactions are excluded, this phase is in internal equilibrium and any infinitesimal change is *reversible*. Therefore

$$(2.1.2),\ (2.1.3) \qquad \mathrm{d}S = \mathrm{d}_e S = \frac{\mathrm{d}q}{T} \quad \text{(reversible process)} \qquad (1)$$

From the First Law

$$(1.5.4),\ (1) \qquad \mathrm{d}U = T\,\mathrm{d}S - p\,\mathrm{d}V \qquad (2)$$

Since $U(S, V)$

$$\mathrm{d}U = \left(\frac{\partial U}{\partial S}\right)_V \mathrm{d}S + \left(\frac{\partial U}{\partial V}\right)_S \mathrm{d}V \qquad (3)$$

$$(2),\ (3) \qquad \left(\frac{\partial U}{\partial S}\right)_V = T, \quad \left(\frac{\partial U}{\partial V}\right)_S = -p \qquad (4)$$

It was because of the particularly simple form assumed by these differential coefficients of U with respect to S and V that Gibbs (1875) (see also Gibbs (1961)) called Eq. (2) the *fundamental equation for a closed phase* for the thermodynamic potential $U(S, V)$. The reason for the name *"thermodynamic potential"* will become clear later in this chapter. It was emphasized in Chapter 1 that a closed phase of constant composition has two degrees of freedom. There is considerable latitude in the choice of these two independent variables. Here we see that when S and V are the chosen independent variables, the differential of U has a particularly simple form.

If we choose U and V as independent variables then

(2)
$$dS = \frac{1}{T}\,dU + \frac{p}{T}\,dV \qquad\qquad (5)$$

which is the fundamental equation for a closed phase for the thermodynamic potential S. Still further choices of independent variables are available. Recall (Section 1.6)

$$H = U + pV \qquad\qquad (6)$$

and define

$$F = U - TS \qquad\qquad (7)$$
$$G = H - TS \qquad\qquad (8)$$

(2), (6)
$$dH = T\,dS + V\,dp \qquad\qquad (9)$$

(2), (7)
$$dF = -S\,dT - p\,dV \qquad\qquad (10)$$

(8), (9)
$$dG = -S\,dT + V\,dp \qquad\qquad (11)$$

Equations (9), (10) and (11) are the fundamental equations for a closed phase for the thermodynamic potentials H, F, and G. $H(S,p)$ is the *enthalpy*, $F(T,V)$ is the *Helmholtz energy* and $G(T,p)$ is the *Gibbs energy*. Which potential is used in any particular application depends on the most convenient choice of independent variables.

Maxwell relations: U, S, F, G and H are functions of the state of the system and their differentials in Eqs. (2), (5), (9), (10) and (11) are complete differentials. (Please review Appendix A1 if necessary.) They are all of the form

$$df = M\,dx + N\,dy \qquad\qquad (12)$$

in which $f(x,y)$ is a function of x and y. The commutative property of df

$$\left(\frac{\partial M}{\partial y}\right)_x = \left(\frac{\partial N}{\partial x}\right)_y \qquad\qquad (13)$$

for these thermodynamic functions, yields the *Maxwell relations*. For example

(10)
$$\left(\frac{\partial S}{\partial V}\right)_T = \left(\frac{\partial p}{\partial T}\right)_V \qquad\qquad (14)$$

(11)
$$\left(\frac{\partial S}{\partial p}\right)_T = -\left(\frac{\partial V}{\partial T}\right)_p \qquad\qquad (15)$$

Exercise 2.2-1

Obtain Maxwell relations from Eqs. (2), (5) and (9).

Answer to Exercise 2.2-1

(13), (2)
$$\left(\frac{\partial T}{\partial V}\right)_S = -\left(\frac{\partial p}{\partial S}\right)_V \qquad (16)$$

(13), (5)
$$\left(\frac{\partial (1/T)}{\partial V}\right)_U = \left(\frac{\partial (p/T)}{\partial U}\right)_V \qquad (17)$$

(13), (9)
$$\left(\frac{\partial T}{\partial p}\right)_S = \left(\frac{\partial V}{\partial S}\right)_p \qquad (18)$$

These Maxwell relations in (16)–(18) are generally not as useful in chemical thermodynamics as those in (14) and (15), in which the independent variables are T and either p or V.

2.3. Calculation of Entropy Changes

From $U(T, V)$

$$dU = \left(\frac{\partial U}{\partial T}\right)_V dT + \left(\frac{\partial U}{\partial V}\right)_T dV \qquad (1)$$

(2.2.5), (1)
$$dS = \frac{1}{T}\left[\left(\frac{\partial U}{\partial T}\right)_V dT + \left(\frac{\partial U}{\partial V}\right)_T dV\right] + \frac{p}{T} dV \qquad (2)$$

(1.6.1), (2)
$$dS = \frac{C_V}{T} dT + \frac{1}{T}\left[\left(\frac{\partial U}{\partial V}\right)_T + p\right] dV \qquad (3)$$

But, for $S(T, V)$

$$dS = \left(\frac{\partial S}{\partial T}\right)_V dT + \left(\frac{\partial S}{\partial V}\right)_T dV \qquad (4)$$

(4), (3), (1.6.1)
$$\left(\frac{\partial S}{\partial T}\right)_V = \frac{C_V}{T} \qquad (5)$$

(4), (3)
$$\left(\frac{\partial S}{\partial V}\right)_T = \frac{1}{T}\left[\left(\frac{\partial U}{\partial V}\right)_T + p\right] \qquad (6)$$

These partial derivatives are less simple than those in (2.2.5). From Problem 1.2

$$\left(\frac{\partial p}{\partial T}\right)_V = \frac{\alpha}{\kappa_T} \tag{7}$$

where α is the expansivity and κ_T the isothermal compressibility,

(6), (2.2.14)
$$\left(\frac{\partial U}{\partial V}\right)_T = T\left(\frac{\partial p}{\partial T}\right)_V - p \tag{8}$$

which is the first *thermodynamic equation of state*, so-called because it expresses the variation of the internal energy U with volume in terms of the variables p, V and T.

Exercise 2.3-1

Show that $U(T, V)$ is necessarily a function of T only, for an ideal gas that obeys the equation of state $pV = nRT$. (**Hint:** Use the first thermodynamic equation of state.)

(4), (5), (2.2.13), (7)
$$dS = \frac{C_V}{T}\,dT + \frac{\alpha}{\kappa_T}\,dV \tag{9}$$

The coefficients are now expressed in terms of measurable quantities and integration of (9) will yield ΔS for prescribed changes in T, V

(9)
$$\Delta S = \int_{T_1}^{T_2} C_V\,d\ln T + \int_{V_1}^{V_2} \frac{\alpha}{\kappa_T}\,dV \tag{10}$$

Exercise 2.3-2

A bulb of volume V filled with an ideal gas at a pressure p_1 is connected via a closed valve to an evacuated bulb of the same volume. Both bulbs are immersed in a thermostat which maintains the system at constant T. On opening the valve, the volume of the ideal gas increases from V to $2V$ without any change in T. Calculate the change in the entropy of the ideal gas as it expands isothermally from V to $2V$. Comment on the sign of ΔS.

Exercise 2.3-3

Derive the equation

$$\left(\frac{\partial U}{\partial V}\right)_T = T^2\left(\frac{\partial(p/T)}{\partial T}\right)_V$$

For condensed matter, T and p are more convenient independent variables than are T and V. For $S(T,p)$

$$dS = \left(\frac{\partial S}{\partial T}\right)_p dT + \left(\frac{\partial S}{\partial p}\right)_T dp \tag{11}$$

(2.2.9)
$$dS = \frac{dH}{T} - \frac{V}{T} dp$$

$$= \frac{1}{T}\left[\left(\frac{\partial H}{\partial T}\right)_p dT + \left(\frac{\partial H}{\partial p}\right)_T dp\right] - \frac{V}{T} dp \tag{12}$$

(12), (1.6.3)
$$= \frac{C_p}{T} dT + \frac{1}{T}\left[\left(\frac{\partial H}{\partial p}\right)_T - V\right] dp \tag{13}$$

(11), (13)
$$\left(\frac{\partial S}{\partial T}\right)_p = \frac{C_p}{T} \tag{14}$$

(11), (13)
$$\left(\frac{\partial S}{\partial p}\right)_T = \frac{1}{T}\left[\left(\frac{\partial H}{\partial p}\right)_T - V\right] \tag{15}$$

(15), (2.2.15)
$$\left(\frac{\partial H}{\partial p}\right)_T = -T\left(\frac{\partial V}{\partial T}\right)_p + V \tag{16}$$

which is a second *thermodynamic equation of state* (*cf.* Eq. (2.3.8)). It expresses the variation of the enthalpy H with pressure in terms of the variables p, V, and T

(13), (16)
$$dS = \frac{C_p}{T} dT - \alpha V\, dp \tag{17}$$

where α is the expansivity

(17)
$$\Delta S = \int_{T_1}^{T_2} C_p\, d\ln T - \int_{p_1}^{p_2} \alpha V\, dp \tag{18}$$

so that the change in S for prescribed changes in T and p may be calculated, using experimental values of C_p and α.

Exercise 2.3-4

What is the final pressure of the ideal gas in the experiment described in Exercise 2.3-1? Calculate the entropy change in the ideal gas from Eq. (18).

Exercise 2.3-5

Starting from the definition of the enthalpy, Eq. (2.2.6), derive the equation

$$C_p - C_V = \frac{\alpha^2 T V}{\kappa_T} \tag{19}$$

Evaluate $C_p - C_V$ for an ideal gas. (**Hint:** Use the thermodynamic equation of state, Eq. (2.3.8). Make a units check on Eq. (19).)

Answers to Exercises 2.3

Exercise 2.3-1

From the first thermodynamic equation of state

$$(8) \qquad \left(\frac{\partial U}{\partial V}\right)_T = T\left(\frac{\partial p}{\partial T}\right)_V - p = 0, \quad \text{if } pV = nRT$$

therefore U, which in general is $U(T, V)$, is a function of T only for an ideal gas obeying $pV = nRT$.

Exercise 2.3-2

For an ideal gas, $V = nRT/p$. Therefore, $\alpha = 1/T$, $\kappa_T = 1/p$ and

$$(10) \qquad \Delta S = nR \int_V^{2V} d\ln V = nR \ln 2 > 0 \tag{20}$$

T is constant and so $\Delta U = 0$, and since the gas expands into a vacuum, $w = 0$. Therefore, $q = 0$ and consequently $\Delta S_0 = 0$. This is a spontaneous process and so $\Delta S_{\text{net}} = \Delta S$, is greater than zero, as expected.

Exercise 2.3-3

$$(8) \qquad \left(\frac{\partial U}{\partial V}\right)_T = T\left(\frac{\partial p}{\partial T}\right)_V - p = T^2\left(\frac{\partial(p/T)}{\partial T}\right)_V$$

Exercise 2.3-4

The ideal gas doubles its volume without any change in T. Therefore, its final pressure $p_2 = p_1/2$

$$(18) \qquad \Delta S = -\int_{p_1}^{p_2} \frac{nR}{p} dp = -nR \ln\frac{1}{2} = nR\ln 2 \tag{21}$$

in agreement with (20).

Exercise 2.3-5

(2.2.6)
$$\left(\frac{\partial H}{\partial T}\right)_p = \left(\frac{\partial U}{\partial T}\right)_p + p\left(\frac{\partial V}{\partial T}\right)_p$$

$$= \left(\frac{\partial U}{\partial T}\right)_V + \left(\frac{\partial U}{\partial V}\right)_T\left(\frac{\partial V}{\partial T}\right)_p + p\left(\frac{\partial V}{\partial T}\right)_p$$

$$C_p - C_V = \left\{T\left(\frac{\partial p}{\partial T}\right)_V - p\right\}\left(\frac{\partial V}{\partial T}\right)_p + p\left(\frac{\partial V}{\partial T}\right)_p$$

$$= -T\left(\frac{\partial p}{\partial V}\right)_T\left(\frac{\partial V}{\partial T}\right)_p\left(\frac{\partial V}{\partial T}\right)_p$$

$$= \frac{\alpha^2 T V}{\kappa_T}$$

The *SI* units of the RS are $\mathrm{N\,m^{-2}\,K^{-2}\,K\,m^3} = \mathrm{J\,K^{-1}}$, which are the *SI* units of heat capacity. For an ideal gas

(19) $$C_p - C_V = n R$$ (22)

2.4. Physical Meaning of the Thermodynamic Potentials

The enthalpy. For a change in state, from state one to state two, of a system in equilibrium with its surroundings at constant pressure ($p_0 = p$), if the only work done is that associated with the reversible change in volume of the system, then

(2.2.6) $$\Delta H = \Delta U + p\,\Delta V = q$$ (1)

Thus the enthalpy change in a closed system is equal to the heat absorbed at constant pressure, provided that the only work done is that due to the reversible change in volume of the system.

The Helmholtz energy. For a change in state of a closed system in equilibrium with its surroundings at constant temperature $T_0 = T$

(2.2.7) $$\Delta F = \Delta U - T\,\Delta S$$ (2)

(2.1.2), (2.1.3), $\Delta F \leq w$ (isothermal process, closed system)   (3)

(3) $\Delta F = w$ (reversible isothermal process, closed system)
                                                                              (4)

In mechanics, the change in the potential energy of a body is equal to the work done on that body and the clear analogy with (4), which states that the work done on a closed system in a reversible isothermal process is equal to the change in the Helmholtz energy, and is the reason for the name *thermodynamic potential*.

Gibbs energy. For a closed system in thermal and mechanical equilibrium with its surroundings $(T = T_0, p = p_0)$

(2.2.8) $$\Delta G = \Delta U + p\Delta V - T\Delta S \qquad (5)$$

(2.1.2), (2.1.3) $$\Delta G \leq w' \quad \text{(constant } T \text{ and } p, \text{ closed system)} \qquad (6)$$

where w' (called the *net work*) is the work done on a system in an isothermal process, excluding the $p\Delta V$ mechanical work associated with the change in volume of the system. If the process is completely reversible $(d_iS = 0)$

(6) $$\Delta G = w' \quad \text{(reversible process, constant } T \text{ and } p, \text{ closed system)} \qquad (7)$$

Phase changes. A crystalline solid at constant pressure is a univariant system, that is, it has one degree of freedom, which we take to be temperature. If the solid is heated at constant pressure, it melts at $T = T_{\text{fus}}$ and T remains constant until the solid has melted completely. The heat absorbed during melting is the enthalpy of fusion, $\Delta_{\text{fus}}H$ and its value per mole of B is

$$\Delta_{\text{fus}}H_{\text{m}}(B) = H_{\text{m}}(B, \text{ liq}) - H_{\text{m}}(B, \text{ cr}) \qquad (8)$$

When the solid has melted completely, the system is bivariant and as it is heated further, the temperature rises until the boiling point, when it again remains constant at T_{vap} until the liquid has evaporated completely. The heat absorbed during vaporization of one mole of B is the molar enthalpy of vaporization

$$\Delta_{\text{vap}}H_{\text{m}}(B) = H_{\text{m}}(B, \text{ g}) - H_{\text{m}}(B, \text{ liq}) \qquad (9)$$

The subscript m may be omitted when it is clear from the content that one mole is intended. Enthalpy has been used here as an example, but the same kind of notation applies to the other thermodynamic potentials.

Exercise 2.4-1

Derive an equation for the difference in heat capacity at constant pressure between the vapour and liquid phases.

Example 2.4-1

The molar enthalpy of vaporization of water at 373.15 K and 101.3 kPa is 40.7 kJ mol^{-1}. Calculate the molar entropy of vaporization of water at this T and p

$$(9) \quad \Delta_{vap}S(H_2O, 373.15\,K) = \frac{\Delta_{vap}H}{T} = \frac{40.7\,\text{kJ mol}^{-1}}{373.15\,K} = 109\,\text{J K}^{-1}\,\text{mol}^{-1}$$

Example 2.4-2

The temperature in a boiler containing superheated water under steam at atmospheric pressure is 110°C. Calculate ΔS and ΔS_0 for the vaporization of one mole of this superheated water

$$\Delta_{vap}H(H_2O, 373.15\,K, 101.3\,\text{kPa}) = 47.3\,\text{kJ mol}^{-1}$$

$$C_p(H_2O(g)) = (30.36 + 9.61 \times 10^{-3}T + 1.18 \times 10^{-6}\,T^2)\,\text{J K}^{-1}\,\text{mol}^{-1}$$

$$C_p(H_2O(liq)) = 75.4\,\text{J K}^{-1}\,\text{mol}^{-1}$$

The system:

The vaporization of water at a temperature 10°C above its normal boiling point is an irreversible process that must be replaced by an alternative (purely conceptual) reversible path.

(1) Cool the liquid water reversibly (i.e. so that T differs only infinitesimally from T_0) from 383.15 K to 373.15 K. The entropy change (see Eq. (2.3.18)) is

$$\Delta S_1 = \int_{383}^{373} C_p(\text{liq})\, \text{d} \ln (T/K)$$

$$= 75.4 \ln (373/383)\,\text{J K}^{-1}\,\text{mol}^{-1} = -1.995\,\text{J K}^{-1}\,\text{mol}^{-1}$$

(2) Vaporize one mole of liquid water reversibly at 373 K and 101.3 kPa

$$\Delta S_2 = \Delta_{vap}H(H_2O, 373\,K, 101.3\,\text{kPa})/T$$

$$= 47.3\,\text{kJ mol}^{-1}/373\,K = 126.8\,\text{J K}^{-1}\,\text{mol}^{-1}$$

(3) Heat the vapour reversibly at constant pressure

$$\Delta S_3 = \int_{373}^{383} C_p(g) \, d\ln(T/K)$$

$$= [30.36 \ln(383/373) + 9.61 \times 10^{-3}(383 - 373)$$

$$+ 5.9 \times 10^{-7}(383^2 - 373^2)] \, J\,K^{-1}\,mol^{-1}$$

$$= 0.903 \, J\,K^{-1}\,mol^{-1}$$

$$\Delta S_1 + \Delta S_2 + \Delta S_3 = (-1.995 + 126.8 + 0.903) \, J\,K^{-1}\,mol^{-1}$$

$$= 125.7 \, J\,K^{-1}\,mol^{-1}$$

The surroundings:

$$\Delta_{vap}C_p = (-45.0 + 9.61 \times 10^{-3}\,T + 1.18 \; 10^{-6}\,T^2)\,J\,K^{-1}\,mol^{-1}$$

For H_2O at $101.3\,kPa$

$$\Delta_{vap}H(383\,K) = \Delta_{vap}H(373\,K) + \int_{373\,K}^{383\,K} \Delta_{vap}C_p\,dT$$

$$= 47\,300 - 45.0(383 - 373) + 4.805 \times 10^{-3}(383^2 - 373^2)$$

$$+ 3.93 \times 10^{-7}(383^3 - 373^3) \, J\,mol^{-1}$$

$$= 46\,890 \, J\,mol^{-1}$$

$$\Delta S_0 = -122.4 \, J\,K^{-1}\,mol^{-1}$$

$$\Delta S + \Delta S_0 = +3.3 \, J\,K^{-1}\,mol^{-1}$$

The net entropy change is positive, as expected for an irreversible process.

Answer to Exercise 2.4-1

Experimental data for C_p may be described by a power series in T

$$C_p = a + bT + cT^2 \tag{10}$$

(Please note that other expressions are also used and a commonly used one will be given in Chapter 8.)

$$\frac{\partial \Delta_{vap}H}{\partial T} = \frac{\partial}{\partial T}(H(g) - H(liq)) = C_p(g) - C_p(liq) = \Delta_{vap}C_p \tag{11}$$

(10), (11) $\quad \Delta_{vap}C_p = \Delta a + \Delta bT + \Delta cT^2 \tag{12}$

The same type of equation applies to other physical and chemical changes. The subscript on Δ is omitted when the process under consideration is not in doubt.

2.5. Conditions for Equilibrium in a Closed System

For a system in thermal equilibrium with its surroundings $(T = T_0)$

$$dS = \frac{dq}{T} + d_i S \tag{1}$$

where $d_i S \geq 0$. For a system in mechanical equilibrium with its surroundings $(p = p_0)$ and with p_0 the only significant force, $dw = -p\,dV$ and so

(1) $$dS = \frac{dU + p\,dV}{T} + d_i S \tag{2}$$

At constant U and V

(2) $$dS = d_i S \geq 0 \tag{3}$$

At constant S and V

(2) $$dU = -T\,d_i S \leq 0 \tag{4}$$

(2), (2.2.6) $$dS = \frac{dH - V\,dp}{T} + d_i S \tag{5}$$

At constant S and p

(5) $$dH = -T\,d_i S \leq 0 \tag{6}$$

(2) $$dS = \frac{dF + T\,dS + S\,dT + p\,dV}{T} + d_i S \tag{7}$$

At constant T and V

(7) $$dF = -T\,d_i S \leq 0 \tag{8}$$

(7) $$dS = \frac{dG + T\,dS + S\,dT - V\,dp}{T} + d_i S \tag{9}$$

At constant T and p

(9) $$dG = -T\,d_i S \leq 0 \tag{10}$$

These Eqs. (3), (4), (6), (8) and (10) state the condition for equilibrium in a closed system in which the only kind of work done on the system is

$$dw = -p\,dV, \quad dw' = 0 \tag{11}$$

To summarize, for a closed system in thermal ($T = T_0$) and mechanical ($p = p_0$) equilibrium with its surroundings, the condition for internal equilibrium is:

$$\left.\begin{array}{l} \text{for prescribed } U \text{ and } V, \text{ that } S(U,V) \text{ is a maximum,} \\ \text{or, for prescribed } S \text{ and } V, \text{ that } U(S,V) \text{ is a minimum,} \\ \text{or, for prescribed } S \text{ and } p, \text{ that } H(S,p) \text{ is a minimum,} \\ \text{or, for prescribed } T \text{ and } V, \text{ that } F(T,V) \text{ is a minimum,} \\ \text{or, for prescribed } T \text{ and } p, \text{ that } G(T,p) \text{ is a minimum.} \end{array}\right\} \tag{12}$$

These statements in (12) justify the name "thermodynamic potentials" for the functions of state S, U, H, F and G. If other forms of work are involved, then the statements in (12) still hold, provided the corresponding variables are prescribed too. Then

$$dw = -p\,dV + dw' = -p\,dV + \sum_i X_i\,dx_i \tag{13}$$

where X_i is a generalized force and dx_i the corresponding displacement. Then

(10) $$dG = -T\,d_iS + dw' \leq 0 \tag{14}$$

and if the system is in internal equilibrium

(14) $$dG = dw' \quad \text{(at constant } T \text{ and } p) \tag{15}$$

Thus the change in the Gibbs energy of a system is equal to the net work (i.e. other than $-p\,dV$) done on the system in a reversible, isothermal process at constant pressure.

Exercise 2.5-1

Equation (2.4.4) states that, when the only kind of work done on a system is that associated with an external pressure, the change in the Helmholtz energy is equal to the work done on the system in a reversible isothermal process. What differences arise when other kinds of work are involved?

Answer to Exercise 2.5-1

The change in internal energy during a reversible process is

$$dU = dq - p \, dV + dw' \tag{16}$$

(16), (7)
$$dS = \frac{dF + T \, dS + S \, dT + p \, dV - dw'}{T} \tag{17}$$

At constant T and V

(17)
$$dF = dw' \quad \text{(at constant T and V)} \tag{18}$$

so that the change in the Helmholtz energy in a reversible isothermal process at constant volume is equal to the net work done on the system. Notice the difference between (18) and (15).

2.6. Entropy of Mixing

For ideal gases, intermolecular forces are negligible and the mixing of two ideal gases at constant pressure takes place without any change in internal energy or temperature, so that the entropy change ΔS is determined only by the larger volume $V_A + V_B$ that is accessible to the molecules of both A and B. Since p and T are constant and $V = n \, R \, T / p$, on using Exercise 2.3-1

(2.3.10)
$$\Delta S = -R[x_A \ln x_A + x_B \ln x_B] \tag{1}$$

Formula (1) is of wider application than just ideal gases. If the molecules of species A and B are very similar, as are isotopes, then the intermolecular forces between the molecules of A and B will be almost the same as those between A and A molecules and those between B and B molecules. Consequently, any change in internal energy on mixing will be negligible and mixing will therefore occur without change in T or p and formula (1) can still be used even though the gases are not strictly ideal. For this reason, Eq. (1) is commonly referred to as the *entropy change on isotopic mixing.* It may also be applied to ideal liquid mixtures and to mixed crystals, when the forces between the particles (atoms, ions or molecules) are sufficiently similar.

2.7. Thermodynamic Potentials for an Open Phase

The internal energy U of an open phase depends on the amount of each substance B present, so that $U(S, V, n_B)$ and

$$dU = \left(\frac{\partial U}{\partial S}\right)_{V,n_B} dS + \left(\frac{\partial U}{\partial V}\right)_{S,n_B} dV + \sum_B \left(\frac{\partial U}{\partial n_B}\right)_{S,V,n_{A \neq B}} dn_B \qquad (1)$$

$(1), (2.2.2)$
$$dU = T\,dS - p\,dV + \sum_B \mu_B\,dn_B \qquad (2)$$

where

$$\mu_B = \left(\frac{\partial U}{\partial n_B}\right)_{S,V,n_{A \neq B}} \qquad (3)$$

is the *chemical potential* of substance B. The sums in (1) and (2) are over all the r components B present in the system. Rearranging (2), gives

$$dS = \frac{1}{T}\,dU + \frac{p}{T}\,dV - \sum_B \frac{\mu_B}{T}\,dn_B \qquad (4)$$

Differentiating $H(= U + pV)$, $F(= U - TS)$ and $G(= U + pV - TS)$ and substituting for dU from (2), gives

$$dH = T\,dS + V\,dp + \sum_B \mu_B\,dn_B \qquad (5)$$

$$dF = -S\,dT - p\,dV + \sum_B \mu_B\,dn_B \qquad (6)$$

$$dG = -S\,dT + V\,dp + \sum_B \mu_B\,dn_B \qquad (7)$$

which are the fundamental equations for open phases. Each term in Eqs. (2) and (4)–(7) is linearly dependent on the amount of each substance in the phase. If a system consists of more than one phase α, then one simply needs to sum the RS of any of these fundamental equations over α, to obtain the infinitesimal change in that thermodynamic potential for the whole system. The same statement applies, of course, to the fundamental equations for a closed phase in Section 2.2.

Exercise 2.7-1

Obtain alternative definitions for the chemical potential of species B from Eqs. (4)–(7). Explain how all four definitions can be valid.

The Gibbs–Duhem equation

Since $U(S, V, n_B)$, and S, V, n_B are extensive properties

$$U(k\,S, k\,V, k\,n_B) = k\,U(S, V, n_B) \qquad (8)$$

Equation (8) shows that U is a homogeneous function of degree 1 in the variables S, V, n_B. Therefore, by Euler's theorem (Section A1)

$$U = \left(\frac{\partial U}{\partial S}\right)_{V, n_B} S + \left(\frac{\partial U}{\partial V}\right)_{S, n_B} V + \sum_B \left(\frac{\partial U}{\partial n_B}\right)_{S, V, n_{A \neq B}} n_B \qquad (9)$$

$$= TS - pV + \sum_B \mu_B\, n_B \qquad (10)$$

(10) $$G = \sum_B \mu_B\, n_B \qquad (11)$$

From (11), the complete differential of G is

$$dG = \sum_B \mu_B\, dn_B + \sum_B n_B\, d\mu_B \qquad (12)$$

(12), (6) $$S\,dT - V\,dp + \sum_B n_B\, d\mu_B = 0 \qquad (13)$$

which is the Gibbs–Duhem equation. It is a relation between infinitesimal variations in T, p and the chemical potentials of all the r components in the system and thus states that only $r + 1$ of the $r + 2$ intensive properties T, p, μ_B are independent. In analogy with (2) and (4)–(7), (13) could be regarded as the fundamental equation for the thermodynamic potential zero, $0(T, p, \mu_B)$.

The partial derivative of any extensive property X with respect to the amount of substance B, when T, p and the amounts of all other substances are held constant, is called the *partial molar X*. Thus an alternative name for the chemical potential of B would be the partial molar Gibbs energy of B. However, for reasons that will become clear later, the name chemical potential is preferred in this instance.

Exercise 2.7-2

Obtain a generalized form of the Gibbs–Duhem equation that is valid for any extensive property $X(T, p, n_B)$.

Exercise 2.7-3

When the independent variables are T, p and μ_B, there is a relation between these intensive variables called the Gibbs–Duhem equation. What relation is there between intensive variables if these are chosen to be T, p, x_B?

Answers to Exercises 2.7

Exercise 2.7-1

From (2)

$$\mu_B = \left(\frac{\partial U}{\partial n_B}\right)_{S,V,n_{A \neq B}} \tag{3}$$

Similarly from (4), (5) and (6)

$$\mu_B = \left(\frac{\partial H}{\partial n_B}\right)_{S,p,n_{A \neq B}} = \left(\frac{\partial F}{\partial n_B}\right)_{T,V,n_{A \neq B}} = \left(\frac{\partial G}{\partial n_B}\right)_{T,p,n_{A \neq B}} \tag{14}$$

The chemical potential is equal to the partial derivative of four different thermodynamic functions with respect to the amount of B, because different pairs of variables from S, T, p, V are held constant during the differentiation with respect to n_B.

Exercise 2.7-2

If X denotes any extensive function, then the corresponding partial molar function X_B is defined by

$$X_B = \left(\frac{\partial X}{\partial n_B}\right)_{T,p,n_{A \neq B}} \tag{14}$$

Partial molar properties are intensive properties with temperature, pressure and composition as independent variables. The complete differential of

$X(T, p, n_B)$ is

$$dX = \left(\frac{\partial X}{\partial T}\right)_{p,n_B} dT + \left(\frac{\partial X}{\partial p}\right)_{T,n_B} dp + \sum_B \left(\frac{\partial X}{\partial n_B}\right)_{T,p,n_{A\neq B}} dn_B \qquad (15)$$

At constant T, p

$$(15),\ (14) \qquad\qquad dX = \sum_B X_B\, dn_B \qquad\qquad (16)$$

Since X is an extensive property and thus a homogeneous function of degree one in the n_B, we have by Euler's theorem

$$X = \sum_B X_B\, n_B \qquad (17)$$

$$(17) \qquad\qquad dX = \sum_B (X_B\, dn_B + n_B\, dX_B) \qquad\qquad (18)$$

$$(18),\ (15) \qquad \sum_B n_B\, dX_B - \left(\frac{\partial X}{\partial T}\right)_{p,n_B} dT - \left(\frac{\partial X}{\partial p}\right)_{T,n_B} dp = 0 \qquad (19)$$

which is the generalized Gibbs–Duhem equation.

Exercise 2.7-3

By definition of the mole fractions, $\sum_B x_B = 1$.

2.8. Other Forms of the Second Law and the Unattainability of $T = 0\,K$

A *heat engine* is a device which, operating in a cycle, absorbs heat q_1 from a reservoir at T_1, performs mechanical work $-w$ and rejects heat $|q_2|$ to a reservoir at T_2 (Figure 2.1a). q_2 is a negative quantity because of our definition of q as the *heat absorbed* by a system. A reservoir is a large system that maintains a constant temperature. The *efficiency* of a heat engine is defined by

$$\eta = \frac{-w}{q_1} = 1 + \frac{q_2}{q_1} \qquad (1)$$

which is ≤ 1, because q_2 is a negative quantity.

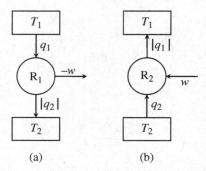

(a) (b)

Figure 2.1.   (a) A heat engine R_1 operates in a reversible cycle in which it absorbs heat q_1 from a reservoir at T_1, it does work $-w = q_1 + q_2$ and rejects heat $|q_2|$ at T_2. R_1 is the thermodynamic system. (b) R_2 is an engine identical with R_1 that operates the same cycle in reverse. R_2 absorbs heat q_2 at T_2, and rejects heat of amount q_1 at T_1. The amount of work done on the system is $w = -(q_1 + q_2)$. (q_1 is now a negative number and q_2 is positive, with $|q_1| > |q_2|$.)

Exercise 2.8-1

Confirm the second equality in Eq. (1). (**Hint**: Equation (1) refers to a system that is operating in a cycle.)

Since U is a function of state, $\Delta U = 0$ for any cyclic process and, therefore, from the First Law, $q + w = 0$. Consequently,

$$\eta = \overline{\overline{q_1}}w = \frac{q_1 + q_2}{q_1} = 1 + \frac{q_2}{q_1} \qquad (1)$$

Remember that q_2 is a negative quantity.

Sadi Carnot (1796–1832) not only realized that cyclic operation was necessary to produce a continuous amount of work, but also that the absorption and rejection of heat and the performance of work had to be carried out reversibly in order to achieve maximum efficiency (Mendoza 1977). For an engine operating in a cycle, each step of which is reversible, the performance of work w on R_2 and the absorption of heat q_2 at T_2 results in the rejection of heat $|q_1|$ to the hot reservoir (Figure 2.1b).

The maximum efficiency of a heat engine (unity) would be obtained if q_2 were zero, but experience shows this to be impossible to achieve. Similarly, when the engine is operated in reverse, it is not possible for w to be zero (Figure 2.1b). Both these idealized situations are impossible and these facts formed the basis for two alternative statements of the Second Law due to Kelvin and to Clausius, respectively, which for brevity we shall call K and C.

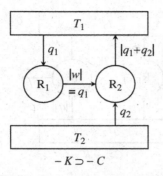

$$-K \supset -C$$

Figure 2.2. This illustrates that $-K \supset -C$. The device $R_1 + R_2$ would absorb heat q_2 at T_2 and reject the same net amount of heat at $T_1 > T_2$, which is $-C$.

K: "It is impossible to construct a device that, operating in a cycle, will produce no effect other than the extraction of heat from a reservoir and the performance of an equivalent amount of work."

C: "It is impossible to construct a device that, operating in a cycle, will produce no effect other than the flow of heat from a colder to a hotter body."

These two statements are equivalent, as we shall now prove, in the manner of Zemansky and Dittman (1997). Suppose that q_2 can be zero for a heat engine, so that K is incorrect (which we denote by $-K$). Let the work produced by the engine R_1 drive a similar engine R_2 in reverse between the same two reservoirs (Figure 2.2). Then the two engines together constitute a device that transfers a net amount of heat q_2 from the cold reservoir to the hot one. Therefore

$$-K \supset -C \tag{2}$$

Now suppose that C is incorrect $(-C)$, which implies that w can be zero for the engine operating the same cycle in reverse, which thus transfers q_2 from T_2 to T_1 (R_1 in Figure 2.3). Let the heat engine R_2 (in Figure 2.3) absorb heat q_1 and reject heat $|q_2|$ so that it does work $-w = q_1 - q_2$. Thus the two engines together constitute a device that absorbs energy $q_1 - q_2$ and converts it completely into work. Thus the failure of C implies the failure of K,

$$-C \supset -K \tag{3}$$

Therefore, C and K are logically equivalent and either both statements are true or both are untrue.

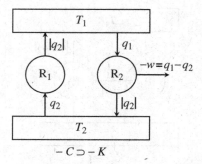

Figure 2.3.   This illustrates that $-C \supset -K$. R_1 absorbs heat q_2 at T_2 and rejects the same amount of heat at T_1, which (if it were possible) would be a statement that $-C$ holds. $R_1 + R_2$ is a device that absorbs heat $q_1 - q_2$ at T_1 and performs an equal amount of work $-w = q_1 - q_2$.

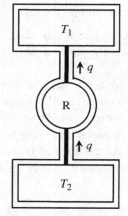

Figure 2.4.   This illustrates the derivation of $+C$ from the statement of the Second Law in Section 2.3. R is a heat engine that operates in a reversible cycle.

Now consider the Clausius statement from the point of view of our statement of the Second Law in Section 2.1. Consider the "thought" experiment illustrated in Figure 2.4, with two large, adiabatically isolated reservoirs at temperatures T_1 and $T_2 < T_1$. (This hypothetical process is the reverse of the natural process described in Section 2.1 (ii).) These two reservoirs comprise the surroundings of the system R, which operates in a reversible cycle. The system R is adiabatically insulated except for being in thermal contact with the two reservoirs. The work done on the system $w = 0$, as in the prescription laid down by Clausius. If this were a feasible process, the entropy change in the system, per cycle, would be

$$\Delta S = \frac{q}{T_2} - \frac{q}{T_1} > 0 \tag{4}$$

which is impossible, since S is a function of the state of the system, and ΔS (system) is zero for any reversible cyclic path. Therefore, the hypothesis that w is zero, is invalid and the Clausius statement follows. But since $C \supset K$, the Kelvin statement also follows from our statement of the Second Law.

The maximum efficiency of a heat engine is obtained by one that operates in a *Carnot cycle*, which consists of the following four steps:

(i) reversible isothermal absorption of heat q_1 at temperature T_1;
(ii) reversible adiabatic change in state to a lower temperature T_2;
(iii) reversible isothermal rejection of heat $\bar{q}_2 = -q_2$ at temperature T_2;
(iv) reversible adiabatic change in state from T_2 to T_1.

The efficiency of the Carnot engine is

(1) $$\eta = \frac{-w}{q_1} = 1 + \frac{q_2}{q_1} \quad (q_2 < 0) \tag{5}$$

Because S is a function of state, the entropy change in the system per cycle is

$$\Delta S = \frac{q_1}{T_1} + \frac{q_2}{T_2} = 0 \tag{6}$$

(5), (6) $$\eta = \frac{-w}{q_1} = 1 - \frac{T_2}{T_1} < 1 \quad (T_2 < T_1) \tag{7}$$

If T_2 could be zero and T_1 finite in (7), then η would be unity, which from (5) requires q_2 also to be zero, in defiance of the Kelvin statement of the Second Law.

Therefore, a *temperature of zero is unattainable* in systems in equilibrium by using a device operating in a Carnot cycle, or indeed in *any* reversible cycle, since the Carnot cycle delivers the maximum possible efficiency.

In Chapter 4 we shall see that even the efficient methods that have been developed for reaching very low temperatures have failed to get to 0 K, in accord with theoretical expectations. Temperatures in the micro-kelvin range, and even lower, have been obtained in systems at equilibrium using the technique of *adiabatic demagnetization* (Chapter 4). So-called negative temperatures refer to systems that are not at equilibrium (Pippard 1966, Zemansky and Dittman 1997).

Answer to Exercise 2.8-1

Since U is a function of state, $\Delta U = 0$ for any cyclic process and therefore, from the First Law, $q + w = 0$. Consequently,

$$\eta = \frac{-w}{q_1} = \frac{q_1 + q_2}{q_1} = 1 + \frac{q_2}{q_1} \tag{1}$$

Remember that q_2 is a negative quantity.

Problems 2

2.1 (a) A gas expands adiabatically, doing work against the external pressure p_0. Is the final temperature of the gas higher or lower than its initial temperature?

   (b) Derive an equation which describes adiabatic, reversible changes in the state of an ideal gas. Hence prove the following relations, which hold for a reversible, adiabatic change in an ideal gas from state 1 to state 2
   (i) $p_1 V_1^\gamma = p_2 V_2^\gamma$; (ii) $T_1 V_1^{\gamma-1} = T_2 V_2^{\gamma-1}$; (iii) $\frac{T_2}{T_1} = \left(\frac{p_2}{p_1}\right)^{(\gamma-1)/\gamma}$
   where $\gamma = (C_p/C_V)$.

   (c) Using 2.1(a)(i), show that if an ideal gas expands (i) isothermally, and (ii) adiabatically from an initial state $p_1 V_1$, then the adiabatic in the pV plane is steeper than the corresponding isothermal (from the same initial state).

2.2 Two moles of an ideal gas expand reversibly and isothermally at 300 K to a state in which the pressure is one tenth of its original value before the expansion. Calculate the change in the Gibbs energy of the gas for this process. If the same change in state were to be carried out irreversibly, would ΔG be larger or smaller than the value found for the reversible expansion? Give reasons for your answer.

2.3 At constant temperature, the internal energy U of an ideal gas is independent of pressure. Does the same statement apply to the enthalpy of an ideal gas?

2.4 (a) For a gas at moderate pressures ($p \leq \frac{1}{2} p_c$) the virial equation

$$pV_m/RT = 1 + B_2(T)/V_m + B_3(T)/V_m^2 + \cdots \tag{1.3.8}$$

can be written in the form

$$pV_m = RT[1 + B'(T)p] \tag{1}$$

The subscript 2 is unnecessary here, and has been dropped for convenience. Show that, to order V_m^{-1}

$$B'(T) = B(T) \tag{2}$$

(b) Derive the equation

$$\left(\frac{\partial U}{\partial p}\right)_T = -T\left(\frac{\partial V}{\partial T}\right)_p - p\left(\frac{\partial V}{\partial p}\right)_T \tag{3}$$

(**Hint:** Use the thermodynamic equation of state.)

(c) Show that, in the pressure range in which (1) applies

$$\left(\frac{\partial U}{\partial p}\right)_T = -T\frac{dB(T)}{dT} \tag{4}$$

(d) The second virial coefficient for oxygen in the temperature range around room temperature is given by

$$B(T) = B_0 - B_1/RT \tag{5}$$

where B_0 is a constant and $B_1 = 0.1515\,\text{J}\,\text{m}^3\,\text{mol}^{-2}$. Calculate the value of $(\partial U/\partial p)_T$ for oxygen at $298\,\text{K}$.

(e) The RS of (4) is the quantity $-c(T)$ which is measured directly in Washburn's experiment (see Eq. (1.6.8)). At $298\,\text{K}$ the value of $-c(T)$ for oxygen is found to be $-6.13 \times 10^{-5}\,\text{m}^3\,\text{mol}^{-1}$. Compare this directly measured value with that calculated in (d) from virial coefficient data. What is the significance of the results of these two experiments with respect to the Second Law of Thermodynamics?

2.5 Complete the following equations for a closed system ($dn_B = 0$, $\forall$ B)

$$\left(\frac{\partial S}{\partial U}\right)_V =; \quad \left(\frac{\partial S}{\partial V}\right)_U =; \quad \left(\frac{\partial H}{\partial S}\right)_p =; \quad \left(\frac{\partial H}{\partial p}\right)_S =;$$

$$\left(\frac{\partial F}{\partial V}\right)_T =; \quad \left(\frac{\partial F}{\partial T}\right)_V =; \quad \left(\frac{\partial G}{\partial p}\right)_T =; \quad \left(\frac{\partial G}{\partial T}\right)_p =$$

2.6 A Carnot engine is one that operates in a cycle, every step of which is reversible. Prove Carnot's theorem that no heat engine operating between two reservoirs can be more efficient that a Carnot engine.

2.7 Calculate η for a Carnot engine that uses an ideal gas as its working substance.

2.8 Prove that all Carnot engines have the same efficiency, irrespective of the working substance. (**Hint**: Consider two reversible engines R_1 and R_2.)

2.9 A heat pump operates like a heat engine in reverse, but with the object of supplying as much heat $-q_1$ as possible to the high-temperature reservoir (e.g. a house) at T_1, while expending as little work w as possible. Calculate the minimum amount of work needed to transfer 1 kJ from outside a house (at $-5°C$) to inside the house at $23°C$.

2.10 Draw a T-S diagram for the Carnot cycle.

2.11 One mole of water at 101.3 kPa is cooled to $-10°C$, when it suddenly freezes. Calculate the net entropy change for the irreversible process

$$H_2O(\text{liq}, 263.16\,\text{K}, 101.3\,\text{kPa}) = H_2O(\text{ice}, 263.16\,\text{K}, 101.3\,\text{kPa})$$

C_p (water) $= 75.3\,\text{J K}^{-1}\text{mol}^{-1}$; C_p (ice) $= 36.9\,\text{J K}^{-1}\text{mol}^{-1}$;
$\Delta H_{\text{fus}}(273.16\,\text{K}, 101.3\,\text{kPa}) = 5\,950\,\text{J mol}^{-1}$.

2.12 The Massieu function $J(T, V)$ and the Planck function $Y(T, p)$ are defined by

$$J = -\frac{F}{T}, \quad Y = -\frac{G}{T} \tag{1}$$

Find the fundamental equations for a closed phase for the potentials J and Y.

3

An Introduction to Statistical Thermodynamics

3.1. Need for a Microscopic Description

Thermodynamics is a macroscopic science. Systems that contain large numbers of particles are described by properties such as temperature, pressure, volume, energy, which are properties of the whole system. The numbers of particles in such systems are so large that it is possible to define small regions of a system which still contain large numbers of particles and to which the Laws of Thermodynamics may be applied. Thus it is perfectly proper to discuss gradients of temperature or concentration or chemical potential in a non-uniform system. Such systems will be dealt with later, but for the present we are concerned with uniform systems containing large numbers of particles. The energies of these particles are distributed over the available quantum states. The details of these distributions depend on the nature of the particles and are different for the kinetic energy of gas molecules (the Maxwell–Boltzmann distribution), the kinetic energy of the electrons in a metal (the Fermi–Dirac distribution), photons and the quasi-particles (phonons) which are quanta of vibrational energy in a crystal (the Bose–Einstein distribution). In the first two chapters, the first three Laws of Thermodynamics (Zeroth, First and Second) were stated. Each of these was associated with the introduction of a thermodynamic property (the temperature T, the internal energy U and the entropy S). From these laws, the fundamental equations for closed and open phases were developed, the thermodynamic potentials $H(S,p)$, $F(T,V)$ and $G(T,p)$ defined and criteria for equilibrium established. It seemed at this stage we were ready to begin the application of thermodynamics to many kinds of physical and chemical systems.

However, new insights are obtained by adopting the *microscopic viewpoint* in which systems are described in terms of the properties of the individual particles of which they are composed. Since the number of these particles is extremely large, a statistical viewpoint is essential, based on the energy states of the individual particles as calculated by the methods of quantum mechanics. This is the approach adopted in *statistical thermodynamics*. In the next section we describe some basic ideas in quantum mechanics, giving only the information required for the present need.

3.2. Postulates of Quantum Mechanics

We state only the minimum necessary at this stage.

I The stationary (time independent) states of a system of N particles are described as completely as possible by the *state function* $\psi(x_i, y_i, z_i), i = 1, 2, \ldots, N$. State functions are *well-behaved*, that is, finite, continuous and single-valued. The way in which ψ describes the state of a system is that the probability that x_i will lie between x_i and $x_i + \mathrm{d}x_i$, y_i between y_i and $y_i + \mathrm{d}y_i$ and z_i between z_i and $z_i + \mathrm{d}z_i$, etc., is $\psi^*\psi\mathrm{d}\tau$, where ψ^* is the complex conjugate of ψ and the volume element

$$\mathrm{d}\tau = \mathrm{d}x_1\mathrm{d}y_1\mathrm{d}z_1\mathrm{d}x_2 \ldots \mathrm{d}z_N = \prod_{i=1}^{N} \mathrm{d}x_i\mathrm{d}y_i\mathrm{d}z_i \tag{1}$$

A state function may be complex, but $\psi^*\psi = |\psi|^2$ is real and this is the reason why the probability of a particle being within $\mathrm{d}\tau$ is defined in terms of $\psi^*\psi$.

II State functions satisfy a partial differential equation called the Schrödinger equation

$$\hat{H}\psi = E\psi \tag{2}$$

where E is the energy of the system and $\hat{H}$ is the *Hamiltonian operator*. For a single particle

$$\hat{H} = -(\hbar^2/2m)\nabla^2 + V \tag{3}$$

where $V(x, y, z)$ is the potential energy of the particle, m is its mass, and

$$\hbar = h/2\pi = 1.0546 \times 10^{-34}\,\mathrm{J\,s} \tag{4}$$

$$h = \text{Planck constant} = 6.6246 \times 10^{-34}\,\mathrm{J\,s} \tag{5}$$

$$\nabla^2 = \frac{\partial^2}{\partial x^2} + \frac{\partial^2}{\partial y^2} + \frac{\partial^2}{\partial z^2} \tag{6}$$

The solution of any problem involving differential equations includes a statement of the *boundary conditions*, that is, the values which ψ is constrained to have for specific values of the variables. It is a property of Eq. (2) that well-behaved solutions of ψ which satisfy the boundary conditions (henceforth often abbreviated to BC) are possible only for certain values of E, called the *eigenvalues* of E. The description "well-behaved" means that acceptable solutions of (2) for ψ must be finite, continuous, and single-valued, and have continuous first and second partial derivatives. The corresponding solutions ψ are called the *eigenfunctions* of the Hamiltonian operator $\hat{H}$.

For a system containing N particles, the potential energy

$$V = V(x_1, y_1, z_1, x_2, \ldots, z_N) \tag{7}$$

But if the particles are only *weakly-interacting*

$$(7) \qquad V = \sum_{i=1}^{N} V_i(x_i, y_i, z_i) \tag{8}$$

Equation (8) states that the potential energy of a system of N weakly-interacting particles is, or can be approximated by, a sum of terms, each of which depends on the coordinates of one particle only. (For example, the first option applies to an ideal gas, the second to a gas at low pressures.) V_i is the potential energy of the ith particle and if it depends only on the coordinates of that particle (x_i, y_i, z_i), then the Hamiltonian is separable into a sum of terms

$$\hat{H} = \sum_{i=1}^{N} \hat{H}_i = \sum_{i=1}^{N} \left\{ \frac{\hbar^2}{2m_i} \nabla_i^2 + V_i \right\} \tag{9}$$

$$(2).\ (9) \qquad \psi = \prod_{i=1}^{N} \psi_i(x_i, y_i, z_i) \tag{10}$$

Thus if the Hamiltonian is separable into a sum of one-particle Hamiltonians, the solution of the Schrödinger equation is the product (10) of one-particle state functions.

If ψ is a solution of (2) then so is $c\psi$, where c is a constant. But if $\psi_1^* \psi_1\, d\tau$ is the probability that particle one lies within the volume element $d\tau_1 = dx_1 dy_1 dz_1$, and similarly, and since all the particles must be somewhere

$$(10) \qquad c^2 \int \psi^* \psi\, d\tau = 1 \tag{11}$$

The integral in (11) is over the volume accessible to the particles. Thus c must be chosen to satisfy (11) and when this has been done, ψ is said to have been *normalized*.

3.3. A Model for the Ideal Gas

Example 3.3-1

Consider a particle of mass m which is free to move in one dimension between two barriers, located at $x = 0$ and $x = a$.

The potential energy of this particle is

$$V = 0, \quad 0 < x < a; \quad V = \infty, \quad x < 0, \quad x > a \tag{1}$$

For $0 < x < a$, the Schrödinger equation

$$(3.2.2),\ (3.2.3) \qquad \frac{-\hbar^2}{2m}\left[\frac{\mathrm{d}^2\psi(x)}{\mathrm{d}x^2}\right] = E\psi \tag{2}$$

is satisfied by both $\sin kx$ and $\cos kx$, so that the general solution of (2) is

$$\psi(x) = A\sin kx + B\cos kx \tag{3}$$

$$(2),\ (3) \qquad \frac{-\hbar^2}{2m}[-k^2 A\sin kx - k^2 B\cos kx] = \frac{\hbar^2 k^2}{2m}\psi(x) = E\psi(x) \tag{4}$$

$$(4) \qquad\qquad E = \frac{\hbar^2 k^2}{2m} \tag{5}$$

Up to this point, there are no restrictions on the value of the wave number k (in three dimensions, the *wave vector* $\boldsymbol{k}$) and therefore of E. Such restrictions, termed *quantization*, arise from the BC. Since the particle cannot penetrate the barriers at $x = 0$ and $x = a$, $\psi(x) = 0$ for $x < 0$ or $x > a$. Therefore, to avoid discontinuities in $\psi(x)$, as required by postulate I

$$\psi(x) = 0, \quad x = 0, \quad x = a \tag{6}$$

$$(3),\ (6) \qquad \text{At } x = 0, \quad \psi(0) = A(0) + B(1) = 0, \quad \Rightarrow B = 0 \tag{7}$$

$$(3),\ (6),\ (7) \qquad \text{At } x = a, \quad \psi(a) = A\sin ka = 0, \quad \Rightarrow k = n\pi/a \tag{8}$$

where n is an integer, called a *quantum number*. Therefore, the eigenvalues are

(5), (8)
$$E = \frac{\hbar^2 k^2}{2m} = \frac{n^2 h^2}{8ma^2} \tag{9}$$

and the corresponding eigenfunctions are

(3), (7), (8)
$$\psi_n(x) = A \sin(n\pi x/a) \tag{10}$$

The constant A is found by normalization

$$1 = \int_0^a A^2 \sin^2(n\pi x/a)\mathrm{d}x = A^2 a/2, \quad A = (2/a)^{1/2} \tag{11}$$

The normalized eigenfunctions are, therefore

(10), (11)
$$\psi_n(x) = (2/a)^{1/2} \sin(n\pi x/a) \tag{12}$$

Orthogonality It is a property of the Schrödinger equation that eigenfunctions belonging to different eigenvalues are orthogonal, that is

$$\int_0^\infty \psi_n * (x)\psi_m(x)\mathrm{d}x = 0, \quad n \neq m \tag{13}$$

The eigenfunctions (12) exhibit this property, since

$$\int_0^a \sin(n\pi x/a) \sin(m\pi x/a)\mathrm{d}x = 0, \quad \text{if } m \neq n \tag{14}$$

Example 3.3-2

Consider a particle confined within a box with sides of lengths a, b and c. Within the box the particle is free, that is, its potential energy

$$V(x, y, z) = 0, \quad 0 < x < a, \quad 0 < y < b, \quad 0 < z < c \tag{15}$$

Outside the box, $V(x, y, z)$ is infinite. The Schrödinger equation is, therefore

(3.2.9)
$$\frac{-\hbar^2}{2m}\nabla^2\psi(x, y, z) = E\psi(x, y, z), \quad 0 < x < a,$$
$$0 < y < b, \quad 0 < z < c \tag{16}$$

Equation (16) is separable into three equations, one for each variable, by the substitution

$$\psi(x, y, z) = X(x)Y(y)Z(z) \tag{17}$$

(16), (17) $$\frac{-\hbar^2}{2m}\left[\frac{1}{X(x)}\frac{\mathrm{d}^2X(x)}{\mathrm{d}x^2}+\frac{1}{Y(y)}\frac{\mathrm{d}^2Y(y)}{\mathrm{d}y^2}+\frac{1}{Z(z)}\frac{\mathrm{d}^2Z(z)}{\mathrm{d}z^2}\right]=E$$

$$(18)$$

Since x, y, z are *independent* variables, each of the three terms on the LS of (18) is equal to a constant. Therefore

(18) $$\frac{-\hbar^2}{2m}\left[\frac{1}{X(x)}\frac{\mathrm{d}^2X(x)}{\mathrm{d}x^2}\right]=E_x \qquad (19)$$

(18) $$\frac{-\hbar^2}{2m}\left[\frac{1}{Y(y)}\frac{\mathrm{d}^2Y(y)}{\mathrm{d}y^2}\right]=E_y \qquad (20)$$

(18) $$\frac{-\hbar^2}{2m}\left[\frac{1}{Z(z)}\frac{\mathrm{d}^2Z(z)}{\mathrm{d}z^2}\right]=E_z \qquad (21)$$

where

(18) $$E_x + E_y + E_z = E \qquad (22)$$

Each of the three equations (19)–(21) is the same as (2) and so their solutions are

$$E_x = \frac{n_x^2 h^2}{8ma^2}; \quad X(x) = (2/a)^{1/2}\sin(n_x\pi x/a) \qquad (23)$$

$$E_y = \frac{n_y^2 h^2}{8mb^2}; \quad Y(y) = (2/b)^{1/2}\sin(n_y\pi y/b) \qquad (24)$$

$$E_z = \frac{n_z^2 h^2}{8mc^2}; \quad Z(z) = (2/c)^{1/2}\sin(n_z\pi z/a) \qquad (25)$$

(17), (23)–(25) $$\psi(x,y,z) = \frac{2^{3/2}}{V}\sin\left(\frac{n_x\pi x}{a}\right)$$

$$\times \sin\left(\frac{n_y\pi y}{b}\right)\sin\left(\frac{n_z\pi z}{c}\right) \qquad (26)$$

where $V = abc$ is the volume of the box.

(9), (22)–(25) $$E = \frac{h^2}{8m}\left(\frac{n_x^2}{a^2}+\frac{n_y^2}{b^2}+\frac{n_z^2}{c^2}\right) \qquad (27)$$

For a cubical box

(27) $$E = \frac{h^2 n^2}{8ma^2}, n^2 = n_x^2 + n_y^2 + n_z^2 \qquad (28)$$

Our model for a monatomic ideal gas, or for the translational motion of a polyatomic ideal gas, is that of N non-interacting particles in a cubical box. Except when $n_x = n_y = n_z$, several different combinations of values of the quantum numbers n_x, n_y, n_z lead to the same energy E. When two or more different eigenfunctions have the same eigenvalue, these two states are said to be *degenerate*. For example the three states specified by $(n_x \; n_y \; n_z) = (2\,1\,1)$ or $(1\,2\,1)$ or $(1\,1\,2)$ all have the energy $6h^2/8ma^2$. The degeneracy of the ith level is denoted by ω_i.

For example,

E_n	n_x	n_y	n_z	ω_n
12	2	2	2	1
14	3	2	1	6

At $E_n = 12$, from (28) the three quantum numbers must each have the same value two and therefore there is only one way in which they can be assigned. When $E_n = 14$, from (28) the quantum numbers must have the values one, two or three. The first one n_x may therefore be chosen in three ways, the second in two ways, which leaves only one choice for the third quantum number. (If still in doubt, write down all six possible choices.)

The energy (kinetic) of the particle is

$$E = \frac{p^2}{2m} = \frac{h^2 k^2}{2m} \tag{29}$$

(29)
$$p = \frac{hk}{2\pi} = \frac{h}{\Lambda} \tag{30}$$

where Λ is the de Broglie wavelength of the particle. If the particles are neutrons obtained from the graphite moderator in an atomic reactor, then their wavelength is that of a slow neutron (about 1 A) and they may be used in neutron diffraction experiments.

3.4. The Number of Translational States

We take as our model for the translational motion of an ideal gas molecule, a particle in a box. We shall use this model to calculate the number of accessible translational energy states in a system containing N molecules and how close these states are together. The average translational energy of a gas molecule at temperature T is $\frac{3}{2}kT$ (as is known from kinetic theory and will be shown in Section 3.7.5). For a hydrogen molecule at 300 K

$$(3.3.28) \qquad n^2 = \frac{3kT}{2} \times \frac{8ma^2}{h^2}$$

$$= \frac{3(1.38 \times 10^{-23}\,\mathrm{J\,K^{-1}})(300\,\mathrm{K})}{2}$$

$$\times \frac{(8)(2.02 \times 1.66 \times 10^{-27}\,\mathrm{kg})(10^{-4}\,\mathrm{m^2})}{(6.625 \times 10^{-34}\,\mathrm{J\ s})^2}$$

$$= 3.8 \times 10^{16}, \quad \text{which gives } n_x \approx 1 \times 10^8 \qquad (1)$$

Approximate values were used for a number of physical quantities in this calculation, since only an order of magnitude was required. For such large numbers, there are a large number of possible combinations of n_x, n_y, n_z that correspond to the same n^2 and so the degeneracy is very large indeed.

Exercise 3.4-1

Show by calculating $\delta E/E$ that the translational energy levels are very closely spaced when E is $\frac{3}{2}kT$.

Exercise 3.4-1 will show that the translational energy states are so closely spaced ($\delta E \ll kT$) that they may be approximated by a continuum of states. To find the number of available energy states with energy $\leq E$, represent each state by a point with coordinates n_x, n_y, n_z. Then all the states with energy $\leq E$ lie in the positive octant of a sphere of radius $(8ma^2E/h^2)^{1/2}$. Each box of unit volume contains 1/8 of eight states, i.e. one state. Therefore the number of states with energy $\leq E$ is given by (1/8)th of the volume of this sphere, so that

$$N(\leq E) = \frac{1}{8}\left(\frac{4\pi}{3}\right)\left(\frac{8ma^2E}{h^2}\right)^{\frac{3}{2}} = \frac{4\pi}{3}\left(\frac{2mE}{h^2}\right)^{\frac{3}{2}}V \qquad (2)$$

(Although the above calculation is for one particle in a set of N non-interacting particles, the same type of calculation holds for a system of N non-interacting particles (Berry *et al.* 2002).) At atmospheric pressure, there would be about 2.5×10^{19} hydrogen molecules in a volume of $1\,\mathrm{cm^3}$. Some of these will have energy greater than their average energy of $\frac{3}{2}kT$

and some less energy but the distribution of energy states is extremely sharp.

Exercise 3.4-2

Calculate the number of translational energy states with energy $\leq \frac{3}{2}kT$ in $1\,\text{cm}^3$ of hydrogen at atmospheric pressure and $300\,\text{K}$.

Exercise 3.4-2 shows that the number of accessible energy states greatly exceeds the number of molecules to be allotted to them. Under these circumstances, we need not worry about the number of molecules that can be accommodated in each state when we come to calculate the *distribution* of the molecules over translational energy states.

We summarize the important results so far. For a gas at ordinary temperatures and pressures:

(i) the number of molecules N is very large;
(ii) the number of translational energy states is very, very large and much greater than N;
(iii) the degeneracy is very large;
(iv) translational energy levels are so closely spaced that a continuum suffices in calculations involving the density of states.

Polyatomic molecules have additional rotational and vibrational degrees of freedom. (A *degree of freedom* is the name used for a variable, or set of variables, that is used to describe a system. Thus, for example, a rigid linear diatomic molecule has two degrees of rotational freedom since two angles must be specified to describe its position with respect to fixed axes.)

It is interesting to note that the translational problem also has a completely different application, that of the phenomenon of *quantum confinement*, where the box is so small that translational levels are sufficiently separated to be observable.

Answers to Exercises 3.4

Exercise 3.4-1

$$(3.3.28) \qquad \frac{\delta E}{E} = \frac{(n_x + 1)^2 - n_x^2}{n_x^2} = \left(1 + \frac{1}{n_x}\right)^2 - 1 \approx \frac{2}{n_x} \approx 2 \times 10^{-8}$$

Thus the translational energy states are very closely spaced so that they may be approximated by a continuum of states.

Exercise 3.4-2

$$N(\leq E) = \frac{4\pi}{3} \left(\frac{3mkT}{h^2} \right)^{\frac{3}{2}} V$$

$$= \frac{4\pi}{3} \left(\frac{\begin{array}{c}(3 \times 2.016 \times 1.66 \times 10^{-27}\,\mathrm{kg}) \\ \times (1.38 \times 10^{-23}\,\mathrm{J\,K^{-1}})(300\,\mathrm{K})\end{array}}{(6.626 \times 10^{-34}\,\mathrm{J\,s})^2} \right)^{\frac{3}{2}} (10^{-4}\,\mathrm{m^3})$$

$$= 3.8 \times 10^{24}$$

You are advised to check the units in this calculation.

3.5. The Boltzmann Distribution

The question we pose is, how are the molecules distributed over the translational energy states given by Eq. (3.3.28)? The number of accessible energy states greatly exceeds the number of molecules to be allotted to them, so that we need not worry about the number of molecules that can be accommodated in each state when we calculate the *distribution* of the molecules over translational energy states. Consider first a set of *distinguishable* particles which are to be assigned in sub-sets to degenerate energy states, such that there are N_i particles in the ith sub-set and so forth. Note that molecules of the same chemical species are *not* distinguishable unless they are *localized* on a lattice, as in a crystal, or in an immobile adsorbed film. The necessary modifications when the molecules are not distinguishable will be considered shortly.

The number of ways of assigning N distinguishable particles to the various energy states is

$$\frac{N!}{\prod_i N_i!} \tag{1}$$

If there are ω_i eigenfunctions with the same energy ε_i, these distinguishable particles may be assigned to the ω_i eigenfunctions in $\omega_i^{N_i}$ ways, provided that there are no restrictions on the number of particles that may be assigned to each eigenfunction (orbital). In fact, such restrictions may exist, but in many situations, including the translational degrees of freedom of molecules, $\omega_i \gg N_i$ and such restrictions become unimportant. (A notable exception is that of electrons in a metal, to which the distribution law we are deriving does not apply.)

The total number of assignments of the N distinguishable particles that is consistent with the particular distribution $\{N_i\}$ is therefore

$$\Omega^D = N! \prod_i \frac{\omega_i^{N_i}}{N_i!} \tag{2}$$

Each such assignment constitutes a quantum state of the macroscopic system. We want to find the *most probable distribution*. To do this requires that we make some assumption about the relative probability of quantum states. It is a fundamental assumption of statistical thermodynamics that each quantum state (assignment) is equally probable. Therefore, the most probable state of the system which it will attain at equilibrium, is the state for which Ω has its maximum value. It is for this reason that Ω is called the *thermodynamic probability*.

We now remove the restriction that the particles be distinguishable. If they are *indistinguishable*, as are the molecules in a gas or in a liquid, then particles may be interchanged without giving a new distribution. Thus our calculated Ω^D is too large by a factor of $N!$ if the particles are indistinguishable, and

$$\Omega^I = \frac{\Omega^D}{N!} = \prod_i \frac{\omega_i^{N_i}}{N_i!} \tag{3}$$

Both Ω^I and Ω^D lead to the same distribution, the *Boltzmann distribution*.

There are two methods of deriving the Boltzmann distribution. The first is the standard procedure, using Lagrange's method of undetermined multipliers (which we shall employ) and another method described by Lorimer (1966). Any state of the system at specified N, E, V must be consistent with the constraints

$$\sum_{i=1}^{r} N_i = N \tag{4}$$

$$\sum_{i=1}^{r} \varepsilon_i N_i = E \tag{5}$$

For Ω (or $\ln \Omega$) to be a maximum, subject to arbitrary variations in the N_i that are consistent with (4) and (5)

$$d \ln \Omega = 0 \tag{6}$$

$$dN = \sum_{i=1}^{r} dN_i = 0 \tag{7}$$

$$dE = \sum_{i=1}^{r} \varepsilon_i dN_i = 0 \tag{8}$$

On using Stirling's approximation for the factorials of large numbers

(3)
$$\ln \Omega^{\mathrm{I}} = \sum_{i=1}^{r} N_i \ln \omega_i - N_i \ln N_i + N_i$$

$$= N + \sum_{i=1}^{r} N_i \ln \frac{\omega_i}{N_i} \tag{9}$$

(9), (7)
$$d \ln \Omega^{\mathrm{I}} = \sum_{i=1}^{r} \ln \frac{\omega_i}{N_i} dN_i \tag{10}$$

Similarly, for Ω^{D}

(2)
$$d \ln \Omega^{\mathrm{D}} = \sum_{i=1}^{r} \ln \frac{\omega_i}{N_i} dN_i \tag{11}$$

Therefore, for either Ω^{I} or Ω^{D}

(11), (12), (6)
$$d \ln \Omega = \sum_{i=1}^{r} \ln \frac{\omega_i}{N_i} dN_i = 0 \tag{12}$$

Because of the two constraints (7) and (8), only $r - 2$ of the r variables $\{N_i\}$ are independently variable. To include the constraints, multiply (7) by α and (8) by $-\beta$ and add the results to (12), giving

$$\sum_{i=1}^{r} \left\{ \ln \frac{\omega_i}{N_i} + \alpha - \beta \varepsilon_i \right\} dN_i = 0 \tag{13}$$

Require α and β to be chosen such that two of the factors in $\{\}$ are zero. Then, since $r - 2$ variations in dN_i can be chosen arbitrarily, the remaining

$r - 2$ factors in $\{\}$ are zero. Therefore, all the r factors in $\{\}$ in (13) are equal to zero

(13)
$$\ln \frac{\omega_i}{N_i} + \alpha - \beta \varepsilon_i = 0, \quad i = 1, \dots, r \tag{14}$$

(14)
$$N_i = \omega_i \exp(\alpha) \exp(-\beta \varepsilon_i) \tag{15}$$

(4), (15)
$$\sum_{i=1}^{r} N_i = N = \exp(\alpha) \sum_{i=1}^{r} \omega_i \exp(-\beta \varepsilon_i) \tag{16}$$

(16)
$$\frac{N_i}{N} = P_i = \frac{\omega_i \exp(-\beta \varepsilon_i)}{q} \tag{17}$$

where

$$q = \sum_{i=1}^{r} \omega_i \exp(-\beta \varepsilon_i) \tag{18}$$

is the molecular *partition function*, a name due to Darwin and Fowler (1922). The Boltzmann distribution may also be written in terms of one-particle *energy states* rather than degenerate *energy levels*, as in (17)

(17)
$$\frac{N_j}{N} = f_j = \frac{\exp(-\beta \varepsilon_j)}{q} \tag{19}$$

where

$$q = \sum_{j=1}^{r} \exp(-\beta \varepsilon_j) \tag{20}$$

Equation (19) has wide applicability; it is obeyed by *all molecules* except *light molecules at low temperatures*. For certain other systems of particles (or quasi-particles) other distribution laws must be used:

Photons, phonons, and light molecules with integral nuclear spin (e.g. ^{4}He) at low temperatures, obey *Bose–Einstein* statistics and are therefore called *bosons*.

Electrons in metals obey Fermi–Dirac statistics and are called *fermions*.

Anticipating a more detailed treatment in Chapter 10, we note that bosons have symmetrical state functions and the number of bosons in a quantum state is unrestricted, whereas fermions have antisymmetrical state functions and the number of fermions in a particle-quantum state is restricted to either zero or one.

The contribution from the translational energy of the gas molecules to the internal energy of the system is the energy calculated from (17), which is the most probable distribution. The energy is actually fluctuating about this most probable value, but because of the very large number of particles, these fluctuations are very small. The molecules in an ideal gas do not all have the same speed. Instead their speeds are distributed about some most probable value.

3.6. Translational Motion in an Ideal Gas

For our model of an ideal gas, that is, N non-interacting particles contained in a box, the translational energy states are given by

$$\text{(3.3.27)} \qquad \varepsilon_t = \frac{h^2}{8m}\left(\frac{n_x^2}{a^2} + \frac{n_y^2}{b^2} + \frac{n_z^2}{c^2}\right) \tag{1}$$

Therefore the translational partition function is

$$\text{(3.5.18)} \qquad q_t = \sum_{n_x=1}^{\infty}\sum_{n_y=1}^{\infty}\sum_{n_z=1}^{\infty} \exp\left\{-\frac{\beta h^2}{8m}\left(\frac{n_x^2}{a^2} + \frac{n_y^2}{b^2} + \frac{n_z^2}{c^2}\right)\right\} \tag{2}$$

Because $\Delta\varepsilon$ is so very small, the translational levels are very closely spaced and therefore the sums may be replaced by integrals and the lower limit replaced by zero. Therefore

$$\text{(2)} \qquad q_t = \int_0^{\infty} \exp\left\{-\frac{\beta h^2}{8m}\left(\frac{n_x^2}{a^2}\right)\right\} dn_x \int_0^{\infty} \exp\left\{-\frac{\beta h^2}{8m}\left(\frac{n_y^2}{b^2}\right)\right\}$$

$$\times\, dn_y \int_0^{\infty} \exp\left\{-\frac{\beta h^2}{8m}\left(\frac{n_z^2}{c^2}\right)\right\} dn_z$$

$$= \left\{\frac{a}{2}\left(\frac{8\pi m}{\beta h^2}\right)^{\frac{1}{2}}\right\}\left\{\frac{b}{2}\left(\frac{8\pi m}{\beta h^2}\right)^{\frac{1}{2}}\right\}\left\{\frac{c}{2}\left(\frac{8\pi m}{\beta h^2}\right)^{\frac{1}{2}}\right\}$$

$$= V\left\{\left(\frac{2\pi m}{\beta h^2}\right)^{\frac{3}{2}}\right\} \tag{3}$$

$$\text{(3)} \qquad \ln q_t = \ln V + \frac{3}{2}\ln\left(\frac{2\pi m}{\beta h^2}\right) \tag{4}$$

For a monatomic ideal gas, the only kind of internal energy is the translational energy of the gas molecules, so that $q = q_t$ and U is equal to $E = E_t$. Therefore, the pressure of the gas is

(2.2.2), (3.5.17) $$p = -\frac{\partial E}{\partial V} = \frac{N \sum_{i=1}^{r} -(\partial \varepsilon_i/\partial V)\omega_i \exp(-\beta\varepsilon_i)}{q} \tag{5}$$

(4) $$= \left(\frac{N}{\beta}\right)\frac{\partial \ln q}{\partial V} = \left(\frac{N}{\beta V}\right) \quad \text{(ideal gas)} \tag{6}$$

But for an ideal gas, $pV = NkT$. Therefore

(6) $$\beta = 1/kT \tag{7}$$

(3.5.18) $$q = \sum_{i=1}^{r} \omega_i \exp(-\varepsilon_i/kT) \tag{8}$$

$$f_i = \frac{\omega_i \exp(-\varepsilon_i/kT)}{q} \tag{9}$$

Alternatively

$$q = \sum_{j=1}^{s} \exp(-\varepsilon_j/kT) \quad \text{(sum over states)} \tag{10}$$

and

$$f_j = \frac{\exp(-\varepsilon_j/kT)}{q} \tag{11}$$

In (8) the sum is over energy levels of energy ε_i and degeneracy ω_i whereas in (10) the sum is over one-particle energy states of energy ε_j

(3), (7) $$q_t = V(2\pi mkT/h^2)^{3/2} = V/\Lambda^3 \tag{12}$$

where

$$\Lambda = h/(2\pi mkT)^{1/2} \tag{13}$$

Λ has the dimensions of length and is known as the *thermal de Broglie wavelength* because it is equal to the Planck constant h divided by the linear momentum

$$\Lambda = h/mv \tag{14}$$

Exercise 3.6-1

Find the SI units of the RS of (13).

Answer to Exercise 3.6-1

The SI units of the RS of (13) are

$$\left(\frac{J^2 s^2}{kg\, J\, K^{-1} K}\right)^{\frac{1}{2}} = m$$

3.7. Some Applications of the Boltzmann Distribution

3.7.1. *Distribution of translational energies*

The number of translational states with energy $\leq \varepsilon$ is

$$(3.4.2) \qquad\qquad N(\leq \varepsilon) = \frac{4\pi}{3}\left(\frac{2m\varepsilon}{h^2}\right)^{\frac{3}{2}} V \qquad\qquad (1)$$

Therefore, the number of translational states between ε and $\varepsilon + d\varepsilon$ is

$$dN = \left(\frac{\partial N(\leq \varepsilon)}{\partial \varepsilon}\right) d\varepsilon = 2\pi\left(\frac{2m}{h^2}\right)^{\frac{3}{2}} \varepsilon^{\frac{1}{2}} V d\varepsilon \qquad (2)$$

The fraction of molecules with energy in this range ε to $\varepsilon + d\varepsilon$ is, therefore,

$$f(\varepsilon)d\varepsilon = \text{number of molecular states}$$

$$\times \text{probability of occupation}$$

$$(2),\ (3.6.11) \qquad = \left\{2\pi\left(\frac{2m}{h^2}\right)^{\frac{3}{2}} \varepsilon^{\frac{1}{2}} V d\varepsilon\right\} \times \frac{\exp(-\varepsilon/kT)}{q}$$

$$(3.6.12) \qquad = \left(\frac{4\varepsilon}{\pi}kT\right)^{\frac{1}{2}} \frac{1}{kT}\exp(-\varepsilon/kT)d\varepsilon \qquad (3)$$

3.7.2. *Distribution of molecular speeds in an ideal gas*

Since the kinetic energy

$$\varepsilon = p^2/2m = \frac{1}{2}mv^2 \qquad\qquad (4)$$

where p is the linear momentum and v the speed of a particle

$$(3) \quad f(v)dv = \left(\frac{2}{\pi^{1/2}}\right)\left(\frac{mv^2}{2kT}\right)^{\frac{1}{2}} \frac{1}{kT}\exp(-mv^2/2kT)\, d\left(\frac{1}{2}mv^2\right) \qquad (5)$$

$$(5) \quad f(v)dv = \left(\frac{m}{2\pi kT}\right)^{\frac{3}{2}} \exp(-mv^2/2kT)4\pi v^2 dv \qquad (6)$$

This is the *Maxwell distribution function* for molecular speeds (Maxwell 1860, 1868). Though Maxwell's first paper on this subject was published in 1860, he had presented that paper (which uses a different derivation to that given here) at a meeting of the British Association in Aberdeen on 21 September 1859. Plots of $f(v)$ for nitrogen at three temperatures will be seen in the solution to Problem 3.4.

3.7.3. The average speed $\langle v \rangle$

$$\langle v \rangle = \frac{\int_0^\infty v f(v) \mathrm{d}v}{\int_0^\infty f(v) \mathrm{d}v} N$$

$$= 4\pi \left(\frac{m}{2\pi kT} \right)^{\frac{3}{2}} \int_0^\infty v^3 \exp(-mv^2/2kT) \mathrm{d}v$$

$$= 4\pi \left(\frac{m}{2\pi kT} \right)^{\frac{3}{2}} \left(\frac{4k^2 T^2}{2m^2} \right)$$

$$= \left(\frac{8kT}{\pi m} \right)^{\frac{1}{2}} \tag{7}$$

3.7.4. The root mean square (RMS) speed

$$\langle v^2 \rangle = \int_0^\infty v^2 f(v) \mathrm{d}v$$

$$= \int_0^\infty v^3 \exp(-mv^2/2kT) \mathrm{d}v$$

$$= 4\pi \left(\frac{m}{2\pi kT} \right)^{\frac{3}{2}} \int_0^\infty v^4 \exp(-mv^2/2kT) \mathrm{d}v$$

$$= 4\pi \left(\frac{m}{2\pi kT} \right)^{\frac{3}{2}} \left(\frac{3\pi^{\frac{1}{2}}}{8} \right) \left(\frac{2kT}{m} \right)^{\frac{5}{2}} = \left(\frac{3kT}{m} \right) \tag{8}$$

(8)
$$\langle v^2 \rangle^{\frac{1}{2}} = \left(\frac{3kT}{m} \right)^{\frac{1}{2}} \tag{9}$$

Exercise 3.7-1

Arrange the RMS speed $\langle v^2 \rangle^{\frac{1}{2}}$, the most probable speed v_p and the average speed $\langle v \rangle$ in numerical order. (**Hint:** See Problem 3.1.)

3.7.5. *Distribution of velocity components*

In one dimension

(3.3.9) $$E_x = \frac{n_x^2 h^2}{8ma^2}, \quad n_x = 1, 2, \ldots$$ (10)

(10) $$n_x = (2a/h)(2mE_x)^{1/2} = (2a/h)|p_x|$$ (11)

The number of states with momentum between 0 and $+p_x$ is therefore

$$\frac{1}{2}\left(\frac{2a}{h}\right) p_x = \left(\frac{a}{h}\right) p_x \tag{12}$$

and the number of states with momentum between $+p_x$ and $+p_x + dp_x$ is $(a/h)dp_x$.

Therefore, the fractional number of molecules with x-component of momentum between $+p_x$ and $+p_x + dp_x$ is

(3.6.11) $$f(+p_x)dp_x = \left(\frac{a}{h}\right) dp_x \times \frac{\exp(-p_x^2/2mkT)}{q_x}$$

$$= \frac{\exp(-p_x^2/2mkT)}{(2\pi mkT)^{1/2}} dp_x \tag{13}$$

and the fractional number of molecules with x-component of velocity between $+v_x$ and $+v_x + dv_x$ is

(13) $$f(+v_x)dv_x = \left(\frac{m}{2\pi kT}\right)^{\frac{1}{2}} \exp(-mv_x^2/2kT)dv_x \tag{14}$$

These results (13) and (14) depend on p_x^2 and v_x^2, as we might have anticipated from the fact that the direction of the x-axis is arbitrary. Consequently, $f(+p_x) = f(-p_x) = f(p_x)$ and $f(+v_x) = f(-v_x) = f(v_x)$.

Exercise 3.7-2

What is the average speed in the x-direction $\langle v_x \rangle$ and the most probable speed in the x-direction $v_{x,p}$?

Consider collisions with a wall normal to the x-axis. All those molecules within a vertical distance $v_x t$ will strike the wall in time t. If there are n molecules per unit volume, the number of collisions per area of wall, per time by molecules with an x-component of velocity between v_x and $v_x + dv_x$, and similarly, is

$$Z_w = n \int_0^\infty v_x f(v_x)\mathrm{d}v_x \int_{-\infty}^\infty f(v_y)\mathrm{d}v_y \int_{-\infty}^\infty f(v_z)\mathrm{d}v_z$$

(14)
$$= n \int_0^\infty \left(\frac{m}{2\pi kT}\right)^{\frac{1}{2}} v_x \exp(-mv_x^2/2kT)\mathrm{d}v_x$$

$$= n \left(\frac{kT}{2\pi m}\right)^{\frac{1}{2}} = n\left(\frac{\langle v \rangle}{4}\right) = \frac{p}{\sqrt{2\pi mkT}} \qquad (15)$$

which is the Hertz–Knudsen equation. This is a useful equation since it may be used to determine the vapour pressure as a function of temperature provided the relative molar mass is known.

Answers to Exercises 3.7

Exercise 3.7-1

Since, from Problem 3.1, $v_p = (2kT/m)^{\frac{1}{2}}$, the numerical order is $\langle v^2 \rangle^{\frac{1}{2}} > \langle v \rangle > v_p$.

Exercise 3.7-2

The RS of equation (14) is a Gaussian function, symmetrical about $x = 0$, with its maximum value at $x = 0$. Therefore, $\langle v_x \rangle$ and $v_{x,p}$ are both zero.

3.8. The Partition Function

3.8.1. *Degrees of freedom*

Because ε appears in the exponent in (3.6.8) and (3.6.10) the partition function factorizes into a *product* whenever ε is a *sum* of several terms. The translational term is always separable (Konig's theorem of classical mechanics). Separation of the other terms is, in many cases, a good approximation. A notable exception is the case of light molecules at low temperatures. When the energy of a molecule is separable into the sum of terms

$$\varepsilon = \varepsilon_n + \varepsilon_e + \varepsilon_t + \varepsilon_r + \varepsilon_v \qquad (1)$$

(1)
$$q = q_n q_e q_t q_r q_v \qquad (2)$$

The subscripts stand for nuclear, electronic, translational, rotational and vibrational, respectively, which are referred to as the *degrees of freedom* of a molecule.

Exercise 3.8-1

Let a and b denote two different particles or two different degrees of freedom of a molecule. Then the energy of the combined system ab is

$$\varepsilon_l = \varepsilon_{ai} + \varepsilon_{bj} \tag{3}$$

As the quantum numbers i, j take on all their allowed values, they enumerate all the quantum states $l = 0, 1, 2, \ldots$ of the combined system ab. Find an expression for the partition function q_{ab}.

Answer to Exercise 3.8-1

$$q_{ab} = q_a q_b = \left(\sum_{i=0}^{r} \omega_i \exp(-\varepsilon_{ai}/kT) \right) \left(\sum_{j=0}^{s} \omega_j \exp(-\varepsilon_{bj}/kT) \right)$$

$$= \sum_{i=0}^{r} \sum_{j=0}^{s} \omega_i \omega_j \exp[-(\varepsilon_{ai} + \varepsilon_{bj})/kT]$$

$$= \sum_{l=0}^{t} \omega_l \exp(-\varepsilon_{abl}/kT)$$

3.8.2. Unexcited degrees of freedom

If $\varepsilon_0 \ll \varepsilon_1$, that is, $\Delta\varepsilon_1 = \varepsilon_1 - \varepsilon_0 \gg kT$, then

$$q = \omega_0 \exp(-\varepsilon_0/kT) \left\{ 1 + \frac{\omega_1}{\omega_0} \exp(-\Delta\varepsilon_1/kT) + \cdots \right\}$$

$$\approx \omega_0 \exp(-\varepsilon_0/kT) = \omega_0 \tag{4}$$

if we choose the energy zero to be ε_0. When all the particles in a system are in the ground state of a particular degree of freedom, it is said to be *unexcited*. The separation between nuclear energy levels is so vast that all nuclei remain in their ground state during normal (terrestrial) chemical processes. If there are several nuclei $A, B, \ldots$ in a molecule

$$q_n = \omega_{n0}^{A} \times \omega_{n0}^{B} \times \cdots = \omega_{n0} \tag{5}$$

Since q_n appears as a constant factor in the partition function for all thermodynamic states, it is usually omitted. An exception occurs in the discussion of the thermodynamic properties of light molecules at low temperatures, when the nuclear and rotational degrees of freedom must be considered together.

Similar considerations often apply to the electronic partition function q_e. If $\Delta\varepsilon_1 \gg kT$, when the energy zero is chosen to be the energy of the state in which all nuclei are in their ground state and the atom or molecule is in its ground electronic state

$$\varepsilon_{n0} + \varepsilon_{e0} = \varepsilon_{n0}^{A} + \varepsilon_{n0}^{B} + \cdots + \varepsilon_{e0} = 0 \tag{6}$$

and consequently $q_e = \omega_{e0}$. If $\Delta\varepsilon_1/kT$ is not $\gg 1$, as may be true for some atoms and molecules, particularly at high temperatures, then

$$q_e = \omega_{e0} + \omega_{e1} \exp(-\Delta\varepsilon_{e1}/kT) + \cdots \tag{7}$$

Usually 1, 2, or 3 terms are sufficient.

3.8.3. *Classical degrees of freedom*

In contrast, when $\Delta\varepsilon_1 \ll kT$, the energy levels are so closely spaced that the energy distribution approximates a continuum. Such degrees of freedom are referred to as *classical*. In such cases, the summation in the partition function may be replaced by an integral

$$q = \sum_{i=0}^{\infty} \omega_i \exp(-\varepsilon_i(n)/kT) \approx \int_0^{\infty} \omega_i \exp(-\varepsilon_i(n)/kT)\mathrm{d}n \tag{8}$$

where the energy of each level depends on the quantum number n. We have already made use of the fact that the translational degree of freedom behaves classically when evaluating q_t in Section 3.6.

3.8.4. *General rules*

The following rules are generally valid and thus act as a good guide.

Degree of freedom	Behaviour
nuclear[a]	always unexcited
electronic	generally unexcited
translational	always classical
rotational[a]	classical, except at very low T
vibrational	use quantum sum, except at high T, when classical and at very low T, when unexcited

[a]not separable for light molecules at very low T.

A non-linear polyatomic molecule containing n atoms has $3n$-6 vibrational degrees of freedom, but a linear polyatomic molecule has $3n$-5 vibrational degrees of freedom. This is because, as explained earlier, a linear molecule has two rototational degrees of freedom. Thus from a knowledge of the energy levels and degeneracies in each degree of freedom (information which is obtained from quantum mechanics and spectroscopy) we can find the partition function of any particular molecule. This is important because the thermodynamic functions can be expressed in terms of Q.

3.8.5. *Partition function for a system of N weakly-interacting particles*

Providing the particles $1, 2, \ldots, N$ are weakly-interacting

$$E = \varepsilon(1) + \varepsilon(2) + \varepsilon(2) + \varepsilon(3) + \cdots + \varepsilon(N) \tag{9}$$

Therefore the partition function for the system of N weakly-interacting distinguishable particles of the same kind is

$$(9) \qquad Q^{\mathrm{D}} = q(1) \times q(2) \times q(3) \times \cdots \times q(N) = q^N \tag{10}$$

But if the particles are indistinguishable, Q^{D} is too large by a factor of $N!$ and

$$Q^{\mathrm{I}} = \frac{q^N}{N!} \tag{11}$$

Recall that the factor $N!$ comes from the number of permutations of the N indistinguishable particles among the N occupied states.

3.9. Evaluation of Thermodynamic Functions from Q

The energy. Because the distribution of particles over the available energy states is extremely sharp, the average value per particle may be replaced by its most probable value, as given by the Boltzmann distribution. Denoting the average value of P by $\langle P \rangle$, the average energy of a N-particle system is

$$\langle E \rangle = N\langle \varepsilon \rangle = \frac{N \sum_{i=1}^{r} \varepsilon_i \omega_i \exp(-\varepsilon_i/kT)}{q} = NkT^2 \frac{\partial \ln q}{\partial T} \tag{1}$$

$$(1), (3.8.11) \qquad\qquad \langle E \rangle = kT^2 \frac{\partial \ln Q}{\partial T} = U \tag{2}$$

The LS of (2) is the average energy $\langle E \rangle$ referred to an energy zero which is the energy of this system of N particles at rest and in their ground nuclear and electronic states. This energy is the quantity called the *internal energy* of the system and denoted in thermodynamics by the symbol U. The First Law dealt only with the *energy change* between two states, so that the energy zero was immaterial and never defined. Statistical thermodynamics has provided us with a clearly defined energy zero. The average translational energy per molecule is

(1), (3.6.12) $$\langle \varepsilon_t \rangle = kT^2 \frac{\partial \ln q_t}{\partial T} = \frac{3}{2} kT \tag{3}$$

There are three translational degrees of freedom, corresponding to the resolution of the velocity of a particle along three mutually perpendicular directions. The orientation of these three directions is arbitrary; therefore, the average translational kinetic energy associated with each of x, y, z is $\frac{1}{2} kT$. This is an example of the *equipartition principle*, which is that the molecular energy associated with each quadratic term contributed by a classical degree of freedom to the Hamiltonian function $H = E_K + V$ is $\frac{1}{2} kT$. Equation (3) demonstrates the validity of the equipartition principle for the translational motion of molecules with kinetic energy

$$E_K = \frac{1}{2} m (v_x^2 + v_y^2 + v_z^2) \tag{4}$$

The pressure

(1), (2) $$p = -\frac{\partial U}{\partial V} = \frac{N \sum_{i=1}^{r} -(\partial \varepsilon_i / \partial V) \omega_i \exp(-\varepsilon_i / kT)}{q} \tag{5}$$

(5) $$p = NkT \frac{\partial \ln q}{\partial V} \tag{6}$$

(6), (3.8.11) $$p = kT \frac{\partial \ln Q}{\partial V} \tag{7}$$

The Helmholtz function and the entropy

$$F = U - TS \tag{8}$$

(8) $$dF = -pdV - SdT \tag{9}$$

(8), (9) $$F = U + T \frac{\partial F}{\partial T} \tag{10}$$

(10)
$$-\frac{U}{T^2} = \frac{1}{T}\frac{\partial F}{\partial T} - \frac{F}{T^2} = \frac{\partial}{\partial T}\left(\frac{F}{T}\right)$$
(11)

(9), (7)
$$\frac{\partial}{\partial V}\left(\frac{F}{T}\right) = -\frac{p}{T} = -k\frac{\partial \ln Q}{\partial V}$$
(12)

(11), (2)
$$\frac{\partial}{\partial T}\left(\frac{F}{T}\right) = -\frac{U}{T^2} = -k\frac{\partial \ln Q}{\partial T}$$
(13)

Integrate (12) and (13)

(12)
$$\frac{F}{T} = -k\ln Q + a(T)$$
(14)

(13)
$$\frac{F}{T} = -k\ln Q + b(V)$$
(15)

From (14) and (15) it follows that

$$a(T) = b(V) = -c$$
(16)

where c is an extensive constant, independent of T and V at constant N. The minus sign on the RS of (16) is purely for convenience; it has no physical significance

(6)
$$S = \frac{U}{T} - \frac{F}{T}$$
(17)

(2), (14)–(17)
$$S = kT\frac{\partial \ln Q}{\partial T} + k\ln Q + c$$
(18)

Thus the entropy contains the undetermined constant c; being independent of T and V, c is independent of the thermodynamic state of the system and so represents that part of the entropy of a system at 0 K which does not change with T and which is not included in Q. For perfect, pure crystals, this is taken to be the entropy which arises from frozen-in metastability, from isotopic mixing and from the degeneracy of the nuclear and electronic ground states. The other degrees of freedom are usually non-degenerate for perfect, pure crystalline solids at 0 K.[a]

---

[a] "Perfect" and "pure" are ideal abstractions which cannot be realized in practice. We could, however, set some practical limits which would have no visible effect on bulk thermodynamic properties.

Since these contributions do not change with T, they are often neglected and are so in defining the *conventional spectroscopic entropy* by

$$S = S_{\text{spect}} = kT\frac{\partial \ln Q}{\partial T} + k \ln Q \quad (c \text{ set } = 0) \tag{19}$$

The description "spectroscopic" is used because the necessary information on energy levels and their degeneracy is usually obtained from spectroscopy. The *conventional calorimetric entropy* is defined by

$$S_{\text{cal}} = S - S_0 = \int_0^T \frac{C_p}{T}\mathrm{d}T + \sum \Delta S_{\text{t}} \tag{20}$$

where $\sum \Delta S_{\text{t}}$ denotes the sum of the entropy changes at each phase transition that occurs between $T = 0\,\text{K}$ and T in (20). The conventional entropies obtained from (19) and (20) usually agree, but sometimes $S_{\text{spect}} > S_{\text{cal}}$ and, when this is so, it indicates that some additional degeneracy in the ground state of the crystal has been neglected, in which case

$$S(0\,\text{K}) = c + c_0 \tag{21}$$

$$c = c_1 + c_2 \tag{22}$$

where

$c_0 =$ entropy due to additional degeneracy of the ground state
$c_1 =$ entropy due to isotopic mixing and nuclear ground states
$c_2 =$ entropy due to frozen-in metastability

The Massieu function and the Planck function

The *Massieu* function J is given by

$$J = S - \frac{U}{T} = -\frac{F}{T} = k \ln Q \tag{23}$$

The *Planck* function Y is given by

$$Y = S - \frac{H}{T} = -\frac{G}{T} = k \ln Q + kV\frac{\partial \ln Q}{\partial V} \tag{24}$$

We now have formulae for U, p, S, J and Y (and therefore F and G) in terms of Q.

It is not necessary to use both the thermodynamic potentials F and J, and likewise the potentials G and Y, but in some contexts J and Y may be preferred because of their closer analogy to S. For example, the equilibrium conditions (2.5.12) may be stated as a closed system is in equilibrium if:

$$\text{for given } U \text{ and } V, S \text{ is a maximum, or}$$
$$\text{for given } T \text{ and } V, J \text{ is a maximum, or} \qquad (25)$$
$$\text{for given } T \text{ and } p, Y \text{ is a maximum.}$$

Absolute activity

The absolute activity λ (Fowler and Guggenheim 1939) is defined by

$$\lambda = \exp(\mu/kT), \quad \mu = kT\ln\lambda \qquad (36)$$

where μ is the chemical potential per molecule. Whether molecular or molar values are intended will always be clear from the context (i.e. the use of k or R). While it is not necessary to use both μ and λ, the absolute activity may be a more convenient function to use in some contexts.

3.10. Internal Degrees of Freedom

3.10.1. *Diatomic molecule*

The two examples studied so far were rather simple because we assumed a system of non-interacting particles and considered only translational energy states. The solution of the Schrödinger equation becomes much more laborious when our system consists of weakly-interacting diatomic or polyatomic molecules. Fortunately, many of the differential equations that occur are standard forms whose solutions are well known in mathematical physics and may therefore be quoted without detailed solution. Nevertheless, the first part of this section may be heavy-going for some readers, in which case they could skip to Eq. (18).

In a system of weakly-interacting diatomic molecules, the potential energy of a single molecule is a function of r, the distance between the two atoms comprising the molecule. The Schrödinger equation is therefore

$$(3.2.2), \ (3.2.9) \qquad \left\{ \frac{\hbar^2}{2m_1}\nabla_1^2 + \frac{\hbar^2}{2m_2}\nabla_2^2 + V(r) \right\}\Psi = E\Psi \qquad (1)$$

where the state function Ψ is a function of the positional coordinates of both particles, $\Psi(x_1, y_1, z_1, x_2, y_2, z_2)$. Transform coordinates from $(x_1, y_1, z_1, x_2, y_2, z_2)$ to the Cartesian coordinates of the centre of mass (X, Y, Z) and the coordinates (x, y, z) of particle two with respect to particle one.

Since V does not depend on X, Y, Z, the Schrödinger equation in this new coordinate system is separable by the substitution

$$\Psi(x_1, y_1, z_1, x_2, y_2, z_2) = \psi_t(X, Y, Z)\psi(x, y, z) \tag{2}$$

into two equations, one of which involves X, Y, Z only and describes the *translational motion of the centre of mass*. The other, which contains x, y, z only, describes the relative motion of the two particles with respect to one another (*vibration*) and the *rotation* of the centre of mass about fixed axes. The first equation is just like that for the translational motion of a free particle of mass $M = m_1 + m_2$. The equation describing vibrational and rotational motion is

$$\frac{-\hbar^2}{2\mu}\nabla^2\psi + V(r)\psi = E\psi \tag{3}$$

where the *reduced mass* μ is given by

$$\frac{1}{\mu} = \frac{1}{m_1} + \frac{1}{m_2} \tag{4}$$

Now transform ∇^2 in (3) from Cartesian coordinates x, y, z to spherical polar coordinates r, θ, ϕ (Stephenson 1961, Margenau and Murphy 1962), giving

$$\frac{-\hbar^2}{2\mu r^2}\left[\frac{\partial}{\partial r}\left(r^2\frac{\partial}{\partial r}\right) + \frac{1}{\sin\theta}\frac{\partial}{\partial\theta}\left(\sin\theta\frac{\partial}{\partial\theta}\right) + \frac{1}{\sin^2\theta}\frac{\partial^2}{\partial\phi^2}\right]\psi + V(r)\psi = E\psi \tag{5}$$

The terms in [] in (5) are the operator ∇^2 in spherical polar coordinates.

Example 3.10-1 A rigid rotator in the xy plane

In this example, r is constant and equal to the equilibrium internuclear separation r_e and θ is constant, equal to $\pi/2$, so that $\sin\theta = 1$. $V(r_e)$ is a constant, which we set equal to zero by our choice of energy zero. Therefore (5)

$$\frac{-\hbar^2}{2\mu r_e^2}\frac{\mathrm{d}^2\psi}{\mathrm{d}\phi^2} = E\psi \tag{6}$$

μr_e^2 is the *moment of inertia*, I. The solution of (6) is

$$\psi = A \exp(im\phi) \tag{7}$$

For ψ to be single-valued, $\psi(\phi + 2\pi)$ must be equal to $\psi(\phi)$, and so

$$m = 0, \pm 1, \pm 2, \ldots \tag{8}$$

(6), (7)
$$E = \frac{\hbar^2 m^2}{2I} \tag{9}$$

Normalization shows that

(7)
$$A = \frac{1}{\sqrt{2\pi}} \tag{10}$$

Exercise 3.10-1

Verify that (7) is the solution of (6), the quantization of E by (8) and that the normalization factor A in (7) is that given by (10). Tabulate the energy of the first four levels, giving also the rotational quantum number m and the degeneracy of each level.

Example 3.10-2 A diatomic molecule with two degrees of rotational freedom

Since V depends on r only, (5) is separable by the substitution

$$\psi = R(r)\Theta(\theta)\Phi(\phi) \tag{11}$$

Substitute from (11) in (5) and multiply the result by $(-2\mu r^2 \sin^2 \theta / \hbar^2 R\Theta\Phi)$, giving

$$\frac{\sin^2 \theta}{R} \frac{\partial}{\partial r}\left(r^2 \frac{\partial R}{\partial r}\right) + \frac{\sin \theta}{\Theta} \frac{\partial}{\partial \theta}\left(\sin \theta \frac{\partial \Theta}{\partial \theta}\right)$$

$$+ \frac{1}{\Phi} \frac{\partial^2 \Phi}{\partial \phi^2} + \left(\frac{2\mu r^2 \sin^2 \theta}{\hbar^2}\right)\{E - V(r)\} = 0 \tag{12}$$

The third term in (12) depends on ϕ only and so may be equated to a constant, which we choose to be $-m^2$. The solution has already been given by (7), (8) and (10). (In the problem of the rigid rotator confined to the xy plane, this was the whole solution. For the diatomic molecule in 3-D, this is just the factor $\Phi(\phi)$ in (11).) The remaining part of (12), which depends on r and θ, must then be equal to $+m^2$. On dividing by $\sin^2\theta$

$$\text{(12)} \quad \frac{1}{R}\frac{\partial}{\partial r}\left(r^2\frac{\partial R}{\partial r}\right) + \frac{1}{\Theta\sin\theta}\frac{\partial}{\partial\theta}\left(\sin\theta\frac{\partial\Theta}{\partial\theta}\right)$$

$$-\frac{m^2}{\sin^2\theta} + \left(\frac{2\mu r^2}{\hbar^2}\right)\{E - V(r)\} = 0 \quad \text{(13)}$$

To complete the separation of variables, equate the second and third terms, which depend on θ only, to a constant $-\beta$ giving

$$\text{(13)} \quad \frac{1}{\sin\theta}\frac{\partial}{\partial\theta}\left(\sin\theta\frac{\partial\Theta}{\partial\theta}\right) - \frac{m^2}{\sin^2\theta} + \beta\Theta = 0 \quad \text{(14)}$$

$$\text{(13)} \quad \frac{1}{R}\frac{\partial}{\partial r}\left(r^2\frac{\partial R}{\partial r}\right) - \beta R + \left(\frac{2\mu r^2}{\hbar^2}\right)\{E - V(r)\} = 0 \quad \text{(15)}$$

When $\beta = J(J+1)$, J integral, the solutions to (14) (which occurs in several different contexts in physics and chemistry) are known to be the *spherical harmonics*. The rotational quantum number is called J here and l in the solution of the Schrödinger equation for the H atom

$$m = 0, \pm 1, \pm 2, \ldots, \pm J \quad \text{(16)}$$

so that the degeneracy of the rotational states is $2J + 1$.

The simplest approximation for $V(r)$ is to assume that, as the two atoms vibrate, the restoring force f is proportional to the displacement $r - r_e$ from the equilibrium position r_e

$$f(r) = k(r - r_e) \quad \text{(17)}$$

r is the instantaneous distance between the centres of the two atoms in the vibrating diatomic molecule and k is the *force constant*. The vibrational motion is then *simple harmonic* with

$$V(r) = \frac{1}{2}k(r - r_e)^2 \quad \text{(18)}$$

Thus (5) has been separated into three equations, each of which contains one variable only. With $V(r)$ given by (8), Eq. (15) has well-behaved solutions only if

$$E = \frac{\hbar^2}{2I}[J(J+1)] + \left(v + \frac{1}{2}\right)hv \quad \text{(19)}$$

where $v = 0, 1, 2, 3, \ldots$ is the vibrational quantum number and the vibrational frequency

$$v = \frac{1}{2\pi} \sqrt{\frac{k}{\mu}} \tag{20}$$

Since E does not depend on m, it follows from (16) that each rotational state is $(2J + 1)$-fold degenerate.

3.10.2. *Polyatomic molecules*

The internal vibrational motion of a polyatomic molecule containing n atoms can be described by the superposition of $3n - 5$ or $3n - 6$ simple harmonic vibrations called *normal modes*. The ambiguity arises because non-linear molecules can rotate about three axes, but a linear molecule has only two components of rotational energy because its mass is concentrated along the molecular axis and therefore makes a negligible contribution to the angular momentum about that axis. Each vibrational degree of freedom contributes two quadratic terms to the classical Hamiltonian, one from kinetic energy and one from the potential energy. The harmonic approximation is a good one for small displacements, but at high temperatures displacements are larger, and anharmonicity corrections may need to be taken into account. One way to do this is to add terms of higher order in $(v + \frac{1}{2})$ to Eq. (18) (Atkins 1986, Davidson 1962).

3.10.3. *Rotational partition function*

Simplify notation by introducing the *rotational constant*

$$B = h/8\pi^2 Ic \tag{21}$$

where c is the speed of light in vacuum. B has dimensions of length^{-1} and experimental results are almost always quoted in cm^{-1}. The rotational energy is

$$(19),\ (21) \qquad E_J = \frac{\hbar^2}{2I}[J(J+1)] = hcB[J(J+1)] \tag{22}$$

$$(22) \qquad \frac{E_J}{kT} = \frac{hcB}{kT}J(J+1) = \frac{\Theta_r}{T}J(J+1) \tag{23}$$

$$(22),\ (23) \qquad \frac{\Theta_r}{T} = \frac{hcB}{kT} = \frac{\hbar^2}{2I\,kT} = 1.4388(B/T)\ \text{cm K} \tag{24}$$

Θ_r is called the *rotational temperature*.

Exercise 3.10-2

Verify the numerical value of hc/k given in (24).

For $T \gg \Theta_r$, the sum in q_r may be replaced by an integral

$$(23), (3.5.18) \qquad q_r = \frac{1}{\sigma} \int_0^\infty (2J+1)\exp[-J(J+1)(\Theta_r/T)]dJ \qquad (25)$$

$$(24) \qquad\qquad = \frac{T}{\sigma\Theta_r} = \frac{8\pi^2 IkT}{\sigma h^2} \qquad (26)$$

σ is a symmetry number, equal to 2 for AA molecules and equal to 1 for AB molecules. If (26) is not sufficiently accurate, it may be replaced by the expansion

$$q_r = \frac{1}{\sigma\eta}\left(1 + \frac{\eta}{3} + \frac{\eta^2}{15}\cdots\right), \qquad \eta = \Theta_r/T \qquad (27)$$

(Fowler and Guggenheim 1939). Linear molecules have only one moment of inertia μr_e^2, which is the same for rotation about either of the two orthogonal axes normal to the molecular axis. For a non-linear polyatomic molecule, the moment of inertia is a second-order symmetric tensor which can always be diagonalized by a principal-axis transformation so that the kinetic energy becomes

$$E_K = \frac{1}{2}I_A\omega_A^2 + \frac{1}{2}I_B\omega_B^2 + \frac{1}{2}I_C\omega_C^2 \qquad (28)$$

where $I_A \leq I_B \leq I_C$ are the *moments of inertia* referred to principal axes and $\omega_A, \omega_B, \omega_C$ are the *angular velocities* about these axes. Notice, as stated previously, that rotation contributes three quadratic terms to the kinetic energy of a polyatomic molecule. To derive the rotational partition function for a polyatomic molecule we need to multiply that for rotation around two axes (26) by that for rotation in a plane. For a non-linear polyatomic molecule, I_A is not necessarily equal to I_B and therefore the partition function for rotation about two of the principal axes A and B is

$$(26) \qquad q_r(A,B) = \frac{8\pi^2 kT}{\sigma h^2}(I_A I_B)^{\frac{1}{2}} = T/\sigma(\Theta_A\Theta_B)^{\frac{1}{2}} \qquad (29)$$

For rotation about the C axis

$$(9) \qquad\qquad E = \frac{\hbar^2 m'^2}{2I_C} \qquad (30)$$

$$(30) \qquad\qquad q_r(C) = \sum_{m'=-\infty}^{\infty} \exp(-\hbar^2 m'^2/2I_C kT) \qquad (31)$$

$$= \int_{-\infty}^{\infty} \exp(-m'^2 \Theta_C / T) \mathrm{d}m' \tag{32}$$

$$= (\pi T / \Theta_C)^{\frac{1}{2}} \tag{33}$$

(29), (33) $$q_\mathrm{r} = \frac{8\pi^2 (2\pi kT)^{\frac{3}{2}}}{\sigma h^3} (I_A I_B I_C)^{\frac{1}{2}} \tag{34}$$

which agrees with formula (326,6) of Fowler and Guggenheim (1939), derived by the methods of classical statistical mechanics. For the moments of inertia of polyatomic molecules see Moore (1983).

3.10.4. *Symmetry numbers*

In a diatomic molecule with two different nuclei, such as HCl, the original configuration is reproduced after a rotation through an angle of 2π about the x or y axes (where, as is usually the case, the z-axis has been chosen along the molecular axis). However, in homonuclear diatomic molecules and, in general, polyatomic molecules with one or more rotational axes of symmetry, rotations about these axes through angles of less than 2π result in a configuration that is indistinguishable from the original configuration. For example, NH_3 has one three-fold axis of symmetry. Rotations through $2\pi/3$ about this axis result in three indistinguishable configurations and the symmetry number is three. In general, σ is the order of the rotation sub-group in the symmetry group of the molecule. The rotation sub-group contains the identity and all the rotation operators, but no other symmetry operators (see, for example, Jacobs 2005, Chapter 2).

Exercise 3.10-3

Determine the rotational symmetry numbers for (a) water (H_2O), (b) methane (CH_4) and (c) benzene (C_6H_6). (***Hints:*** The shape of the methane molecule is tetrahedral and that of benzene is a hexagon in the xy plane.)

3.10.5. *Pauli exclusion principle*

The state function of a set of particles with spin angular momentum of magnitude $\frac{1}{2}\hbar$ (fermions) is *antisymmetric* with respect to the exchange of indistinguishable particles. This statement is known as the *Pauli exclusion principle*; it is a fundamental law of nature, that is, a description of the way things are. Consider a system of N indistinguishable particles and let P_{12}

be the operator that interchanges the positions and spins of two of these particles, labelled 1 and 2. Then

(postulate I) $$|P_{12}\psi(1,2)|^2 = |\psi(1,2)|^2 \qquad (35)$$

(35) $$P_{12}\psi(1,2) = \exp(i\gamma)\psi(1,2) \qquad (36)$$

where $\exp(i\gamma)$ is a complex number of modulus unity, called a *phase factor*. On repeating the interchange

(36) $$P_{12}P_{12}\psi(1,2) = \exp(2i\gamma)\psi(1,2) = \psi(1,2) \qquad (37)$$

(36), (37) $$P_{12}\psi(1,2) = \pm\psi(1,2) \qquad (38)$$

Equation (38) shows that ψ could be either symmetric or antisymmetric with respect to the interchange of indistinguishable particles, but in fact, for *fermions*, ψ is *antisymmetric* and

(38) $$P_{12}\psi(1,2) = -\psi(1,2) \qquad (39)$$

The Pauli principle (39) applies to the *total state function* ψ, which is often a product of several factors and one or more of these factors may be symmetric under P_{12}. Designate the nuclear state functions with $m_S = \pm\frac{1}{2}$ (in the usual angular momentum units of $\hbar$), by α and β respectively. The four two-proton state functions that are eigenfunctions of the interchange operator P_{12} are given in Table 3.1. The first three functions are symmetric with respect to the interchange of indistinguishable particles but the fourth one is antisymmetric.

Interchanging the two nuclei is equivalent to a rotation through π about an axis normal to z, that is, an increase in θ by π. Both $\sin\theta$ and $\cos\theta$ change sign under this operation. Therefore the rotational state functions with $J = 0, 2, 4, \ldots$ are even under P_{12}, but those with J odd, $J = 1, 3, 5, \ldots$,

Table 3.1. Nuclear state functions ψ_n of $^1H_2$ which are eigenfunctions of the interchange operator P_{12}. S is the total spin angular momentum of two protons (each with spin 1/2) and m_S is its projection along the z axis; EV = eigenvalue.

State function ψ_n	ev of P_{12}	m_S	S
$\alpha(1)\alpha(2)$	1	1	
$\alpha(1)\alpha(2) + \alpha(2)\beta(1)$	1	0	1
$\alpha(1)\beta(2)$	1	-1	
$\alpha(1)\beta(2) - \alpha(2)\beta(1)$	-1	0	0

change sign under P_{12}. Since $\psi = \psi_n \psi_r$ must change sign under P_{12}, there are two kinds of molecular hydrogen:

ortho-hydrogen, with parallel nuclear spins, $S = 1$,

and $J = 1, 3, 5, \ldots$.

para-hydrogen, with anti-parallel nuclear spins, $S = 0$,

and $J = 0, 2, 4, \ldots$.

The *average* molecular rotational partition function of hydrogen is therefore

$$q_r = \frac{1}{4} \left\{ \sum_{J \text{ even}} (2J + 1) \exp[-J(J + 1)\Theta_r/T] \right.$$

$$\left. + 3 \sum_{J \text{ odd}} (2J + 1) \exp[-J(J + 1)\Theta_r/T] \right\} \tag{40}$$

In fact, this differs little from

$$q_r \approx \frac{1}{2} \left\{ \sum_{\text{all } J} (2J + 1) \exp[-J(J + 1)\Theta_r/T] \right\} \tag{41}$$

The approximation (41) becomes more satisfactory at high temperatures and is even better for homonuclear diatomics that don't contain hydrogen (because of their lower rotational temperatures).

3.10.6. *Vibrational partition function*

In the vibrational ground state, the vibrational energy (*the zero-point energy*) of a harmonic oscillator is

(19) $$E_o = \frac{1}{2} \frac{h v}{kT} \tag{42}$$

Define the characteristic vibrational temperature Θ_v by

$$\Theta_v = h v / k = 1.4388 \,\text{cm K} \,\breve{v} \tag{43}$$

where $\breve{v}$ is v/c in cm^{-1}, so that

$$\frac{E_v - E_0}{kT} = \frac{\Theta_v}{T} \tag{44}$$

The zero-point energy E_0 is present at any temperature. Because of the magnitude of Θ_v (Table 3.2) at room temperature most molecules will be in their ground vibrational state. In (44), vibrational energies are

Table 3.2. Rotational and vibrational temperatures, bond lengths and dissociation energies for some diatomic molecules.

Gas	Θ_v/K	Θ_r/K	r_e/A	D_0/eV
H_2	6215	85.4	0.742	4.47
D_2	4394	42.7	0.742	4.55
N_2	3374	2.88	1.098	9.76
O_2	2256	2.07	1.207	5.08
CO	3070	2.77	1.128	9.14
NO	2690	2.42	1.150	6.48
HCl	4140	15.2	1.275	4.43
HBr	3787	12.02	1.408	3.75
HI	3266	9.06	1.604	2.75
Cl_2	808	0.351	1.988	2.48
Br_2	463	0.116	2.284	1.97
I_2	308	0.0537	2.667	1.54

being measured *with respect to the lowest quantum state*. Therefore, the vibrational partition function is

$$q_v = \exp(-\Theta_v/2T) \sum_{v=0}^{\infty} \exp(-v\Theta_v/T) = \frac{\exp(-\Theta_v/2T)}{1 - \exp(-\Theta_v/T)} \qquad (45)$$

Exercise 3.10-4

Verify the second equality in (45). (***Hint:*** What kind of series occurs in q_v?)

Exercise 3.10-5

Why is the energy of a harmonic oscillator not zero in its lowest quantum state?

$$(3.9.1), (45) \qquad U_v = Nk\left(\frac{\Theta_v}{2} + \frac{\Theta_v}{\exp(\Theta_v/T) - 1}\right) \qquad (46)$$

(the first term in brackets in (46) being the *zero-point energy*) and

$$(46) \qquad C_{V,v} = Nk\left(\frac{\Theta_v}{2}\right)^2 \left(\frac{\exp(\Theta_v/T)}{(\exp(\Theta_v/T) - 1)^2}\right) \qquad (47)$$

$$(47) \qquad = Nk\left(\frac{\Theta_v/2T}{\sinh \Theta_v/T}\right)^2 \qquad (48)$$

are the contributions to the internal energy and heat capacity from a single vibrational mode. For polyatomic molecules, add $3n - 6$ or 5 terms like those in (46) and (47). Vibrational and rotational temperatures for several diatomic molecules are given in Table 3.2.

Exercise 3.10-6

Verify the formula given in Eq. (47) for the vibrational contribution to the heat capacity of a gas composed of diatomic molecules. (**Hint:** Start from Eq. (46).)

3.10.7. *Monatomic crystal*

Each atom in the crystal contributes three vibrational modes. Therefore, in the high-temperature limit $(T \gg \Theta_{\mathrm{v}})$ Eq. (39) yields for the molar vibrational energy

$$(46) \qquad\qquad\qquad U_{\mathrm{m,v}} = 3RT \qquad\qquad\qquad (49)$$

in agreement with the classical equipartition principle

$$(49) \qquad\qquad\qquad C_{\mathrm{V,m,v}} = 3R \qquad\qquad\qquad (50)$$

which explains the empirical law of Dulong and Petit for the heat capacity of monatomic crystals at elevated temperatures. However, the heat capacity of crystalline solids is strongly temperature dependent, tending to zero as $T \to 0$. One of the earliest applications of Planck's quantum hypothesis was Einstein's treatment of this problem in 1905. With the simple *ad hoc* assumption that the actual vibrational frequencies in a crystal could be replaced by a single average value v_{E}, the molar heat capacity of a monatomic crystal would be given by

$$(47) \qquad\qquad C_{V,\mathrm{v}} = Nk \left(\frac{\Theta_{\mathrm{E}}}{T}\right)^2 \left(\frac{\exp(\Theta_{\mathrm{E}}/T)}{(\exp(\Theta_{\mathrm{E}}/T) - 1)^2}\right) \qquad\qquad (51)$$

The Einstein function (28) shows the right sort of strong variation of $C_{V,\mathrm{v}}$ with T, but comparison with experimental data shows that the approximation of a single frequency may be too drastic. Let $g(v)\, dv$ be the number of frequencies between v and $v + dv$. Debye assumed that the crystal vibrations could be approximated by those of a three-dimensional elastic solid, for which the frequency distribution $g(v)$ is proportional to

v^2, up to a high-frequency cut-off v_D. This frequency distribution yields an expression for $C_{V,v}$ that cannot be evaluated in a closed form except at low temperatures, where

$$C_{V,v} = aT^3, \quad T \ll \Theta_D = hv_D/k, \quad a = \text{constant} \tag{52}$$

At higher temperatures, tables must be used. Equation (52) is known as the *Debye T^3-law*. Alternatively, the simpler Einstein formula (51) can be used with an empirical adjustment of the Debye temperature determined from (52) (Guggenheim 1967).

Answers to Exercises 3.10

Exercise 3.10-1

The solution to (6) is $\psi = A\exp(im\phi)$ because $d^2\psi/d\phi^2 = -m^2\psi$ and therefore (6) is satisfied with $E = \hbar^2 m^2/2I$. The normalization condition is given by $\int_0^{2\pi} A^2 d\phi = 1$ so that $A = 1/\sqrt{2\pi}$. For the first four levels:

m	ω	$E/(\hbar^2/2I)$
0	1	0
±1	2	1
±2	2	4
±3	2	9

Exercise 3.10-2

From Cohen *et al.* (2007) p. 111, $hc = 1.986446 \times 10^{-25}$ J m, $k = 1.38065 \times 10^{-23}$ J K^{-1} so that

$$\frac{hc}{k} = \frac{1.986446 \times 10^{-23} \text{ J cm}}{1.38065 \times 10^{-23} \text{ J K}^{-1}} = 1.4388 \text{ cm K}$$

Exercise 3.10-3

(a) The water molecule has one two-fold axis of symmetry. The rotational sub-group contains just two operators, the identity and the rotation through π about the two-fold axis. Therefore, $\sigma = 2$. (b) The easiest way to visualize the tetrahedral shape of methane is to draw a cube, place the carbon atom at the centre and a hydrogen atom at alternate corners of the cube. If you don't want to be bothered to do this yourself, tables of all the symmetry operators for the point groups are given in virtually every book on group theory (e.g. Cotton 1990, Jacobs 2005). From your sketch, or the literature, observe that the point symmetry group of methane is T_d, the rotational symmetry operators are $3C_2$ and $4C_3$ and the number of indistinguishable configurations is $1 + 3 + 8 = 12$. The identity is, of course, counted only once. (c) The shape of benzene is that of a hexagon. Though the point group is D_{6h}, the rotational sub-group is C_6, with a C_6 axis and coincident C_3 and C_2 axes normal to the plane of the molecule, and two sets of three C_2 axes (one set through the atoms and one set bisecting the bonds). The number of indistinguishable configurations achieved through rotations is therefore $1 + (2 + 2 + 1) + 3 + 3 = 12$. (It would make no difference if we ignored the coincident C_3 and C_2 axes and instead counted five configurations from the C_6 axis, giving $1 + 5 + 3 + 3 = 12$ for σ.)

Exercise 3.10-4

The series in (22) is a *geometric progression*, with first term 1 and common ratio $\exp(-\Theta_v/T)$. Its sum is therefore

$$(45) \qquad q_v = \sum_{v=0}^{\infty} \exp(-v\Theta_v/T) = \frac{1}{1 - \exp(-\Theta_v/T)} \qquad (46)$$

Exercise 3.10-5

If the energy of a harmonic oscillator could be zero, its position r would be exactly at $r = r_e$, so that both energy and position would be known precisely, which is not allowed according to the Heisenberg uncertainty principle.

Exercise 3.10-6

$$U_v = Nk \left(\frac{\Theta_v}{2} + \frac{\Theta_v}{\exp(\Theta_v/T) - 1} \right) \qquad (47)$$

Differentiating U_v with respect to T at constant V, the vibrational contribution to the molar heat capacity at constant volume is

$$(47) \qquad C_{V,m,v} = R \left(\frac{-\Theta_v(-\Theta_v/T^2)[\exp(\Theta_v/T)]}{[\exp(\Theta_v/T) - 1]^2} \right)$$

$$C_{V,m,v} = R \left(\frac{\Theta_v}{2} \right)^2 \left(\frac{\exp(\Theta_v/T)}{(\exp(\Theta_v/T) - 1)^2} \right) \qquad (48)$$

3.11. Thermodynamic Functions for a Monatomic Ideal Gas

This collection of structureless particles is one of the simplest systems that can be imagined and yet is still of practical use (the rare gases, for example). It is used here for illustrative purposes. Referred to an energy zero, U_0, which is that of one mole of substance, with the molecules at rest in their ground nuclear and electronic states

$$\text{(3.6.12), (3.7.11), (3.8.2)} \qquad U_{\mathrm{m}} = U_0 + kT^2 \frac{\partial \ln Q}{\partial T} = U_0 + \frac{3}{2}RT \qquad (1)$$

$$H_{\mathrm{m}} = U_0 + \frac{3}{2}RT + pV_{\mathrm{m}} \qquad (2)$$

With $c = 0$ (see Section 3.8)

$$\text{(3.9.27)} \qquad J_{\mathrm{m}} = -\frac{F_{\mathrm{m}}}{T} = R \ln Q = \ln[(2\pi mkT/h^2)^{3/2}V/N] \qquad (3)$$

$$\text{(3.8.19)} \qquad S_m/R = \ln[(2\pi m/h^2)^{3/2}(kT)^{5/2}/p] + 5/2 \qquad (4)$$

In the event that the ground state is degenerate, a further term $\ln \omega_0$ must be added to the RS of (4). This formula for the entropy is independent of the energy zero but these are still conventional entropies unless the entropy of the pure, perfect crystal is zero at $0\,\mathrm{K}$, which may be assumed to be so for the rare gases, but may not be so when these formulae (which are the contributions from the translational partition function) are used for more complicated materials. The *ab initio* calculation of a thermodynamic property is an exciting event and was rightly regarded as such when first done by Sackur (1911, 1912) and by Tetrode (1912a, b; 1915). The *standard entropy* of a substance is the entropy of one mole of the substance in its standard state, that is, its normal physical state at the standard temperature $T^{\ominus}$ of 298.15 K and standard pressure $p^{\ominus}$ of 10^5 Pa $(=1\,\mathrm{bar})$.[b]

For numerical calculation, (4) may be written as

$$\text{(4)} \qquad \frac{S_{\mathrm{m}}(T,p)}{R} = \frac{3}{2} \ln M_r + \frac{5}{2} \ln \frac{T}{T^{\ominus}} - \ln \frac{p}{p^{\ominus}}$$

$$+ \frac{5}{2} + \ln[(2\pi m_u/h^2)^{3/2}(kT^{\ominus})^{5/2}/p^{\ominus}] \qquad (5)$$

---

[b]There is more to standard states than the reader might infer from this statement. Necessary qualifications will be introduced in the appropriate places.

where M_r is the *relative molar mass* of the substance (that is the mass of one molecule divided by one twelfth of the mass of one atom of ^{12}C) and m_u is the *atomic mass constant*, the value of which is 1.66054×10^{-27} kg.

Example 3.11-1

Calculate the standard entropy of argon. Verify that the argument of ln in the last term in (5) is a pure number.

From Eq. (3.10.5), the standard entropy is

$$\frac{S_m^{\ominus}}{R} = \frac{3}{2}\ln M_r + \frac{5}{2} + \ln[(2\pi m_u/h^2)^{3/2}(kT^{\ominus})^{5/2}/p^{\ominus}]$$

$$= \frac{3}{2}\ln 39.948 + \frac{5}{2} + \ln\left[\frac{2\pi(1.66054 \times 10^{-27}\,\mathrm{kg})}{(6.626075 \times 10^{-34}\,\mathrm{J\,s})^2}\right]^{3/2}$$

$$\times \left[\frac{(1.38066 \times 10^{-23} \times 298.16\,\mathrm{J})^{5/2}}{10^5\,\mathrm{Pa}}\right]$$

$$= 5.53136 + 2.5 + 1.30062 + 9.29402$$

$$= 18.626, \quad \mathrm{or}, \quad S_m^{\ominus} = 154.86\,\mathrm{J\,K^{-1}mol^{-1}}$$

The units of the argument of ln in this term are

$$\frac{\mathrm{kg}^{3/2}\,\mathrm{J}^{5/2}}{\mathrm{J}^3\,\mathrm{s}^3\mathrm{N}\,\mathrm{m}^{-2}} = \frac{\mathrm{kg}^{3/2}}{\mathrm{s}^3\,\mathrm{m}^{-3}\,\mathrm{J}^{3/2}} = 1,$$

so the argument is dimensionless, as expected.

Problems 3

3.1 Show that the linear momentum of the free particle constrained to move in one dimension between two barriers is $p = \hbar k$. (**Hint**: "Free" means that the potential energy is zero between the barriers.) Find the momentum $\boldsymbol{p}$ for the particle in a box.

3.2 Calculate the wavelength of slow neutrons of energy 0.1 eV.

3.3 Find the most probable speed v_p of the molecules in an ideal gas. (**Hint**: Use the Maxwell distribution of speeds.)

3.4 Using Mathcad or any other mathematical package of your choice, plot $f(v)$ against v for nitrogen at $T = 300\,\mathrm{K}$, $500\,\mathrm{K}$ and $1000\,\mathrm{K}$.

3.5 Find the rate of effusion through a small hole and show how this equation can be used to calculate the vapour pressure of a solid by measuring the rate of effusion.

3.6 Find the fundamental equations for an open phase for the thermodynamic potentials J and Y.

3.7 x, y, and z are independent variables and

$$Ax + By + Cz = 0 \qquad (P3.7.1)$$

What can be said about the values of A, B and C?

3.8 Derive an expression for the chemical potential of a pure, ideal gas in terms of the molecular partition function q.

3.9 Calculate the contributions from translation, rotation and vibration to the thermodynamic functions U, H and S for 1 mole of N_2 at 101.325 kPa and (a) 300 K (b) 500 K (c) 1000 K. Calculate also the values of F and G at these three temperatures.

4

The Third Law of Thermodynamics

4.1. Entropy and Probability

An isolated system changes spontaneously until it reaches a state of equilibrium, unless some external constraint exists to prevent this change. For a natural process in this isolated system, $dS > 0$, so that S increases until it reaches the maximum value consistent with the prescribed values of (U, V, N). If the independent variables are changed to new values, then the system changes to a new equilibrium state in which S is a maximum, consistent with the new (U, V, N). Each thermodynamic state has a certain probability of occurrence. (A quantitative measure of the probability of a thermodynamic state will be defined below, after Eq. (1).) A natural process is a transformation from a state of lower probability to a state of higher probability. The equilibrium state is a state of maximum entropy S; it is also the most probable state. We therefore expect a correlation between S and the thermodynamic probability Ω.

Let A and B denote two separate systems. Since S is an extensive property, the entropy of the combined system AB is

$$S_{AB} = S_A + S_B \tag{1}$$

Recall (Section 3.1) that each assignment of the molecules in A to the available energy states constitutes a quantum state of the macroscopic system A and that since each such assignment has the same *a priori* probability, the number of such assignments is the *thermodynamic probability* Ω of that macroscopic state. But, since any of the Ω_A assignments in A may be combined with any one of the Ω_B assignments in B, probabilities are multiplicative so that

$$\Omega_{AB} = \Omega_A \Omega_B \tag{2}$$

Define the relation between S and Ω by

$$S = f(\Omega) + c \tag{3}$$

where c is an extensive constant that is independent of the thermodynamic state of the system but dependent on the nature (chemical composition) of the system

$$(1),\ (2),(3) \qquad f(\Omega_A\Omega_B) + c_{AB} = f(\Omega_A) + c_A + f(\Omega_B) + c_B \qquad (4)$$

$$(4) \qquad\qquad\qquad\qquad c_{AB} = c_A + c_B \qquad\qquad\qquad\qquad (5)$$

$$(4) \qquad\qquad\qquad\quad f(\Omega_A\Omega_B) = f(\Omega_A) + f(\Omega_B) \qquad\qquad\quad (6)$$

Equation (6) is satisfied if f is the logarithmic function, so that

$$(1),\ (5),\ (6) \qquad\qquad\quad {}^{\bullet}\ \ S = k' \ln \Omega + c \qquad\qquad\qquad (7)$$

where k' is a universal constant and c is a constant that is independent of the thermodynamic state of the system but dependent on the nature (chemical composition) of the system. In the next section we shall find k' by considering a particular system, namely an ideal gas.

4.2. The Boltzmann Entropy Equation

Consider an adiabatically isolated system that consists of two sub-systems which are bulbs of volumes V_1 and $V_2 = V_1$. V_1 is filled with one mole of ideal gas and V_2 is evacuated. The two sub-systems are connected via a valve that is initially closed, so that the ideal gas occupies V_1 only. On opening the valve, the gas will occupy the whole system. The entropy change per mole accompanying this process is

$$(2.3.19) \qquad\qquad\qquad \Delta S_m = R \ln 2 > 0 \qquad\qquad\qquad (1)$$

Alternatively, we might say that the system has changed from a state one of lower probability, in which all the gas molecules are in V_1 at the instant the valve is opened, to a state of higher probability in which they occupy the whole volume $V_1 + V_2$. The probability of a single molecule being in the volume $V_1 + V_2$ is twice that of it being in the volume V_1 and so the probability of N molecules being in $V_1 + V_2$ is 2^N times that of them being in V_1. In the initial state one all N molecules fill the total available volume V_1 and the relative probability of this state is one. Therefore

$$\Omega_2/\Omega_1 = 2^N \qquad\qquad\qquad\qquad (2)$$

$$(4.1.7),\ (2) \qquad\qquad \Delta S_m = k' \ln(\Omega_2/\Omega_1) = k' N \ln 2 \qquad\qquad (3)$$

(1), (3) $$k' = k \tag{4}$$

(4.1.7), (4) $$S = k \ln \Omega + c \tag{5}$$

Equation (5) is the Boltzmann equation connecting the entropy S with the thermodynamic probability Ω; k is the Boltzmann constant introduced in (1.3.6) and c is a constant which is independent of the thermodynamic state of the system but dependent on the nature (chemical composition) of the system.

Example 4.2-1 The random alloy

Consider the mixing of two crystals, one of atoms of A, the other composed of B atoms, by diffusion to form a random alloy. Since the N_A atoms of A and N_B atoms of B are distributed randomly over the $N_A + N_B$ sites

$$\Omega_{AB} = \frac{(N_A + N_B)!}{N_A! \, N_B!} \tag{6}$$

(7), (6) $$\ln \Omega_{AB} = -N_A \ln \left(\frac{N_A}{N_A + N_B} \right) - N_B \ln \left(\frac{N_B}{N_A + N_B} \right) \tag{7}$$

(5), (8) $$S_{AB} = k(N_A + N_B)(-x_A \ln x_A - x_B \ln x_B) + c_{AB} \tag{8}$$

In the pure, perfect A crystal, there is only one possible arrangement of the N_A indistinguishable atoms on the N_A sites, so that $\Omega_A = 1$ and

(5) $$S_A = c_A \tag{9}$$

Similarly, $\Omega_B = 1$ and

(5) $$S_B = c_B \tag{10}$$

Since c in Eq. (5) is an extensive constant that is independent of the thermodynamic state

$$c_{AB} = c_A + c_B \tag{11}$$

(9), (10), (11), (12) $$S_m = R(-x_A \ln x_A - x_B \ln x_B) \tag{12}$$

which is identical with Eq. (2.6.1) for the entropy of mixing. Since c is that part of the entropy which is due to nuclear degrees of freedom and to isotopic mixing, it does not change with T or with any terrestrial thermodynamic process (excepting nuclear reactions), and so c is usually ignored in Eq. (6) and the Boltzmann entropy equation written as

$$(6) \qquad\qquad\qquad S = k\ln\Omega \qquad\qquad\qquad (13)$$

as inscribed on Boltzmann's tombstone in the Zentralfriedhof in Vienna.

4.3. The Third Law of Thermodynamics

For a large number of isothermal processes I $\rightarrow$ II, it is found experimentally that

$$S_{\text{II}} - S_{\text{I}} = \Delta S \rightarrow 0 \text{ as } T \rightarrow 0 \text{ K} \qquad\qquad (1)$$

(Nernst 1906). Examples are:

 (i) a phase transition between two crystalline phases α, β, for example, S(monoclinic) $\rightarrow$ S(rhombic);
 (ii) the phase transition between liquid and solid He;
 (iii) a chemical reaction between two crystalline phases, e.g. $Pb + I_2 = PbI_2$, at specified T, p;
 (iv) the transformation of a crystal at $(T, x) \rightarrow (T, x')$, where x is p, V or B.

This result that

$$\lim_{T \rightarrow 0 \text{ K}} \Delta S = 0 \qquad\qquad (2)$$

where ΔS is the entropy change in an *isothermal* process, is presumed to be of general validity for pure phases in internal equilibrium, or in frozen metastable equilibrium, if this is not disturbed. It was originally called the *Nernst heat theorem* but has since come to be known as the **Third Law of Thermodynamics**. For any such process I $\rightarrow$ II

$$\Delta S = S_{\text{II}} - S_{\text{I}}$$
$$= (k\ln\Omega_{\text{II}} + c_{\text{II}}) - (k\ln\Omega_{\text{I}} + c_{\text{I}})$$
$$= k\Delta(\ln\Omega) \qquad\qquad (3)$$

since c, being independent of the thermodynamic state, vanishes

$$(2), (3) \qquad\qquad\qquad \Delta S = k\Delta(\ln\Omega_0) = 0 \qquad\qquad (4)$$

which is a re-statement of the Third Law in terms of Ω. Since (4) is true for a variety of processes and materials, the question naturally arises, "is

$\ln \Omega_0 = 0$?", i.e. "is $\Omega_0 = 1$?". At very low temperatures (close to 0 K) all systems are in their lowest energy state and as long as we limit consideration to the electronic, vibrational, rotational and translational degrees of freedom, the ground state of perfect, pure crystals and of liquid helium is non-degenerate, i.e. $\Omega_0 = 1$. If we accept that, with certain well-understood exceptions (to be discussed in detail in Section 4.4), $\Omega_0 = 1$ as $T \to 0$ K then, with S_0 the value of S as $T \to 0$ K

$$S_0 = k \ln \Omega_0 + c \tag{5}$$

(5) $$S_0 = c \quad \text{(when } \Omega_0 = 1) \tag{6}$$

Therefore, if $\Omega_0 = 1$ and if the convention is adopted that contributions to S_0 from nuclear degrees of freedom and isotopic mixing (which are in c) are ignored

(6) $$S_0 = 0 \quad (\Omega_0 = 1, c \text{ excluded}) \tag{7}$$

This is the form in which the Third Law is commonly stated (Planck 1945), though the conditions are not always clarified. The contributions from c are normally unaffected by terrestrial physical and chemical changes and so their neglect by setting $c = 0$ is inconsequential. Any frozen-in metastability contributes to Ω_0.

$$S - S_0 = S - k \ln \Omega_0 - c = S_{\text{cal}} \tag{8}$$

is called the calorimetric entropy because it is the entropy calculated by extrapolation using calorimetric data. At temperatures low enough for all substances (except liquid He under its own vapour pressure) to be a crystalline solid

(2.3.14) $$S(T) - S_0 = \int_0^T (C_p/T) \mathrm{d}T \tag{9}$$

There is a lower limit, say T_1, below which C_p cannot be measured, but at low temperatures $T < T_1$

$$C_p \approx C_V = aT^3 \tag{10}$$

may be used (for all crystalline substances) to extrapolate to $T = 0$ K. The dependence of C_V on T in (10) is known as the Debye T^3-law (3.10.52). Equations (9) and (10) allow the extrapolation of $S - S_0$ down to $T = 0$ K. We shall adopt the usual convention of ignoring c (when it is possible to do so). Then the experimental evidence is that $S - S_0$ is zero at $0\,\mathrm{K}$ and, when there is no reason to believe that Ω_0 is not 1, S is zero at $0\,\mathrm{K}$.

4.4. Experimental Observations Concerning the Third Law

4.4.1. *Calorimetric entropy of rhombic and monoclinic sulphur*

The heat capacities of the rhombic and monoclinic forms of sulphur can be measured down to low temperatures, by taking advantage of the slow rate of conversion of S(monoclinic) to the stable rhombic form. The entropy of transition at the transition temperature of $368.5\,\mathrm{K}$ (from $\Delta H/T$) is

$$S_\mathrm{m}(\mathrm{monocl}, 368.5\,\mathrm{K}) - S_\mathrm{m}(\mathrm{rh}, 368.5\,\mathrm{K}) = 1.09\,\mathrm{J\ K^{-1}mol^{-1}} \tag{1}$$

The calorimetric entropies of the two forms are

$$S_\mathrm{m}(\mathrm{rh}, 368.5\,\mathrm{K}) - S_\mathrm{m}(\mathrm{rh}, 0\,\mathrm{K}) = (36.86 \pm 0.20)\mathrm{J\ K^{-1}mol^{-1}} \tag{2}$$

$$S_\mathrm{m}(\mathrm{monocl}, 368.5\,\mathrm{K}) - S_\mathrm{m}(\mathrm{monocl}, 0\,\mathrm{K}) = (37.82 \pm 0.40)\mathrm{J\ K^{-1}mol^{-1}} \tag{3}$$

$$(1),\ (2),\ (3)\quad S_\mathrm{m}(\mathrm{monocl}, 0\,\mathrm{K}) - S_\mathrm{m}(\mathrm{rh}, 0\,\mathrm{K}) = (0.13 \pm 0.60)\mathrm{J\ K^{-1}mol^{-1}} \tag{4}$$

The entropy difference between the two crystalline forms of S at $0\,\mathrm{K}$ is therefore zero, within experimental error, confirming the Nernst statement of the Third Law Eq. (4.3.1).

4.4.2. *Vapour pressure of liquid helium*

We shall apply the Maxwell relation

$$(2.2.14) \qquad \left(\frac{\partial S}{\partial V}\right)_T = \left(\frac{\partial p}{\partial T}\right)_V \tag{5}$$

to liquid helium. Because solid He is unstable under its own vapour pressure, He remains a liquid down to temperatures in the micro-kelvin range (unless the pressure is increased). Since the system consists of two phases, it has only one degree of freedom and p depends only on T. Therefore, the partial derivative on the RS of (5) may be replaced by the total derivative dp/dT. At constant T and S, V can only change when an amount dn is transferred between the two phases. Since S and V are both extensive properties, the partial derivative on the LS of (5) is independent of dn and so may be replaced by $\Delta S/\Delta V$. Therefore

(5)
$$\frac{dp}{dT} = \frac{\Delta S}{\Delta V}$$
(6)

which is known as the Clapeyron equation. ΔS in (6) is the entropy change when a particular amount is transferred from the condensed phase to the vapour phase and ΔV is the corresponding volume change that accompanies the same process. It was shown by Simon and Swenson (1950) that, at temperatures below $1.4\,\mathrm{K}$

$$dp/dT = 0.425\,T^7$$
(7)

and thus becomes extremely small as T is lowered further. Therefore from (6), the entropy of vaporization at constant T also tends to zero as $T \to 0\,\mathrm{K}$, in accordance with the Third Law.

4.4.3. *Thermal expansion at low temperatures*

Exercise 4.4-1

What can be predicted about the coefficient of thermal expansion at low temperatures? (**Hint**: Use the Third Law and a Maxwell relation.)

4.4.4. *Frozen-in configurational entropy*

In the CO crystal, a random mixture of CO and OC (defined in relation to a particular crystal direction) is frozen-in and the rate of conversion to the more stable state of perfectly aligned molecules is negligibly small. It is possible to calculate the entropy of such crystals by applying the statistical-thermodynamical formulae in Chapter 3. The results are called spectroscopic entropies S_{spect}, because energy levels determined by spectroscopic methods are used in the calculation. Usually, S_{spect} and S_{cal} agree, but when they do not, this is a clear indication of a degenerate ground state, $\Omega_0 \neq 1$, for then

$$S_{\mathrm{spect}} - S_{\mathrm{cal}} = k \ln \Omega_0$$
(8)

Data for four such crystals are given in Table 4.1. For a random alignment of CO or NNO molecules, $\Omega_0 = 2^N$ and $k \ln \Omega_0 = R \ln 2 = 0.69\,R$, for one mole. Table 4.1 shows that the random alignment of CO and NNO molecules is not quite complete. In the NO crystal, the molecular unit is a pair of oppositely

Table 4.1.   Ground state degeneracy and configurational molar entropy in CO, NNO, NO and $FClO_3$. The 10–20% discrepancy between the third and fourth columns is due to some ordering having occurred as the crystals are cooled.

	Ω_0	$(1/N_A) \ln \Omega_0$	$(S_{m,spect} - S_{m,cal})/R$
CO	2^{N_A}	0.69	0.55
NNO	2^{N_A}	0.69	0.55
NO	$2^{N_A/2}$	0.35	0.33
$FClO_3$	4	1.38	1.22
Ice	$(3/2)^{N_A}$	0.405	0.40

Arrangement of $(NO)_2$ units in the NO crystal:

NO

ON

It is the random arrangement of $N_A/2$ of these units that gives rise to the low-temperature configurational entropy of NO.

oriented molecules, as shown in Table 4.1. There are $N/2$ randomly oriented pairs, so that $\Omega_0 = 2^{N/2}$, as confirmed by the data in Table 4.1.

In ice, the HOH molecules are oriented such that each O atom is surrounded by the two H atoms of the HOH molecule ("short" bonds) and two more H atoms that form hydrogen bonds ("long" O...H...O bonds) (see Barnes *et al.* 1979). Each O atom has four adjacent O atoms at the corners of a tetrahedron. The two H atoms of the central molecule are oriented towards two of the four adjacent O atoms. Each of these has only two of the four tetrahedral directions available to form H bonds (the other two being occupied by H atoms in the short bonds). Ω_0 is therefore $3/2$.

Exercise 4.4-2

Suggest a reason why the configurational entropy of SCO is zero whereas that of NNO and CO is non-zero.

4.4.5. *Hydrogen and deuterium*

In Section 4.3 we stated that it was usual to ignore nuclear spin states and isotopic mixing in calculating the conventional entropy S, by including them in a constant c, which was then ignored. This is a satisfactory method of simplifying the calculation of S since, for most substances, c is independent of the thermodynamic state of the substance. However, hydrogen and its isotope deuterium are exceptions to this general rule. Because of their low mass, correspondingly small moments of inertia, and comparatively low intermolecular forces, these molecules rotate at very low temperatures, even in the crystalline state. Ordinary hydrogen is a mixture (in the ratio 3:1) of *ortho*-hydrogen in which the nuclear spins are parallel, with a statistical weight $\omega = (2 \times 1) + 1 = 3$, and *para*-hydrogen, in which the nuclear spins are anti-parallel, with a statistical weight $\omega = (2 \times 0) + 1 = 1$. Therefore, the contribution (per mole) to S_0 from nuclear spin degeneracy is

$$S_0 = k \ln \Omega_0 (\text{nuclear spin}) = \frac{3}{4} R \ln 3 + \frac{1}{4} R \ln 1 = \frac{3}{4} R \ln 3 \qquad (1)$$

The contribution to S_0 from the mixing of o-H_2 and p-H_2 is

$$S_o(o, p \text{ mixing}) = -R \left[\frac{3}{4} \ln \frac{3}{4} + \frac{1}{4} \ln \frac{1}{4} \right] \qquad (2)$$

but this is often neglected, in accordance with the usual convention, since it is constant, provided the natural mole fractions of o and p are not interfered with. Below $12\,K$, the heat capacity of hydrogen is anomalously larger than that expected from the T^3-law. This is attributed to the ordering of the molecules of *ortho*-hydrogen.

It is possible to carry out the catalytic conversion of the o-H_2 in ordinary hydrogen to p-H_2. This conversion is accompanied by the decrease to zero of the entropy of ordinary hydrogen at $0\,K$, which is the sum of the contributions from nuclear spins and o, p mixing and which amounts to

$$(1), (2) \qquad S_0 = \frac{3}{4} R \ln 3 - R \left[\frac{3}{4} \ln \frac{3}{4} + \frac{1}{4} \ln \frac{1}{4} \right] = R \ln 4 \qquad (3)$$

Alternatively, if the usual convention of neglecting c had been applied to hydrogen, then the entropy of pure p-H_2 on this scale would be $-R \ln 4$ (*cf.* Guggenheim 1967). The analysis for deuterium is similar.

Answers to Exercises 4.4

Exercise 4.4-1

From the Maxwell relation

$$(2.2.11) \qquad \left(\frac{\partial V}{\partial T} \right)_p = - \left(\frac{\partial S}{\partial p} \right)_T \qquad (9)$$

$$(9) \qquad \Delta S = S(T, p) - S(T, 0) = - \int_0^p \left(\frac{\partial V}{\partial T} \right)_T dp \qquad (10)$$

From the Third Law, $\Delta S \to 0$ as $T \to 0\,\mathrm{K}$, and so, therefore, does the expansivity. This has been verified by experimental measurements, *inter alia*, on copper, silver and aluminium.

Exercise 4.4-2

CO and NNO are linear and the intermolecular forces for the two orientations are sufficiently similar for molecules to be rotated through $180°$ with little change in the total energy. Therefore, the molecules adopt a random orientation as the crystal forms. The same is not true for SCO, which therefore crystallizes with $\Omega_0 = 1$.

4.5. The Thomsen and Berthelot Rule

At constant T, p

$$(2.2.8) \qquad \Delta G = \Delta H - T\Delta S \qquad (1)$$

Since ΔS for an isothermal process does not become infinite as $T \to 0\,\mathrm{K}$

$$(1) \qquad \Delta G = \Delta H (T \to 0\,\mathrm{K}) \qquad (2)$$

Equation (2) is a formal statement of the empirical rule of Thomsen and Berthelot (TB) that the equilibrium state is the one which satisfies (2). It was Nernst's attempt to find a theoretical basis for the TB rule that led him to postulate the general validity of (4.3.1). Actually, the TB principle was that the equilibrium state to which a system evolves is the one which results in the production of the most heat, or, in modern terminology, maximizes $-\Delta H$ (the heat evolved at constant pressure (see (2.2.5)) and therefore minimizes G, in agreement with (2.5.12).

The TB principle actually works at moderate temperatures well above zero. This implies that not only are ΔH and ΔG equal at zero temperature, but that their gradients $(\partial \Delta H/\partial T)_p$ and $(\partial \Delta G/\partial T)_p$ are also nearly equal over quite a substantial temperature range. Now, at $T = 0\,\mathrm{K}$

$$(1), (3.3.1) \qquad \left(\frac{\partial \Delta G}{\partial T}\right)_p = \Delta\left(\frac{\partial G}{\partial T}\right)_p = -\Delta S = 0 \qquad (3)$$

$$\left(\frac{\partial \Delta H}{\partial T}\right)_p = \Delta\left(\frac{\partial H}{\partial T}\right)_p = \Delta C_p = 0 \qquad (4)$$

since the heat capacity at constant pressure vanishes at zero temperature, as shown by Eq. (4.3.10). From (2), (3) and (4), we can expect ΔH and

ΔG to be approximately equal over (at least) a moderate range of T, as found by Thomsen and Berthelot.

4.6. Unattainability of $T = 0\,\mathrm{K}$

4.6.1. *Paramagnetism*

The most efficient method of obtaining really low temperatures is by *adiabatic demagnetization*. The third postulate of quantum mechanics (which was not needed until now) is:

III Electrons possess intrinsic angular momentum (called "spin angular momentum") described by the spin vector s and a magnetic moment

$$m = -g_e s(\mu_B/\hbar) = -g_e s(e/2m_e) \tag{1}$$

μ_B is the atomic unit of magnetic moment called the *Bohr magneton*

$$(1) \qquad \mu_B = e\hbar/2m_e = 0.9274 \times 10^{-23}\,\mathrm{J\,T^{-1}} \tag{2}$$

For free electrons

$$g_e = 2.00232 \tag{3}$$

The only allowed value of the spin quantum number s is $1/2$, so that $m_s = \pm\frac{1}{2}$. These numerical values of s and g_e have been chosen to agree with experimental results. In particular, the Stern–Gerlach experiment shows that $s = 1/2$ and that $g_e \approx 2$.

(**Note:** Readers not greatly concerned with the thermodynamics of adiabatic demagnetization, could omit the following explanations and rejoin the text at Section 4.6.2.)

We shall restrict consideration to atoms up to the first two rows of d-block elements, to which Russell–Saunders or LS coupling applies. In atoms with several unpaired electrons, the orbital angular momenta couple to give a total orbital angular momentum L, the spin angular momenta couple to give a total spin angular momentum S, and then L and S add to give the total angular momentum J.

Compounds containing unpaired electrons (notably salts of transition metals) have magnetic moments which tend to line up along the direction of the applied magnetic field (paramagnetism). The work done during the *isothermal* magnetization of a system is

$$(A2.2.18) \qquad dw = \int dV \mu_0 H \cdot dM = \int dV\, \mathbf{B} \cdot d\mathbf{M} \tag{4}$$

B is understood to mean the *magnetic induction associated with the external field*, since H is the external field (see Appendix A2, Section A2.2). Equation (4) is the contribution to the Helmholtz energy of a system in a magnetic field that is due to the interaction of the external magnetic field with the material system placed in that field. Therefore the fundamental equation for a closed phase for $F(T, M)$ is

$$\mathrm{d}F(T, M) = -S\mathrm{d}T + \int \mathrm{d}V\, \mathbf{B} \cdot d\mathbf{M} \tag{5}$$

For solid samples any $p\,\mathrm{d}V$ contribution is negligible and that is why V does not appear as an independent variable in (5). However, B is a more useful independent variable than M. Define the Helmholtz function $F(T, B)$ by

$$F(T, B) = F(T, M) - \int \mathrm{d}V \mathbf{M} \cdot \mathbf{B} \tag{6}$$

(5), (6) $\qquad\qquad \mathrm{d}F(T, B) = -S\mathrm{d}T - \int \mathrm{d}V \mathbf{M} \cdot \mathrm{d}\mathbf{B} \tag{7}$

With a suitable experimental arrangement (a small spherical sample at the centre of a solenoid whose length is sufficiently greater than the sample dimensions for end effects to be negligible) the magnetic field is uniform and

(7) $\qquad\qquad \mathrm{d}F(T, B) = -S\mathrm{d}T - \mathbf{m} \cdot \mathrm{d}\mathbf{B} \tag{8}$

where

$$m = \int \mathbf{M}\mathrm{d}V \tag{9}$$

is the *magnetic moment* of the system. Molecules with no unpaired electrons exhibit *diamagnetism* because an applied magnetic field induces circulation currents which give rise to a magnetic field which opposes the applied field. Diamagnetism involves the ground state orbitals and is a property of all matter, though in a system containing unpaired electrons it is masked by the larger paramagnetism. Induced paramagnetism may also occur, but requires the presence of low-lying excited state orbitals. Induced paramagnetism may be distinguished from the paramagnetism due to the angular momentum of unpaired electrons by being temperature independent, whereas the paramagnetism due to the magnetic moments of unpaired electrons is temperature dependent.

Consider a system in which each molecule contains an atom with unpaired $3d$ or $4f$ electrons and total angular momentum quantum number J. The magnitude of the total angular momentum is $\sqrt{J(J+1)}\hbar$. There are $2J+1$ states with the total angular momentum vector $\boldsymbol{J}$ aligned such that its projection along the direction of the magnetic field (chosen to be the z-axis) is $M_J\hbar$, where $M_J = -J, -J+1, \ldots, 0, 1, \ldots, +J$. The partition function of this system is

$$q = \sum_{M_J=-J}^{J} \exp(g_J M_J \mu_B B/kT) \tag{10}$$

where g_J is the Landé splitting factor

$$g_J = 1 + \frac{J(J+1) + S(S+1) - L(L+1)}{2J(J+1)} \tag{11}$$

Define

$$x = (g_J J \mu_B B/kT) \tag{12}$$

so that

$$(10), (12) \quad q = \sum_{M_J=-J}^{J} \exp(-x M_J/J) = \exp(x) \sum_{n=0}^{2J} \exp[(-x/J)^n] \tag{13}$$

The sum in (13) is a geometric progression with common ratio $\exp(x/J)$, so that

$$\begin{aligned}
q &= \exp(x) \left[\frac{1 - \{\exp(-x/J)\}^{2J+1}}{1 - \exp(x/J)} \right] \\
&= \left[\frac{\exp[x(2J+1)/2J] - \exp[-x(2J+1)/2J]}{\exp(x/2J) - \exp(-x/2J)} \right] \\
&= \frac{\sinh[x(2J+1)/2J]}{\sinh[x/2J]} = \frac{\sinh[(2J+1)g_J\mu_B B/2kT]}{\sinh[(g_J\mu_B B/2kT)]}
\end{aligned} \tag{14}$$

The magnetic contribution to the molar Helmholtz function is therefore

$$F_m(T, \boldsymbol{B}) = -kT \ln Q = -RT \ln \frac{\sinh[(2J+1 g_J\mu_B B/2kT)]}{\sinh[(g_J\mu_B B/2kT)]} \tag{15}$$

The molar magnetic moment of the system is

$$(8), (15) \quad \boldsymbol{m} = -(\partial F_m/\partial \boldsymbol{B})_T \tag{16}$$

and the magnetization of the system

$$\boldsymbol{M} = \frac{1}{V_m} \left(\frac{\partial F_m}{\partial \boldsymbol{B}} \right)_T \tag{17}$$

This is perhaps an opportune time to remind the reader that $\boldsymbol{B}$ is the magnetic induction (or magnetic flux density) of *the external field* (see Eq. (4) *et seq.*) and that B is the magnitude of this vector quantity $\boldsymbol{B}$.

For large values of B, sinh x may be replaced by $\frac{1}{2}\exp x$, so that

$$(13),\ (16) \qquad F_m(T, \boldsymbol{B}) = -N_A J g_J \mu_B B, \quad \boldsymbol{M} = N_A J g_J \mu_B / V_m$$

$$(\mu_B B \gg kT) \qquad (18)$$

Equation (18) describes the saturation of the magnetic moment at high fields. At the other extreme of low B, expand each sinh term, retaining only the first two terms. Then expand the ln function, again retaining only the first two terms. This gives for the molar Helmholtz function

$$(15) \qquad F_m(T, \boldsymbol{B}) = -RT \ln(2J + 1) - N_A J(J + 1)(g_J \mu_B B)^2 / 6kT$$

$$(\mu_B B \ll kT) \qquad (19)$$

$$(19) \qquad V_m \boldsymbol{M} = N_A J(J + 1)(g_J \mu_B)^2 \boldsymbol{B} / 3kT \quad (\mu_B B \ll kT) \qquad (20)$$

Equation (20) states that $\boldsymbol{M}$ is proportional to $\boldsymbol{B}$ and inversely proportional to T, in accordance with *Curie's law*, which has been verified experimentally for a variety of substances.

4.6.2. *Adiabatic demagnetization*

For low fields, the molar entropy is

$$(19) \quad S_m / R = \ln(2J + 1) - (1/3)J(J + 1)(N_A g_J \mu_B B / RT)^2 \quad (\mu_B B \ll kT)$$
$$(21)$$

Since the system is of constant composition and volume changes are negligible, the entropy $S(T, \boldsymbol{B})$ remains constant during a reversible adiabatic change. Therefore, at sufficiently small fields

$$(20) \qquad\qquad B \propto T \quad \text{(adiabatic change)} \qquad\qquad (22)$$

Thus if a paramagnetic substance is magnetized at a low temperature of about 1 K and the magnetic field reduced adiabatically, really low temperatures in the micro- and even nano-kelvin range (using nuclear, instead of electronic moments) can be obtained by this process of *adiabatic demagnetization*. It might appear that T could be reduced to zero, but in fact this does not occur because all substances eventually cease to be paramagnetic and become either diamagnetic or ferromagnetic. This loss of paramagnetism is accompanied by a reduction in the molar entropy by $R \ln$

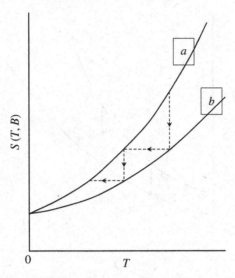

Figure 4.1. Temperature dependence of the molar entropy at very low temperatures for (a) unmagnetized material and (b) the magnetized material. The dashed lines show the path of a thermodynamic process in which the material is magnetized isothermally and then demagnetized adiabatically. The curves meet as T → 0 K, in accordance with Nernst statement of the Third Law.

(2S+1) so that S_m is reduced to zero when the field is zero and T is small but finite (Eq. (20)), in accordance with the Third Law, Eq. (4.3.2). Consider the plot of $S_m(T)$ shown in Figure 4.1. Curve a is for the unmagnetized material and curve b for magnetized material. Vertical and horizontal dashed lines show the entropy and temperature changes during isothermal magnetization and adiabatic demagnetization. Figure 4.1 shows that as T → 0 K, the two curves I and II meet, in accordance with the Nernst statement that

$$\lim_{T \to 0\,\mathrm{K}} \Delta S = 0 \quad (\Delta T = 0) \tag{23}$$

the isothermal process here being

$$S^{\mathrm{I}}(T, 0) \to S^{\mathrm{II}}(T, \boldsymbol{B}) \quad (\Delta T = 0) \tag{24}$$

Figure 4.2 shows that if

$$\lim_{T \to 0\,\mathrm{K}} |\Delta S| > 0 \tag{25}$$

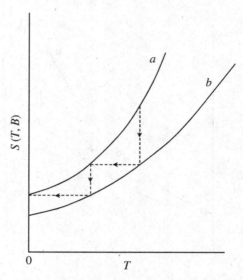

Figure 4.2.   Hypothetical temperature dependence of the molar entropy for a material that remains paramagnetic at 0 K. Since no such material exists, it is impossible to attain a temperature of zero on the thermodynamic scale.

and if a material could be found that remained paramagnetic, then it would be possible to reach $T = 0\,\mathrm{K}$ in a finite number of steps. But no such material exists.

Problems 5

4.1 Nernst had doubts about whether the Third Law was a new law or whether it could be deduced from the Second Law. What do you think? (**Hint:** Consider the inaccessibility of a temperature of $0\,\mathrm{K}$.)

4.2 Below 1 K, the entropy of $CoSO_4 \cdot 7H_2O$ is derived almost entirely from electron spin. Given that the ground state of Co(II) in this material is a quartet, calculate the electron-spin contribution to the entropy in this temperature range.

4.3 The mole fractions of the four isotopes in naturally-occurring Pb are: $x(^{204}Pb) = 0.015$, $x(^{206}Pb) = 0.236$, $x(^{207}Pb) = 0.226$, $x(^{208}Pb) = 0.523$. Calculate the molar entropy of isotopic mixing in natural lead. Is this term part of the Third Law entropy of Pb?

4.4 Experimental values of the heat capacity of Ag at a constant pressure of 101.3 kPa are given in Table P4.2. Calculate the entropy of silver at

Table P4.2. Molar heat capacity of silver as a function of temperature.

T/K	15	30	50	70	90	110	130	150
$C_{p,m}/\mathrm{J\,K^{-1}mol^{-1}}$	0.67	4.77	11.65	16.33	19.13	20.96	22.13	22.97
T/K		170	190	210	230	250	270	290	300
$C_{p,m}/\mathrm{J\,K^{-1}mol^{-1}}$	23.61	24.09	24.42	24.73	25.03	25.31	25.44	25.50

298.15 K by numerical integration. Explain any conventions involved. (**Hint**: Use Eqs. (4.3.9) and (4.3.10).)

4.5 Verify the value of $\Omega_0 = 1.5$ for ice and calculate the configurational entropy of ice. (**Hint**: The easiest way to visualize the tetrahedral arrangement of O atoms around a water molecule is to draw a cube. Place a HOH molecule at the centre and O atoms at alternate corners of the cube.)

4.6 Ordinary deuterium is a mixture, in the ratio 2:1, of *ortho*-deuterium with $\Omega = 1$ and *para*-deuterium with $\Omega = 3$. What value of the entropy of deuterium is to be expected on extrapolating crystal data to 0 K? State clearly the origin of the contributions included. If ordinary deuterium is converted to pure $o\text{-}D_2$, what is S_0 for $o\text{-}D_2$?

4.7 The conventional entropies of liquid and gaseous water at $T = 298.15$ K and $p = 101.325$ kPa are:

	$H_2O(\mathrm{liq})$	$H_2O(\mathrm{g})$
$S(298.15\,\mathrm{K},\ 1\ \mathrm{atm.})/\mathrm{J\ K^{-1}mol^{-1}}$	69.91	188.83

Calculate the entropy of vaporization under these conditions. Do you think it reasonable that your result is positive and comparatively large? (**Hint**: Compare the entropy change you have calculated with that in Eq. (4.4.1).)

4.8 The rotational temperature of H_2 is 85.4 K. Calculate the rotational contributions at 140 K to the entropy and the heat capacity of $o\text{-}H_2$ and of $p\text{-}H_2$.

4.9 The coefficient of thermal expansion of crystalline solids tends to zero as $T \to 0\,\mathrm{K}$. Show that in this limit the entropy of a crystalline solid is independent of pressure. What are the consequences with respect to the Third Law of Thermodynamics?

5

Systems of One Component

5.1. Real Gases

An ideal gas is one which obeys the equation of state

$$pV_m = RT \quad \text{(ideal gas)} \tag{1}$$

at all pressures. All gases (called *real gases*, when it is necessary to distinguish them from this hypothetical ideal gas) behave ideally at sufficiently low pressures but as the pressure increases, deviations from ideality occur, as described in Section 1.3.

It follows from (1) and the first thermodynamic equation of state (2.3.8) that the internal energy U and of an ideal gas is a function of T only (Ex. 2.3-1)

5.1.1. *The critical state*

The densities of the liquid ρ^L and saturated vapour ρ^G are equal at the critical temperature T_c. T_c may be found from the intersection of the linear plot of the mean density $\bar{\rho}$ (that is, the average of the vapour and liquid densities) against T, with the $\rho(T)$ curve, which gives T_c and $V_c = M/\rho_c$ (see Figure 5.1).

5.1.2. *Principle of corresponding states*

In any equation of state, the molar volume V_m depends on p, T and on two or more parameters which depend on the nature of the gas (e.g. van der Waals equation (P1.11.1) and the virial equation (1.3.8)). However, if we use the *reduced variables*

$$V_{m,r} = V_m/V_{m,c}, \quad p_r = p/p_c, \quad T_r = T/T_c \tag{2}$$

then, for many gases, to a good approximation

$$V_{m,r} = V_m(p_r, T_r) \tag{3}$$

with *no parameters* that are specific to any particular gas.

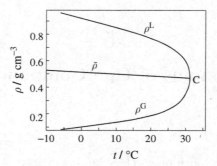

Figure 5.1. Density of liquid carbon dioxide ρ^L and of the saturated vapour in equilibrium with the liquid, ρ^G, as functions of temperature. $\bar{\rho}$ is the mean of ρ^L and ρ^G (after Cailletet and Mathias 1886). C marks the critical point.

Table 5.1. Critical constants for several gases. The relative molecular mass M_r is the ratio of the mass m of one molecule of a substance to the atomic mass constant, $m_u = m_a(^{12}C)/12 = 1.660\,538\,782(83) \times 10^{-27}$ kg, i.e. the mass of one atom of ^{12}C divided by exactly 12. The values of the reduced Boyle temperature for He and H_2 are abnormally high, due to quantum effects (data from Zemansky and Dittman 1997).

	M_r	p_c/MPa	$V_{m,c}$/cm^{3}mol^{-1}	T_c/K	$\dfrac{p_c V_{m,c}}{RT_c}$	T_B/K	T_B/T_c
He	4.003	0.229	57.8	5.21	0.306	22.64	4.35
Ne	20.18	2.73	41.7	44.8	0.306	121	2.70
Ar	39.95	4.86	75.5	150.7	0.293	411.5	2.73
Kr	83.80	5.48	92.1	209.4	0.290	575	2.75
Xe	131.30	5.84	118.8	289.8	0.288	768	2.65
H_2	2.016	1.297	65.0	33.23	0.305	110	3.31
N_2	28.013	3.39	90.2	126.0	0.292	327	2.59
O_2	31.99	5.04	74.5	154.3	0.292	406	2.63
CO	28.011	3.50	93.2	133.0	0.294	345	2.59
CO_2	44.010	7.38	94.2	304.2	0.288	715	2.35
CH_4	16.032	4.63	98.8	190.3	0.289	491	2.58

Exercise 5.1-1

Values of the critical constants p_c, V_c and T_c for several gases are shown in Table 5.1. Verify the value of the ratio $p_c V_c / RT_c$ for Ne

$$(1.3.1) \qquad Z = \frac{pV_m}{RT} = \frac{p_r V_{m,r}}{T_r} \times \frac{p_c V_{m,c}}{RT_c} \qquad (4)$$

The compression factor

$$(4) \qquad Z = Z(p_r, T_r) \qquad (5)$$

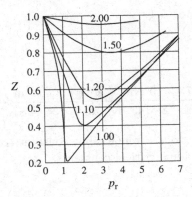

Figure 5.2. Reduced compressibility factor Z_r as a function of reduced pressure p_r for various values of the reduced temperature T_r (after Gouq-Jen Su 1946). Curves are labelled by values of the reduced temperature. (Individual points on the curves are not shown, but data for the following gases were used: nitrogen, methane, ethane, ethylene, propane, n-butane, isopentane, n-heptane, carbon dioxide and water.)

is independent of the nature of the gas. Plots of $Z(p_r, T_r)$ at fixed T_r (e.g. Figure 5.2) may be used to estimate $V_m(T,p)$ for any gas when its critical constants are known. This is a demonstration of the *Principle of Corresponding States* (PCS) that the properties of gases, when expressed as functions of reduced variables p_r, T_r are, to a good approximation, independent of the nature of the gas.

Exercise 5.1-2

Table 5.1 shows the ratio $p_c V_{m,c}/RT_c$ to be remarkably constant. Explain why this is so.

Exercise 5.1-3

The critical temperature of NH_3 is 405.4 K and its critical pressure is 11.3 MPa. Estimate the molar volume of NH_3 at 486.5 K and 22.6 MPa.

The values of the Boyle temperature in Table 5.1 (excluding the two lightest gases, helium and hydrogen) range from 122 K to 710 K, and yet the ratio of the Boyle temperature to the critical temperature (excluding He and H_2) is approximately constant, a good example of the principle of corresponding states.

Exercise 5.1-4

A particular edition of a text on physical chemistry does not list the Boyle temperature for CO. Estimate its value.

Answers to Exercises 5.1

Exercise 5.1-1

For neon

$$Z_c = \frac{p_c V_{m,c}}{RT_c} = \frac{2.73 \times 10^6 \mathrm{Nm}^{-2} \times 41.7 \times 10^{-6} \mathrm{m}^3 \mathrm{mol}^{-1}}{8.3145 \, \mathrm{JK}^{-1} \mathrm{mol}^{-1} \times 44.8 \, \mathrm{K}} = 0.306 \qquad (6)$$

The values of this ratio are remarkably constant, considering the wide variations in the individual critical constants.

Exercise 5.1-2

From Eqs. (3) and (4)

$$Z = \frac{p_r V_r}{T_r} \frac{p_c V_{m,c}}{RT_c} = f(p_r, T_r) \frac{p_c V_{m,c}}{RT_c}$$

And since (5) shows that $Z = Z(p_r, T_r)$, we expect that the ratio $p_c V_{m,c}/RT_c$ will be approximately the same for all substances, as illustrated in Table 5.1.

Exercise 5.1-3

$$p_r = \frac{22.6 \, \mathrm{MPa}}{11.3 \, \mathrm{MPa}} = 2.00, \quad T_r = \frac{486.5 \, \mathrm{K}}{405.4 \, \mathrm{K}} = 1.200 \qquad (7)$$

By interpolation of plots of $Z(p_r)$ for various values of the parameter T_r, we find

$$Z = 0.595 = \frac{22.6 \, \mathrm{MPa} \times V_m}{8.3145 \mathrm{J \, K}^{-1} \, \mathrm{mol}^{-1} \times 486.5 \, \mathrm{K}} \qquad (8)$$

$$(8) \qquad\qquad V_m = 106.5 \, \mathrm{cm}^3 \, \mathrm{mol}^{-1} \qquad (9)$$

This result (9) is a much better approximation than assuming ideal behaviour, which would give

$$(5.1.1) \qquad V_m = \frac{8.3145 \, \mathrm{J \, K}^{-1} \mathrm{mol}^{-1} \times 486.5 \, \mathrm{K}}{22.6 \, \mathrm{MPa}} = 179 \, \mathrm{cm}^3 \, \mathrm{mol}^{-1} \qquad (10)$$

Exercise 5.1-4

Excluding He, H_2 and CO, the mean value of the reduced Boyle temperature $T_{B,r} = T_B/T_c$ for the gases in Table 5.1 is 2.62. For CO

$$T_B = T_{B,r} \times T_c = 2.62 \times 133\,\mathrm{K} = 348\,\mathrm{K}$$

which is less than 1% higher than the experimental value quoted in Table 5.1.

5.2. Intermolecular Potentials

Practical applications of the PCS have been illustrated in Section 5.1.2, but we have not yet faced the question as to why the principle appears to hold so widely for gases and also, as we shall see, for liquids and solids as well. This implies that it is associated with the *intermolecular forces* that are responsible for cohesion in condensed matter and for deviations from ideal behaviour in gases. The two essential features of the intermolecular potential energy for non-polar molecules are a strong repulsion at close distances of approach, due to the finite size of the molecules, and a longer-range attraction which is due to *dispersion forces*. The electron distribution in a molecule is not static but in constant motion, which results in all molecules possessing a fluctuating dipole moment. This fluctuating dipole moment induces a dipole moment in nearby molecules and this dipole-induced dipole interaction results in an attractive force (the London dispersion force) that leads to a potential energy of interaction that varies as the inverse sixth power of the distance between molecular centres. This is usually called the van der Waals term in the intermolecular potential, since it is responsible for the attractive force between molecules envisaged by van der Waals. If we assume that the rapidly-moving, rotating molecules can be represented by hard-spheres, so that the potential energy rises steeply to infinity at $r = \sigma$ (Figure 5.3) then this hard-sphere repulsion, combined with the inverse sixth power van der Waals attraction, would yield an ∞, 6 model for the intermolecular interaction.

When the charge distributions in two molecules overlap, their energy rises steeply but not vertically, as assumed in the hard-sphere potential. From studies of the cohesive energy of rare-gas crystals, John Lennard-Jones concluded that this strong repulsion could be represented by an inverse twelfth power of the distance r between molecular centres, which when combined with the inverse sixth power van der Waals term, gives the

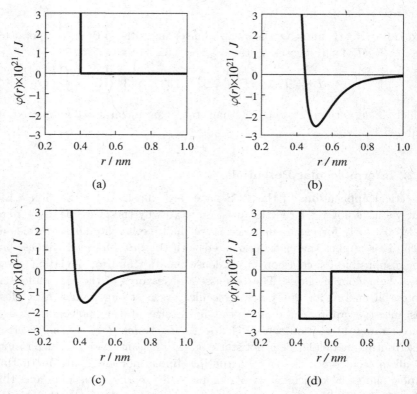

Figure 5.3. Intermolecular potentials: (a) hard-sphere; (b) Lennard-Jones 12,6 (c) exp, 6, (d) square well. The Lennard-Jones parameters used are those for CO_2 (Moore 1983). The exp, 6 parameters used are for the short-range $Br^- - Br^-$ interaction in RbBr (Jacobs and Vernon 1998).

Lennard-Jones 12,6 potential

$$\varphi(r) = 4\varepsilon \left[\left(\frac{\sigma}{r}\right)^{12} - \left(\frac{\sigma}{r}\right)^{6} \right] \qquad (1)$$

(Figure 5.3). Alternatively, the overlap repulsion may be represented by an exponential term, giving

$$\varphi(r) = A \exp(-r/\rho) - C/r^6 \qquad (2)$$

which is referred to as the exp, 6 potential. The forms of these potentials are sketched in Figure 5.3.

The PCS implies that the intermolecular potential energy should, to a good approximation, be a universal function

$$\varphi_r\left(\frac{r}{r_c}\right), \varphi_r = \frac{\varphi}{kT_c}, \quad \frac{4}{3}\pi r_c^3 = V_c/N_A \tag{3}$$

For example, for the Lennard-Jones 12,6 potential

$$(1) \quad \varphi_r\left(\frac{r}{r_c}\right) = 4\varepsilon_r\left[\left(\frac{\sigma_r}{r_r}\right)^{12} - \left(\frac{\sigma_r}{r_r}\right)^6\right] = 4\varepsilon_r\left[\left(\frac{\sigma}{r}\right)^{12} - \left(\frac{\sigma}{r}\right)^6\right] \tag{4}$$

where

$$\varepsilon_r = \left(\frac{\varepsilon}{kT_c}\right), \quad \sigma_r = \left(\frac{\sigma}{r_c}\right), \quad r_r = \left(\frac{r}{r_c}\right) \tag{5}$$

When only pair-interactions are allowed for, the second virial coefficient is given by (e.g. Fowler and Guggenheim 1939)

$$B(T) = \frac{N_A}{2}\int_0^\infty (1 - \exp[-\varphi(r)/kT])4\pi r^2 dr \tag{6}$$

Use of the appropriate intermolecular potential $\varphi(r)$ in (6) will yield the corresponding expression for the second virial coefficient.

Exercise 5.2-1

Calculate the second virial coefficient for a gas composed of perfectly hard-spheres of diameter σ.

For the Lennard-Jones potential, break the range of integration into two sections. When $r < \sigma$, $\varphi(r)$ is positive and increases rapidly as r decreases. The exponential term can therefore be neglected and this section may be approximated by

$$I_1 = \int_0^\sigma 4\pi r^2 dr = 4\pi\sigma^3/3 \tag{7}$$

When $r > \sigma$, in the temperature range where $\varphi(r) \ll kT$, expand the exponential retaining only the first two terms, so that

$$I_2 = \int_\sigma^\infty 16\pi\varepsilon\sigma^6\, r^{-4}dr = 16\pi\varepsilon\sigma^3/3kT \tag{8}$$

(6), (7), (8)
$$B(T) = (2\pi N_A\sigma^3/3)[1 - 8\varepsilon/3kT] \tag{9}$$

Exercise 5.2-2

What is the Boyle temperature of a gas of molecules that interact according to the Lennard-Jones intermolecular potential?

Answers to Exercises 5.2

Exercise 5.2-1

With r the distance between the centres of two molecules

$$B(T) = \frac{N_A}{2} \int_0^\infty (1 - \exp[-\varphi(r)/kT])4\pi r^2 dr$$

For $r > \sigma$, $\varphi(r) = 0$, and there is no contribution to B.

Therefore
$$B(T) = \frac{N_A}{2} \int_0^\sigma 4\pi \, r^2 dr = \frac{2\pi\sigma^3 N_A}{3} \tag{10}$$

Exercise 5.2-2

The Boyle temperature is the temperature at which the second virial coefficient changes sign. Therefore, for a gas with a Lennard-Jones intermolecular potential

$$(9) \qquad\qquad T_B = 8\varepsilon/3k \tag{11}$$

5.3. The Joule–Thomson Effect

The Joule–Thomson (J–T) effect is observed when a steady stream of gas at p_1. T_1 is passed through a porous plug in a thermally insulated tube. When steady state is attained, it is observed that the pressure of the gas that has passed through the plug, p_2, is always less than p_1 but that its temperature T_2 may be either less than or greater than T_1. Consider the transfer of a definite mass of gas through the plug. Since the expansion is adiabatic, the change in internal energy is

$$U_2 - U_1 = -\int_{V_1}^0 p_1 \, dV - \int_0^{V_2} p_2 \, dV$$

$$= p_1 V_1 - p_2 V_2 \tag{1}$$

$$(1) \qquad\qquad U_2 + p_2 V_2 = U_1 + p_1 V_1 \tag{2}$$

$$(2) \qquad\qquad H_2 = H_1 \quad \text{(J–T expansion)} \tag{3}$$

Since a J–T expansion occurs at constant enthalpy $H(T, p)$

$$\left(\frac{\partial H}{\partial T}\right)_p \left(\frac{\partial T}{\partial p}\right)_H = -\left(\frac{\partial H}{\partial p}\right)_T \tag{4}$$

The change in T with p at constant H is called the *Joule–Thomson coefficient*

(4), (2.3.16)
$$\left(\frac{\partial T}{\partial p}\right)_H = -\frac{1}{C_p}\left(\frac{\partial H}{\partial p}\right)_T = \frac{V}{C_p}(\alpha T - 1) \tag{5}$$

where α is the expansivity of the gas and C_p its heat capacity at constant pressure. The adiabatic expansion of a gas against an external pressure always leads to cooling because of the decrease in internal energy due to the performance of work. But in the J–T experiment either heating, when $(\partial T/\partial p)_H < 0$, or cooling, if $(\partial T/\partial p)_H > 0$, can occur. Figure 5.4 shows isenthalpic curves for nitrogen. The dashed curve, called the J–T *inversion curve*, is the locus of the maxima in these curves where the J–T coefficient changes sign. The J–T effect is of practical use in the liquefaction of gases, although initial cooling by adiabatic expansion may be required to ensure that the gas is below its upper inversion temperature.

Exercise 5.3-1

Can an ideal gas be cooled by a Joule–Thomson expansion?
From the virial expansion to order V_m^{-2}

(1.3.11), (1.3.14)
$$pV_m/RT = 1 + B_2/V_m + B_3/V_m^2$$
$$= 1 + B_2'p/RT + B_3'(p/RT)^2 \tag{6}$$

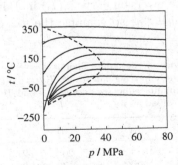

Figure 5.4.   Joule–Thomson inversion curve for nitrogen (dashed curve) which is the locus of the maxima in the isenthalpic curves. Circle o shows the critical point.

(6) $C_{p,m} \left(\dfrac{\partial T}{\partial p} \right)_H = T(\partial V_m / \partial T) - V_m$

$$= \left(T \dfrac{dB_2'}{dT} - B_2' \right) + \dfrac{p}{R} \left(\dfrac{dB_3'}{dT} - \dfrac{2B_3'}{T} \right) \qquad (7)$$

At moderate pressures, the expansion may be terminated at the term of order p giving

(7), Ex.1.3-3 $\left(\dfrac{\partial T}{\partial p} \right)_H = \dfrac{1}{C_{p,m}} \left(T \dfrac{dB_2}{dT} - B_2 \right) \qquad (8)$

Therefore, at moderate pressures, the inversion temperature T_i is the temperature at which the tangent to $B_2(T)$ through the origin meets the $B_2(T)$ curve. Some values of the Joule–Thomson inversion temperature T_i are given in Table 5.2. Despite the wide range of values for T_i, the ratio T_i/T_c is roughly constant (except for H_2 and He, in which quantum effects are significant). This is one further example of the Principle of Corresponding States. (Note that the virial equation of state is a classical equation and therefore deviations are to be expected for light molecules, especially at low temperatures. Notice the relatively low critical temperatures of helium and hydrogen.)

Exercise 5.3-2

Estimate a value for the upper inversion temperature of xenon.

To utilize Eq. (7) requires measurements of $B_2(T)$ and $B_3(T)$ over wide ranges of p and T. Such data are often not available. We shall therefore determine the inversion curve $p_i(T_i)$ in reduced coordinates from an equation of state, using as an example the van der Waals equation of

Table 5.2.   Maximum values of the Joule–Thomson inversion temperature T_i. Above the maximum value of T_i, only heating can result. $T_{i,r} = T_i/T_c$ is the reduced inversion temperature, and its maximum value (excluding H_2 and He) is reasonably constant considering the wide range of values for T_i.

	He	Ne	Ar	Kr	Xe	H_2	N_2	O_2	CO_2	CH_4
T_i/K	40	231	723	1090		202	621	724	1500	968
T_c/K	5.21	44.44	150.7	209.4	289.8	33.23	126.3	154.8	304.2	190.6
$T_{i,r}$/K	7.68	5.20	4.80	5.21		6.08	4.92	4.68	4.93	5.08

state (P1.12.1), which may be rewritten in the form

$$pV_m/RT = (1-y)^{-1} - ay/bRT, \quad y = b/V_m \tag{9}$$

Differentiate (9) with respect to T at constant p and replace $(\partial V_m/\partial T)_p$ by V_m/T, which is the inversion condition (see Eq. (5)). This gives

$$\left(\frac{2a}{bR}\right)\frac{1}{T_i} = (1-y_i)^{-2} \tag{10}$$

in which subscript i refers to the inversion temperature and V_m to the corresponding molar volume. Eliminating y_i between (10) and (9) (with $T = T_i$) yields

$$p_i = -27 + 24\sqrt{3T_i} - 12T_i \tag{11}$$

Figure 5.5 shows experimental data for nitrogen (Roebuck and Osterberg 1935), as well as the inversion curve calculated from (11). Please notice that in Figure 5.5, $p_i(T_i)$, whereas Figure 5.4 shows the more common $T_i(p_i)$ plot.[a] The van der Waals equation of state certainly predicts an inversion curve of the right form, though agreement with experiment is not quantitative. If there were no attractive forces, the van der Waals equation would reduce to a rather poor equation of state for hard-spheres. This suggests that a better representation of the inversion curve would be obtained by replacing the van der Waals hard-sphere term by a better

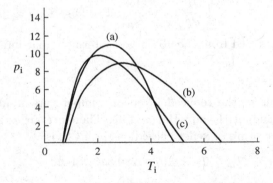

Figure 5.5. Inversion curves: (a) experiment, (b) calculated from van der Waals' equation of state, (c) calculated from the modified van der Waals equation (data from Zemansky and Dittman 1997).

---

[a]I am indebted to Professor A.R. Allnatt for these calculations of the inversion curve and for much helpful discussion.

expression for the pressure of a system composed of hard-spheres. This indeed proves to be the case, as shown in Figure 5.5, where curve c is the inversion curve obtained from the modified van der Waals equation of state

$$Z = Z_{hs}(\text{CS}) - \frac{a}{RTV_{\text{m}}} \tag{12}$$

where $Z_{hs}(\text{CS})$ employs the Carnahan–Starling approximation for the hard-sphere pressure (Carnahan and Starling 1969)

$$Z_{hs}(\text{CS}) = \frac{1 + \eta + \eta^2 - \eta^3}{(1 - \eta)^3}, \quad \eta = b/4V_{\text{m}} \tag{13}$$

As Figure 5.5 shows, the agreement with experiment is improved considerably on the low temperature side of the maximum in the $p_i(T_i)$ curve, which is the region where repulsive interactions are most important.

Answers to Exercises 5.3

Exercise 5.3-1

For an ideal gas, the coefficient of expansion is

$$\alpha \equiv \frac{1}{V_{\text{m}}} \left(\frac{\partial V_{\text{m}}}{\partial T} \right)_p = \frac{1}{T}$$

Therefore the Joule–Thomson coefficient for an ideal gas is

$$(5) \qquad \left(\frac{\partial T}{\partial p} \right)_H = \frac{V}{C_p}(\alpha T - 1) = 0$$

as would be expected from the absence of intermolecular interactions.

Exercise 5.3-2

The mean value of the reduced inversion temperature from the data in Table 5.2 (excluding H_2 and He) is 4.97. Therefore an estimate for the inversion temperature of xenon, based on the PCS, is

$$T_{\text{i,r}} \times T_{\text{c}} = 4.97 \times 289.8 \,\text{K} = 1440 \,\text{K}$$

5.4. Thermodynamic Functions for Gases

In this section we are concerned with real gases that are only moderately non-ideal, so that only terms up to the second virial coefficient B_2 need be included. For a pure substance, the chemical potential is the Gibbs energy per mole, so that

(2.2.11) $$d\mu = dG_m = -S_m dT + V_m dp \tag{1}$$

(1.3.12), (1.3.15) $$V_m = \frac{RT}{p} + B_2 \quad \text{(moderate pressures)} \tag{2}$$

The expansivity and compressibility are

(2) $$\alpha = \frac{1}{V_m}\left(\frac{\partial V_m}{\partial T}\right) = \left(\frac{R}{p} + \frac{dB_2}{dT}\right)\frac{1}{V_m} \tag{3}$$

(1.3.12), (1.3.15) $$\kappa_T = -\frac{1}{V_m}\left(\frac{\partial V_m}{\partial p}\right) = \left(\frac{RT}{pV_m}\right)\frac{1}{p} \tag{4}$$

(1), (2) $$\mu = \mu^{\ominus}(T) + RT\ln(p/p^{\ominus}) + B_2 p \tag{5}$$

where $p^{\ominus}$ is the agreed standard pressure of 10^5 Pa (Mills *et al.* 1988, Cohen *et al.* 2007, *cf.* also McNaught and Wilkinson 1997)

(5) $$S_m = -[d\mu^{\ominus}(T)/dT] - R\ln(p/p^{\ominus}) - (dB_2/dT)p \tag{6}$$

(5), (6) $$H_m = \mu^{\ominus}(T) - T[d\mu^{\ominus}(T)/dT] + [B_2 - T(dB_2/dT)]p \tag{7}$$

Exercise 5.4-1

Write down the corresponding formulae for V_m, μ, S_m and H_m for an ideal gas.

The last three formulae all contain the chemical potential of the substance, $\mu^{\ominus}(T)$ at the standard pressure $p^{\ominus}$ and/or its derivative with respect to temperature. $\mu^{\ominus}(T)$ contains a constant term which we denoted originally by U^0 (see Eq. (3.11.1) and associated text). For polyatomic molecules there are additional contributions to $U(T)$ from rotational and vibrational energy and these appear also in $H(T)$ and $\mu^{\ominus}(T)$. It is not necessary (and would frequently be impossible) to calculate these temperature dependent terms. Instead, this term is fixed at one temperature by setting $H(T)$ equal to zero for each element at $T = 298.15$ K and $p = p^{\ominus}$. This convention also fixes $H(298.15 \text{ K}, p^{\ominus})$ of a compound as the enthalpy change $\Delta_f H$ accompanying the formation of that compound from its elements at the standard temperature and pressure, as will be discussed in more detail in Chapter 8. $H(T)$ at other temperatures may be calculated by using experimental values of the heat capacity at constant pressure. This is expressed as a power series in T of the form

$$C_p = a + \underline{b}(T/\mathrm{K}) + c(\mathrm{K}^2/T^2) + d(T^2/\mathrm{K}^2) \tag{8}$$

As many terms as are necessary may be used in (8), but three terms are often sufficient. Notice that the coefficients a, b, c, d, all have the units of heat capacity, namely $\mathrm{J\,K^{-1}\,mol^{-1}}$. There is also a constant term in $-[d\mu^{\ominus}(T)/dT]$ in S which was called $S^{\ominus}$ and which is taken to be zero at $0\,\mathrm{K}$ for all pure elements. The exceptional cases of hydrogen and deuterium were discussed in Section 4.4.5. To clarify that these definitions are in use, the chemical potential $\mu^{\ominus}$ is called the *standard chemical potential* and $S^{\ominus}$ the *standard entropy*. (Since these are conventions, the alternative descriptor *conventional* has been proposed, but *standard* is the agreed term (Cohen *et al.* 2007).)

Example 5.4-1

Calculate the enthalpy change $\Delta_{\mathrm{vap}}H$ at constant pressure $p^{\ominus}$ as 1 mole of liquid water at $293\,\mathrm{K}$ is transformed into steam at $383\,\mathrm{K}$, given that $C_p(\mathrm{H_2O(liq)}) = 75.0\,\mathrm{J\,K^{-1}\,mol^{-1}}$; $C_p(\mathrm{H_2O(gas)}) = 35.4\,\mathrm{J\,K^{-1}\,mol^{-1}}$; $\Delta_{\mathrm{vap}}H(\mathrm{H_2O},\ 373\,\mathrm{K}) = 47.3\,\mathrm{kJ\,mol^{-1}}$.

   Since the heat capacities are assumed to be independent of temperature,

$$H(\mathrm{H_2O(liq)}373\,\mathrm{K}) - H(\mathrm{H_2O(liq)}293\,\mathrm{K}) = 75.0\,\mathrm{JK^{-1}\,mol^{-1}} \times 80\,\mathrm{K}$$
$$= 6.0\,\mathrm{KJ\,mol^{-1}}\Delta_{\mathrm{vap}}H(\mathrm{H_2O},\ 373\,\mathrm{K}) = 47.3\,\mathrm{kJ\,mol^{-1}}$$
$$H(\mathrm{H_2O(gas)}383\,\mathrm{K}) - H(\mathrm{H_2O(gas)},\ 373\,\mathrm{K})$$
$$= 35.4\,\mathrm{JK^{-1}\,mol^{-1}} \times 10\,\mathrm{K} = 0.35\,\mathrm{kJ\,mol^{-1}}$$

Summing, $H(\mathrm{H_2O(gas)}383\,\mathrm{K}) - H(\mathrm{H_2O(liq)}273\,\mathrm{K}) = 53.7\,\mathrm{kJ\,mol^{-1}}$.

   If thermodynamic properties at higher pressures are required, then the virial expansion in (2) can be extended, but an alternative approach is described in the next section.

Answer to Exercise 5.4-1

For an ideal gas

$$V_{\mathrm{m}} = \frac{RT}{p} \tag{9}$$

(1), (9) $\mu = \mu^{\ominus}(T) + RT\ln(p/p^{\ominus}) \tag{10}$

(5) $S_{\mathrm{m}} = -[d\mu^{\ominus}(T)/dT] - R\ln(p/p^{\ominus}) \tag{11}$

(10), (11) $H_{\mathrm{m}} = \mu^{\ominus}(T) - T[d\mu^{\ominus}(T)/dT] \tag{12}$

5.5. Gases at High Pressures

For an ideal gas

(5.4.10) $$\mu = \mu^{\ominus}(T) + RT\ln(p/p^{\ominus}) \tag{1}$$

and for a real gas at moderate pressures

(5.4.5) $$\mu(T,p) = \mu^{\ominus}(T) + RT\ln(p/p^{\ominus}) + B_2 p \tag{2}$$

Adding additional terms from the virial expansion results in increasingly complicated expressions for $\mu(T,p)$. In order to retain the formal simplicity of the ideal gas formula (1) for the chemical potential for *all gases at all pressures*, a new quantity, called the *fugacity* $\tilde{p}$, is defined by

(1) $$d\mu = RT\,d\ln\tilde{p} = V_m dp \quad \text{(any gas, constant } T) \tag{3}$$

$$\lim_{p\to 0}\left(\frac{\tilde{p}}{p}\right) = 1 \tag{4}$$

Subtract $RT\,d\ln p$ from each side of (3), divide each side by RT, and integrate from p' to p, where p' is some low pressure

$$\ln\left(\frac{\tilde{p}}{p}\right) - \left[\ln\left(\frac{\tilde{p}}{p}\right)\right]_{p=p'} = \int_{p'}^{p}\left(\frac{V_m}{RT} - \frac{1}{p}\right)dp \tag{5}$$

In the limit $p' \to 0$

(5), (4) $$\ln\left(\frac{\tilde{p}}{p}\right) = \int_{0}^{p}\left(\frac{Z-1}{p}\right)dp \tag{6}$$

There are three methods for finding $\tilde{p}$ from Eq. (6):

(i) by numerical integration, using experimental values of $Z(p)$;
(ii) from a suitable equation of state;
(iii) from the Principle of Corresponding States.

(6) $$\ln\left(\frac{\tilde{p}}{p}\right) = \ln\phi = \int_{0}^{p_r}\frac{Z-1}{p_r}dp_r \tag{7}$$

where ϕ is the *fugacity coefficient*. The RS of (7) is independent of the nature of the gas and has been evaluated from known $Z(p_r)$, so that $\tilde{p}/p$ is known as a function of the reduced pressure p_r at various T_r. Thus $\tilde{p}$ may be evaluated from published graphs (Newton 1935, Gouq-Jen Su 1946) for any gas whose critical constants are known.

5.6. Liquids

5.6.1. *Thermodynamic properties*

The compressibility of an ideal gas is p^{-1} and this relation is still roughly true for real gases (Eq. (5.4.4)).

Exercise 5.6-1

Confirm the above statement for a gas described by the virial equation of state up to the second coefficient B_2.

In contrast, the compressibility of a liquid is almost constant. This observation and the fact that the density of a liquid is much higher than that of a gas, except near the critical point (Figure 5.1), suggests a hard-sphere model for a liquid. In a crystalline solid at low temperatures, the molecules occupy sites dictated by the crystal structure (for example, cubic close-packed in the case of argon) and virtually every site is occupied. As the temperature is increased, defects in this perfect crystal structure arise as some molecules leave their normal lattice sites and take up new positions on the surface of the crystal (Schottky defects). When a crystal melts, the molecules are no longer on lattice sites but in a constant state of motion with mean free path (the average path length travelled between collisions) much less than that in a gas. Since

$$\kappa_T = -\frac{1}{V}\left(\frac{\partial V}{\partial p}\right)_T \tag{1}$$

is constant and of the order of $10^{-3}\,\mathrm{MPa^{-1}}$, integrating (1) at constant T yields

$$V_\mathrm{m} = V_\mathrm{m}^{\ominus}\exp[-\kappa_T(p - p^{\ominus})] \approx V_\mathrm{m}^{\ominus}[1 - \kappa_T(p - p^{\ominus})] \tag{2}$$

$V_\mathrm{m}^{\ominus}$ is the molar volume at the standard pressure of 10^5 Pa. Integrating the fundamental equation

$$d\mu = V_\mathrm{m}dp - S_\mathrm{m}dT \tag{3}$$

at constant temperature gives

$$(2),\ (3) \qquad \mu - \mu^{\ominus} = V_\mathrm{m}^{\ominus}[(p - p^{\ominus}) - \frac{1}{2}\kappa_T(p - p^{\ominus})^2] \tag{4}$$

$$= \frac{1}{2}(p - p^{\ominus})(V_\mathrm{m} + V_\mathrm{m}^{\ominus}) \tag{5}$$

Exercise 5.6-2

Complete the derivation of (5) from (4), explaining any approximations made.

Differentiating (5) with respect to T at constant p gives

$$S_m = -\frac{d\mu^\ominus}{dT} - \frac{1}{2}(p - p^\ominus)\left[\left(\frac{\partial V_m}{\partial T}\right)_p + \left(\frac{\partial V_m^\ominus}{\partial T}\right)_p\right]$$

$$= -\frac{d\mu^\ominus}{dT} - \frac{1}{2}(p - p^\ominus)[\alpha V_m + \alpha^\ominus V_m^\ominus] \tag{6}$$

(5), (6) $\quad H_m = \mu^\ominus - T\frac{d\mu^\ominus}{dT} + \frac{1}{2}(p - p^\ominus)[V_m(1 - \alpha T) + V_m^\ominus(1 - \alpha^\ominus T)$

$$\tag{7}$$

It is an empirical fact that the heat capacity C of a liquid is approximately independent of temperature. Assuming that this is so, the standard enthalpy will be of the form

$$H^\ominus = A + CT \tag{8}$$

and the standard entropy

(2.3.14) $$\qquad S^\ominus = \int_{T_1}^{T} \frac{C}{T}dT = C\ln T + B \tag{9}$$

Exercise 5.6-3

Why is the constant of integration $B = S^\ominus(T_1)$ in Eq. (9) not zero?

The *thermal pressure coefficient*

$$\beta = \left(\frac{\partial p}{\partial T}\right)_V = -\frac{(\partial V/\partial T)_p}{(\partial V/\partial p)_T} = \frac{\alpha}{\kappa_T} \tag{10}$$

β determines the volume dependence of the internal energy through the first thermodynamic equation of state, Eq. (2.3.8).

5.6.2. Theoretical description

Our first approach is to try and describe a liquid in the same terms as those used for dense gases. The virial expansion was introduced in Chapter 1 (Eq. (1.3.8)) to provide an empirical description of deviations from ideal gas behaviour. It may readily be derived from a Taylor expansion of the compression factor in terms of the inverse molar volume

$$Z = \frac{pV_m}{RT} = 1 + \frac{1}{V_m} \left[\frac{\partial Z}{\partial(1/V_m)} \right]_{T,(1/V_m)=0}$$

$$+ \frac{1}{2V_m^2} \left[\frac{\partial^2 Z}{\partial(1/V_m)^2} \right]_{T,(1/V_m)=0} + \cdots \tag{11}$$

(11), (1.3.8) $$B_{n+1}(T) = \frac{1}{n!} \left(\frac{1}{V_m} \right)^n \left[\frac{\partial^n Z}{\partial(1/V_m)^n} \right]_{T,(1/V_m)=0},$$

$$n = 1, 2, 3 \ldots \tag{12}$$

It can be shown (Fowler and Guggenheim 1939, Moore 1983, p. 683) that the second virial coefficient B_2 is

$$B_2(T) = \frac{N_A}{2} \int_0^\infty (1 - \exp[-\varphi(r)/kT]) \, 4\pi r^2 dr \tag{13}$$

where $\varphi(r)$ is the potential energy of a pair of molecules separated by a distance r. Equation (13) can be evaluated for various forms of the intermolecular potential $\varphi(r)$. Increasingly complicated formulae exist for the third and higher virial coefficients (Hill 1962, Waldram 1985). The virial expansion works well at low and moderate densities but it fails to converge at high densities so that other approaches must be used.

When argon melts at its triple point, its volume expands by 15%. The structure of the solid is cubic close-packed, or face-centred cubic, in which each Ar atom has 12 nearest neighbours, but in the liquid the atoms are no longer fixed on a lattice and each Ar atom has on average, eight to nine nearest neighbours. The number of molecules at a distance between r and $r + dr$ from a given molecule (O) is $Cg(r)4\pi r^2 dr$, where $C = N/V$ is the number density and $g(r)$ is the *radial distribution function*. We have no reason to suppose that the distribution of molecules around O is other than spherically symmetric. Essentially, $g(r)$ contains information on how the local number density differs from the average density. If there were no attractive intermolecular forces and the volume occupied by the molecules were negligibly small, then the distribution of other molecules around O would be random and $g(r)$ would be 1, independent of r. It is 1 at large distances from O (where the intermolecular interactions with O are negligible), but close to O, $g(r)$ is zero because of the strong repulsion between molecules that are sufficiently close for their charge distributions

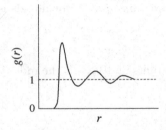

Figure 5.6. Radial distribution function $g(r)$ for a liquid such as argon. The first peak occurs near the minimum r_m in the curve representing the intermolecular potential. At close distances of approach, $g(r) = 0$ because of the finite size of the molecules.

to overlap. At distances larger than one molecular diameter the molecules attract one another and so $g(r)$ becomes greater than one (see Figure 5.6). The first large peak occurs at a value of r near the minimum r_m in the intermolecular potential (Figure 5.3b and c). This first shell is diffuse because the molecules in a liquid are in constant motion. Nevertheless, its presence is enough to attract a second (even more diffuse) second shell and then a third shell. At larger distances any structure becomes indistinct.

$g(r)$ can be obtained from X-ray or neutron scattering experiments or from molecular dynamics (MD) calculations. In MD simulations (e.g. Jacobs and Rycerz 1997) an array of a few hundred or a few thousand molecules is set up on a lattice and the molecules are assigned random velocities. These molecules interact according to a realistic intermolecular potential (see, for example, Catlow and Mackrodt 1982). The Newtonian equations of motion are then solved repetitively at intervals Δt (the 'time step') which is of the order of ps. The positions and velocities of all the particles are stored for subsequent analysis.

The internal energy may be expressed in terms of $g(r)$. The average number of molecules at a distance between r and $r + dr$ from one O is $Cg(r)4\pi r^2 dr$ and the average potential energy of interaction between these molecules in the shell between r and $r + dr$ and the one at O is $\varphi(r)Cg(r)4\pi r^2 dr$. Integrating over r would give the potential energy of interaction of one molecule with the other molecules in the liquid and therefore the molar internal energy is

$$U_m(r) = \frac{CN_A}{2} \int_V \varphi(r)g(r)4\pi r^2 dr \qquad (14)$$

The factor $1/2$ prevents counting each interaction twice. Similarly the pressure p can be calculated from $g(r)$.

A commonly used *equation of state* for liquids is the Tait equation

$$\frac{V_{m,0} - V_m}{pV_{m,0}} = \frac{a}{b+p} \tag{15}$$

a and b are positive constants and $V_{m,0}$ is the molar volume at limiting low pressures.

5.6.3. *Vapour pressure*

Figure 5.1 showed plots of the densities of liquid and saturated vapour for carbon dioxide. Figure 5.7 shows a similar plot, in reduced coordinates, that fits data for several substances. This shows that the principle of corresponding states applies to liquids as well as to gases. The curve in Figure 5.7 was calculated from the empirical formulae

$$\rho^L/\rho_c = 1 + \frac{3}{4}\gamma^3 + \frac{7}{4}\gamma \tag{16}$$

$$\rho^G/\rho_c = 1 + \frac{3}{4}\gamma^3 - \frac{7}{4}\gamma \tag{17}$$

$$\gamma = \left[1 - \frac{T}{T_c}\right]^{1/3} \tag{18}$$

and fits data for Ne, Ar, Kr, Xe, N_2, O_2, CO and CH_4 (Guggenheim 1945).

Exercise 5.6-4

Find expressions for $(\rho^L - \rho^G)/\rho_c$, $(\rho^L + \rho^G)/2\rho_c$.

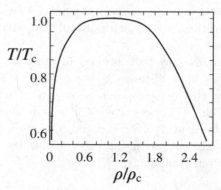

Figure 5.7.   Reduced densities of coexisting liquid and vapour phases, showing that the principle of corresponding states applies to liquids as well as to gases. The curve shown was calculated from the empirical formulae (5.6.16) and (5.6.17) and provides a good fit to experimental data for Ne, Ar, Kr, Xe, O_2, N_2, CO and CH_4 (after Guggenheim 1945).

The formula (16) should only be used for T less than $0.65\,T_C$, although the formula for $(\rho^L - \rho^G)/\rho_c$ (Exercise 5.6-4) is still accurate above $T = 0.65\,T_c$. This system is univariant (two phases, one component) and the dependence of vapour pressure on temperature is given by the Clapeyron equation

$$\frac{dp}{dT} = \frac{\Delta_{vap}S_m}{\Delta_{vap}V_m} \tag{19}$$

In (19), $\Delta_{vap}S_m = \Delta_{vap}H_m/T$, $\Delta_{vap}V_m = V_m^G - V_m^L$, where $V_m^G \gg V_m^L$, so that

(19)
$$\frac{d\ln p}{dT} = \frac{\Delta_{vap}H_m}{RT^2} \tag{20}$$

Equation (20) is the Clausius–Clapeyron (CC) equation. It holds best at low temperatures where the three approximations made, namely that the vapour behaves as an ideal gas, that $V_m^G \gg V_m^L$ and that $\Delta_{vap}H_m$ is independent of temperature, are generally valid.

Exercise 5.6-5

The condition for equilibrium when a system consists of a single component distributed between two phases α and β is

$$\mu^\alpha = \mu^\beta \tag{21}$$

Derive the general form of the Clapeyron equation

$$\frac{dp}{dT} = \frac{\Delta_{vap}S_m}{\Delta_{vap}V_m} \tag{22}$$

(22)
$$\frac{d\ln p}{d(1/T)} = -\frac{\Delta_{vap}H_m}{R} \tag{23}$$

from which the molar heat of vaporization $\Delta_{vap}H_m$ may be determined by plotting $\log p$ against $1/T$. If we do not assume that $\Delta_{vap}H_m$ is constant, then

(2.2.9), (2.3.16)
$$\frac{d\Delta H}{dT} = \Delta C_p + \left(\Delta V - T\frac{\partial \Delta V}{\partial T}\right)\frac{dp}{dT} \tag{24}$$

(22)
$$\frac{d\Delta H}{dT} = \Delta C_p + \left(\Delta V - T\frac{\partial \Delta V}{\partial T}\right)\frac{\Delta H}{T\Delta V} \tag{25}$$

(25)
$$\frac{d\Delta H}{dT} = \Delta C_p + \frac{\Delta H}{T} - \Delta H\left(\frac{\partial \ln \Delta V}{\partial T}\right)_p \tag{26}$$

Formula (26) is due to Planck (1945). Equation (26) is quite general and applies to any phase transition $\alpha \to \beta$. When α is a condensed phase, that is, either a solid or a liquid, and β is the corresponding vapour, the molar volume of the condensed phase can be neglected in comparison with that of the vapour. At low pressures, the vapour may be assumed to behave like an ideal gas. If these two approximations are valid

$$(26) \qquad \frac{\mathrm{d}\Delta H}{\mathrm{d}T} = \Delta C_p, \quad V_{\mathrm{m}}^{L\ \mathrm{or}\ S} \ll V_{\mathrm{m}}^{G} = RT/p \qquad (27)$$

$$(27) \qquad \Delta H(T) - \Delta H(0\,\mathrm{K}) = \int_0^T \Delta C_p \mathrm{d}T \qquad (28)$$

$$(20),\ (28) \qquad \frac{\mathrm{d}\ln p}{\mathrm{d}T} = \frac{\Delta_{\mathrm{vap}}H(0\,\mathrm{K})}{RT^2} + \frac{1}{RT^2}\int_0^T \Delta C_p \,\mathrm{d}T \qquad (29)$$

$$(29) \qquad \ln p = -\frac{\Delta_{\mathrm{vap}}H(0\,\mathrm{K})}{RT} + \frac{\Delta C_p \ln T}{R} + a \qquad (30)$$

In this derivation of (30), ΔC_p has been assumed to be independent of temperature. For the corresponding result when ΔC_p is temperature dependent, see Section 517 of Fowler and Guggenheim (1939). An equation of the form of (30) fits a large amount of experimental data (see Moelwyn-Hughes 1961, p. 699) provided (because of the approximations made) the coefficient of $(RT)^{-1}$ is taken to be an empirical constant c and not interpreted as the limiting value of the enthalpy of vaporization as $T \to 0\,\mathrm{K}$. Similarly, the coefficient of $\ln T$ is another empirical constant b and not $\Delta C_p/R$. If the temperature dependence of C_p were accounted for and the integration performed exactly, the constant a would be the *vapour pressure constant* (Fowler and Guggenheim 1939).

Applying the PCS to (20) leads us to expect that

$$\ln(p/p_{\mathrm{C}}) = A - B(T_{\mathrm{C}}/T) \qquad (31)$$

Guggenheim (1966b) has shown that (31) is a remarkably accurate representation of vapour pressure data for Ar, Kr, Xe, N_2, O_2, CO and CH_4 from the triple point to the critical point. This is understandable at low reduced pressures where the two approximations made ($\Delta_{\mathrm{vap}}H_{\mathrm{m}}$ independent of T, vapour behaves as an ideal gas) are valid. That it continues to hold over such a wide temperature range indicates some compensation between the effects of the two approximations.

Answers to Exercises 5.6

Exercise 5.6-1

The virial equation in terms of the pressure is, to first order

$$\frac{pV_m}{RT} = 1 + B_2'p \quad \text{or} \quad V_m = \frac{RT}{p} + RTB_2'$$

$$\kappa = -\frac{1}{V_m}\left(\frac{\partial V_m}{\partial p}\right)_T = \frac{1}{V_m}\frac{RT}{p^2} = \frac{1}{p}\left(\frac{RT}{pV_m}\right) = \frac{1}{p}(1 - B_2'p + \cdots)$$

$$= p^{-1} - B_2/RT \approx p^{-1}$$

since B_2 is small in comparison with the molar volume.

Exercise 5.6-2

(2), (3)

$$\mu - \mu^\ominus = V_m^\ominus[(p - p^\ominus) - \frac{1}{2}\kappa_T(p - p^\ominus)^2] \tag{4}$$

$$\kappa_T = -\frac{1}{V_m^\ominus}\frac{(V_m - V_m^\ominus)}{p - p^\ominus}$$

gives

$$\mu - \mu^\ominus = V_m^\ominus(p - p^\ominus)\left(1 + \frac{1}{2}\frac{(V_m - V_m^\ominus)}{V_m^\ominus}\right) = \frac{1}{2}(p - p^\ominus)(V_m^\ominus + V_m) \tag{5}$$

The replacement for κ_T is an approximation, but a valid one since κ_T is small.

Exercise 5.6-3

$$\rho^L/\rho_C = 1 + \frac{3}{4}\gamma^3 + \frac{7}{4}\gamma \tag{16}$$

$$\rho^G/\rho_C = 1 + \frac{3}{4}\gamma^3 - \frac{7}{4}\gamma \tag{17}$$

(16), (17)

$$\frac{\rho^L + \rho^G}{2\rho_C} = 1 + \frac{3}{4}\gamma^3, \quad \frac{\rho^L - \rho^G}{\rho_C} = \frac{7}{2}\gamma$$

Exercise 5.6-4

If the temperature T of the system is varied by dT, and consequent changes in the chemical potential of the two phases are $d\mu^\alpha$, $d\mu^\beta$, the system adopts a new equilibrium state in which

$$\mu^\alpha + \mathrm{d}\mu^\alpha = \mu^\beta + \mathrm{d}\mu^\beta \tag{32}$$

(21), (32)
$$\mathrm{d}\mu^\alpha = \mathrm{d}\mu^\beta \tag{33}$$

(33)
$$(\partial\mu^\alpha/\partial T)\mathrm{d}T + (\partial\mu^\alpha/\partial p)\mathrm{d}p$$
$$= (\partial\mu^\beta/\partial T)\mathrm{d}T + (\partial\mu^\beta/\partial p)\mathrm{d}p \tag{34}$$

(34)
$$-S_\mathrm{m}^\alpha\mathrm{d}T + V_\mathrm{m}^\alpha\mathrm{d}p = -S_\mathrm{m}^\beta\mathrm{d}T + V_\mathrm{m}^\beta\mathrm{d}p \tag{35}$$

(35)
$$(V_\mathrm{m}^\beta - V_\mathrm{m}^\alpha)\mathrm{d}p = (S_\mathrm{m}^\beta - S_\mathrm{m}^\alpha)\mathrm{d}T \tag{36}$$

(36)
$$\frac{\mathrm{d}p}{\mathrm{d}T} = \frac{S_\mathrm{m}^\beta - S_\mathrm{m}^\alpha}{V_\mathrm{m}^\beta - V_\mathrm{m}^\alpha} = \frac{\Delta_{\alpha\to\beta}S_\mathrm{m}}{\Delta_{\alpha\to\beta}V_\mathrm{m}} = \frac{\Delta_{\alpha\to\beta}H_\mathrm{m}}{T(\Delta_{\alpha\to\beta}V_\mathrm{m})} \tag{37}$$

The operator $\Delta_{\alpha\to\beta}$ means the change in the molar entropy or molar volume or molar enthalpy (or any other thermodynamic property) that accompanies the transfer of one mole of substance from phase α to phase β.

Exercise 5.6-5

Because B is the entropy of one mole of the pure liquid at the reference temperature T_1 and not the entropy of the pure crystal at $0\,\mathrm{K}$.

5.7. Solids

5.7.1. *Thermodynamic functions*

It is an empirical observation (which will be justified in Section 8.3) that the heat capacity of many solids at constant volume varies as T^3 at low temperatures (the Debye-T^3 law). Therefore

$$C_p \approx C_\mathrm{V} = aT^3 \tag{1}$$

(1)
$$H_\mathrm{m}(T) = H_\mathrm{m}(0\,\mathrm{K}) + \frac{1}{4}aT^4 \tag{2}$$

(1)
$$S_\mathrm{m}(T) = S_\mathrm{m}(0\,\mathrm{K}) + \frac{1}{3}aT^3 \tag{3}$$

(2), (3)
$$\mu(T) = H_\mathrm{m}(0\,\mathrm{K}) - TS_\mathrm{m}(0\,\mathrm{K}) - \frac{1}{12}aT^4 \tag{4}$$

$H_\mathrm{m}(0\,\mathrm{K})$ and $S_\mathrm{m}(0\,\mathrm{K})$ are the molar enthalpy and entropy extrapolated to $T = 0\,\mathrm{K}$. All these formulae (1)–(4) hold at low temperatures, that is, up to approximately $T = 15\,\mathrm{K}$, though this upper limit varies with the material under consideration.

Exercise 5.7-1

Are Eqs. (2) and (3) consistent with the fundamental equation for dH?

At higher temperatures one may use formulae based on the Einstein approximation for the vibrational partition function

(3.10.44), (3.10.45)
$$q_v = \sum_{v=0}^{\infty} \exp(-v\Theta_v/T) = \frac{1}{1 - \exp(-\Theta_v/T)}$$

$$F_m = U_0 + 3RT \ln\{1 - \exp(\Theta_E/T)\} \tag{5}$$

$$U_m = U_0 + \frac{3R\Theta_E}{\exp(\Theta_E/T) - 1} \tag{6}$$

(3.10.22)
$$C_V = Nk \left(\frac{\Theta_E}{2}\right)^2 \left(\frac{\exp(\Theta_E/T)}{(\exp(\Theta_E/T) - 1)^2}\right) \tag{7}$$

(3.10.22), (6)
$$S_m = S_0 - 3R \ln\{1 - \exp(\Theta_E/T)\}$$
$$- \frac{3R\Theta_E}{T \exp(\Theta_E/T) - 1} \tag{8}$$

When T is greater than $\Theta_D/3$, there is little difference ($<2\%$) between the energy and entropy calculated from (6) and (8) and the values calculated from Debye's formulae, provided Θ_E is chosen to be 0.73 Θ_D for energies and 0.71 Θ_D for entropies (Guggenheim 1967).

The data in Table 5.3 show that the PCS applies to solids as well as to liquids and gases. This is only to be expected as a consequence of the dependence of the PCS on intermolecular forces.

Table 5.3. These data illustrate that the principle of corresponding states applies to solids. T_{tp} is the temperature of the triple point, $V_{m,tp}$ the molar volume at the triple point, p_{tp} the pressure of the vapour at T_{tp} and $\Delta_{fus}H$ the molar enthalpy change that accompanies the melting process.

	Ne	Ar	Kr	Xe
T_C/K	44.8	150.7	209.4	289.8
T_{tp}/K	24.6	84.5	116.0	161.3
T_{tp}/T_C	0.549	0.547	0.553	0.557
$p_C/10^2$ kPa	27.3	48.6	54.8	59.0
p_{tp}/kPa	43.1	69.1	73.1	82.1
$10^2 \, p_{tp}/p_C$	1.58	1.42	1.33	1.39
$V_{m,tp}^L/cm^3 \, mol^{-1}$		28.14	34.13	42.68
$V_{m,tp}^S/cm^3 \, mol^{-1}$		24.61	29.65	37.09
$V_{m,tp}^L/V_{m,tp}^s$		1.143	1.151	1.151
$(\Delta_{fus}H/R)K^{-1}$	40.3	141.3	196.2	276
$\Delta_{fus}H/RT_C$	0.990	0.938	0.937	0.952

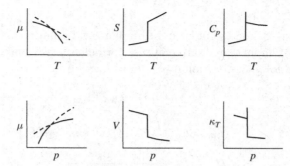

Figure 5.8. Typical form of the chemical potential and of its first and second derivatives with respect to T and p, as functions of T and p.

5.7.2. *Phase transitions in solids*

For an allotropic phase change from one crystalline phase to another, the chemical potential is continuous, but its first derivative with respect to either T or p is discontinuous (Figure 5.8). Phase transitions in which μ is continuous, but its first-order derivatives with respect to T and p are discontinuous, are called *First-Order phase transitions*.

Exercise 5.7-2

Find expressions for the second partial derivatives of μ with respect to T and p.

Example 5.7-1

Ammonium perchlorate undergoes a phase transition from orthorhombic to cubic forms at 513 K. The enthalpy of transition is $10.0\,\mathrm{kJ\,mol^{-1}}$. The densities of the orthorhombic and cubic structures are 1950 and $1760\,\mathrm{kg\,m^{-3}}$.

Calculate (a) the change of the transition point with pressure and (b) the entropy change accompanying the transition

$$dT_{\mathrm{tr}}/dp = \frac{T_{\mathrm{tr}}(V^{\mathrm{cubic}} - V^{\mathrm{ortho}})}{\Delta_{\mathrm{tr}}H}$$

$$= \frac{513 \times 0.1175 \left(\frac{1}{1760} - \frac{1}{1950}\right) \mathrm{K\,kg\,mol^{-1}m^3\,kg^{-1}}}{10\,000\,\mathrm{N\,m\,mol^{-1}}}$$

$$= 3.34 \times 10^{-7}\,\mathrm{K\,Pa^{-1}} \text{ or } 3.38 \times 10^{-2}\,\mathrm{K\,atm^{-1}}$$

$$\Delta S_{\mathrm{tr}} = \frac{\Delta_{\mathrm{tr}}H}{T_{\mathrm{tr}}} = \frac{10\,000\,\mathrm{J\,mol^{-1}}}{513\,\mathrm{K}} = 19.5\,\mathrm{J\,K^{-1}\,mol^{-1}}$$

This relatively large entropy change is associated with the onset of free rotation of the tetrahedral perchlorate anions. NMR measurements show that the tetrahedral ammonium ions rotate freely down to very low temperatures.

Similarly, the change in melting point with pressure can be calculated from the heat of fusion and the densities of the liquid and solid phases. Independent measurements of dT_{fus}/dp, $\Delta_{fus}V$ and $\Delta_{fus}H$ provide confirmation of the Clapeyron equation and hence of the Second Law of Thermodynamics.

Answers to Exercises 5.7

Exercise 5.7-1

At constant pressure the fundamental equation for dH reduces to $dH = TdS$. From (2) and (3), $dH = aT^3 = TdS$.

Exercise 5.7-2

From the fundamental equation for $d\mu = dG/n$

$$\left(\frac{\partial \mu}{\partial T}\right)_p = -S_m, \quad \left(\frac{\partial \mu}{\partial p}\right)_T = V_m$$

$$\left(\frac{\partial^2 \mu}{\partial T^2}\right)_p = -\left(\frac{\partial S_m}{\partial T}\right)_p = -\left(\frac{C_{p,m}}{T}\right), \quad \left(\frac{\partial^2 \mu}{\partial p^2}\right)_T = \left(\frac{\partial V_m}{\partial p}\right)_T = -V_m \kappa$$

5.8. Triple Point

The condition for equilibrium in a system comprising a single component distributed between two phases α and β is

$$\mu^\alpha(T,p) = \mu^\beta(T,p) \tag{1}$$

The presence of a second phase reduces the number of degrees of freedom from two for a single phase to one, which implies the existence of a relation between T and p. This relation is the Clapeyron equation. Thermodynamic states are thus represented by points on a line (recall Section 1.1). If a third phase γ of the same component is added, then the equilibrium condition becomes

$$\mu^\alpha(T,p) = \mu^\beta(T,p) = \mu^\gamma(T,p) \tag{2}$$

and the relation between T and p is now satisfied only at a single point, the *triple point*. The most famous example (already encountered in the

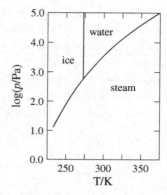

Figure 5.9. Phase diagram $p(T)$ for H_2O showing the triple point where ice is in equilibrium with water and water vapour and no other substances are present. T, p at the triple point are (273.16 K, 611 Pa) and at the critical point are (647 K, 22.1 MPa).

definition of the thermodynamic scale for measuring temperature) is the triple point of water shown in Figure 5.9. If the substance exhibits a phase transition between two crystalline phases, then more than one triple point may exist. A notable example is ice, which exists in several allotropic forms.

5.9. Higher-order Phase Transitions

There exists also the possibility of phase transitions in which both μ and its first derivatives are continuous, but its second partial derivatives with respect to T and p are discontinuous. Such transitions are called *second-order phase transitions*. This classification is due to P. Ehrenfest. The only well-documented example appears to be the superconducting-conducting transition in metals such as tin, in which a finite discontinuity in the heat capacity has been observed. Other examples have been slow to emerge, despite the large research effort devoted to this subject (Tilley and Tilley 1990). In other suspected examples, such as the He I $\rightarrow$ He II transition in liquid helium, which occurs at 2.17 K under its own equilibrium vapour pressure, the heat capacity becomes infinite. Figure 5.10 shows the discontinuity in the heat capacity curve. The $C_p(T)$ curve is reminiscent of the Greek letter lambda and for this reason such transitions are often referred to as λ-*transitions*. Similar discontinuities in C_p occur in α-iron, in quartz near 600°C, in ammonium chloride and in solid methane and hydrogen halides.

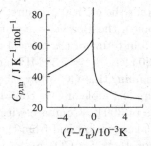

Figure 5.10. Molar heat capacity at constant pressure as a function of T, near the λ-transition in liquid helium.

Problems 5

5.1 The molar volume of carbon dioxide as a function of pressure, at 333 K is given in Table P5.4.

Table P5.4. Molar volume V_m of CO_2 as a function of pressure p at 333 K.

p/kPA	1318	3589	5436	7567	8648
V_m/cm^{3}mol^{-1}	2000	666.7	400	250	200

Calculate the fugacity of CO_2 at 4000 kPa and 333 K.

5.2 Show that the isothermal compressibility, to order (B_2/V_m), is given by the equivalent formulae

$$\kappa^{-1} = p\left(1 + \frac{B_2}{V_m}\right); \quad \kappa^{-1} = p\left(1 + \frac{B_2 p}{RT}\right)$$

Calculate the molar volume and isothermal compressibility of nitrogen at $p = 10$ MPa and $T = 250$ K and 500 K. The values of B_2 in cm^3 mol^{-1} at these temperatures are: -16.2 and $+16.9$ (Dymond and Smith 1980, p. 188).

5.3 Find the contribution to the chemical potential of a gas obeying the virial equation (1.3.12). Calculate this contribution $\Delta\mu_{\text{real}}$ for nitrogen at 273 K and 4 MPa, given that $B_2' = -0.453$ (MPa)$^{-1}$ and $B_3' = 2.25 \times 10^{-6}$ (MPa)$^{-2}$.

5.4 Prove that $k_S/k_T = C_V/C_p$.

5.5 (a) The Berthelot equation of state for one mole of gas is

$$p = \frac{RT}{V - b} - \frac{a}{TV^2} \tag{1}$$

Find expressions for the critical pressure, volume and temperature for a gas that obeys the Berthelot equation. Hence evaluate the constants a and b in terms of the critical constants.

(b) For CO_2, $T_c = 304.2\,\text{K}$, $p_c = 7.38\,\text{MPa}$ and $V_c = 94.2\,\text{cm}^3\,\text{mol}^{-1}$. Find a and b. Assuming that CO_2 obeys the Berthelot equation, calculate the pressure of 1 mole of CO_2 confined in a volume of $10\,\text{dm}^3$ at a temperature of $300\,\text{K}$. Compare with the pressure calculated from van der Waals equation and from the ideal gas equation. The van der Waals constants for CO_2 are $a = 0.364\,\text{m}^6\,\text{Pa}\,\text{mol}^{-2}$ and $b = 42.7\,\text{cm}^3\,\text{mol}^{-1}$.

5.6 Find expressions for $\left(\frac{\partial U}{\partial V}\right)_T$ for a gas that obeys (a) van der Waals equation of state and (b) one that obeys the Berthelot equation of state. Find equations for the change in entropy that accompanies the isothermal change in volume of a gas from V_1 to V_2 if the gas obeys (c) van der Waals equation of state, and (d) one that obeys the Berthelot equation of state.

5.7 The square-well potential (Figure 5.3) is defined by:

$$\varphi(r) = \infty, \quad \text{for } r < \sigma_1$$
$$\varphi(r) = -\varepsilon, \quad \text{for } \sigma_1 < r < \sigma_2$$
$$\varphi(r) = 0, \quad \text{for } r > \sigma_2$$

Find an expression for the second virial coefficient for a gas in which the intermolecular interactions can be approximated by the square well potential. Find an expression for $B_2(T)$ if $\sigma_1 = \sigma$, $\sigma_2 = 1.611\sigma$ and

$$\frac{2}{3}\pi N_A \sigma^3 = 0.440 V_c, \quad \varepsilon = 0.75\,kT_c$$

Table P5.5. The dependence of the vapour pressure of mercury on temperature.

$t/°\text{C}$	$\log(p/\text{kPa})$	p/kPa
0	−4.608	0.00002466
40	−3.100	0.000796
80	−1.946	0.01124
120	−1.007	0.0985
160	−0.272	0.5346
200	0.358	2.280
240	0.762	5.786
280	1.334	21.6

5.8 The vapour pressure of mercury in the temperature range 0–280°C is given in Table P5.5.

Fit these data to the equation

$$\log(p/\text{kPa}) = -\frac{c}{T} + b\log T + a \tag{1}$$

Discuss the physical interpretation of the constants b and c.

6

Systems of More Than One Component

6.1. The Phase Rule

Consider a system of ϕ phases and c components. Suppose that T and p are uniform and constant and that a small quantity dn_B of a component B is transferred from phase α to phase I. Then

$$dG = -\mu^{\alpha} dn_B + \mu^{\beta} dn_B \quad (T, p \text{ constant}) \tag{1}$$

At constant T and p, the condition for equilibrium is $dG = 0$, so that

$$(1) \qquad\qquad \mu_B^{\alpha} = \mu_B^{\beta} \qquad\qquad (2)$$

If the system is not in equilibrium with respect to component B then, subject to any possible constraints, substance B migrates from the phase in which its chemical potential is higher to that in which its chemical potential is lower, until the condition (2) is established. The analogy between this behaviour and that of a system subject to a gravitational potential difference or an electrical potential difference points to the suitability of the name "chemical potential". The same argument establishes the equality of the chemical potential of each of the c components in every one of the ϕ phases

$$(2) \qquad\qquad \mu_B^{\alpha} = \mu_B^{\beta} = \cdots = \mu_B^{\varphi} \qquad\qquad (3)$$

Equation (3) shows that there are $\phi - 1$ equations like (2) for each component and therefore $c(\phi - 1)$ *equations of phase equilibrium* in all, for this system of c components distributed between ϕ phases.

This system would certainly be described completely if values were ascribed to the $c\phi + 2$ variables T, p, μ_B^{α}. But not all these variables are independent. Besides the $c(\phi - 1)$ equations of phase equilibrium there are ϕ Gibbs–Duhem relations, one for each phase. Therefore, the total number of degrees of freedom, or *variance*, v is

$$v = c\phi + 2 - c(\phi - 1) - \phi = c - \phi + 2 \qquad (4)$$

This is the Phase Rule (see Gibbs 1961). If $v = 0$, the system is invariant, if $v = 1$, it is univariant and if $v = 2$, it is bivariant. It was tacitly assumed that the only thermodynamic forces acting on the system were the temperature and pressure of the surroundings. If, for example, the system contains unpaired electrons and is subjected to a magnetic field, then

$$v = c - \phi + 3 \qquad (5)$$

Exercise 6.1-1

What is the variance in a system such as that discussed above (Eq. (2)), but in which the pressure is held constant?

Phase equilibria are commonly displayed in *phase diagrams*, of which we saw an example in Figure 5.10.

Answer to Exercise 6.1-1

$$v = c - \phi + 1 \qquad (6)$$

6.2. Partial Molar Properties

Before reading this section, the reader is advised to review Section 2.7 on partial molar properties. It is convenient to divide the discussion of two-component systems, and also that of three or more components, into *mixtures*, in which the mole fraction of any one component may lie in the range from zero to one, and *solutions* in which one component is always in excess. In the latter case, the component always in excess (and designated component A) is called the *solvent* and the other $r - 1$ components are called the *solutes*. We begin with some further discussion of partial molar properties in mixtures. Using V as an example of an extensive property

$$(2.7.17) \qquad\qquad V = \sum_B V_B n_B \qquad\qquad (1)$$

where V_B is the partial molar volume of substance B. From the Gibbs–Duhem equation (2.7.19), for a mixture of two components, at constant T and p

$$n_A dV_A + n_B dV_B = 0 \qquad (2)$$

Dividing by $n_A + n_B$ expresses the amounts of substances A and B in terms of mole fractions. Since the sum of the mole fractions is unity

$$x_A + x_B = 1 \tag{3}$$

(2), (3) $\qquad (1 - x_B)dV_A + x_B dV_B = 0 \quad$ (constant T, p) $\tag{4}$

The independent variables are T, p and either x_B or x_A. Choosing the former alternative, and maintaining T and p constant

(4) $\qquad (1 - x_B)\left(\dfrac{\partial V_A}{\partial x_B}\right) + x_B \left(\dfrac{\partial V_B}{\partial x_B}\right) = 0 \tag{5}$

There are three methods of determining partial molar properties from experimental data.

6.2.1. *Direct method*

V is measured as a function of n_B while n_A is held constant and the results plotted. The gradient of the curve $V(n_B)$ with T, p, n_A constant is the partial molar volume of B.

6.2.2. *From apparent molar properties*

For a mixture containing n_A moles of A and n_B moles of B, the *apparent molar volume* of B is defined by

$$V_B^{\text{app}} = \frac{V - n_A V_{m,A}^*}{n_B} \tag{6}$$

The superscript * means the pure component, so $n_A V_{m,A}^*$ is the volume that would be occupied by the same amount of component A in the pure state. This is not the same in the mixture because the intermolecular forces between A and B differ from the intermolecular forces between A and A molecules

(6) $\qquad n_B V_B^{\text{app}} = V - n_A V_{m,A}^* \tag{7}$

(7) $\qquad n_B \left(\dfrac{\partial V_B^{\text{app}}}{\partial n_B}\right)_{T,p,n_A} + V_B^{\text{app}} = \left(\dfrac{\partial V}{\partial n_B}\right)_{T,p,n_A} \tag{8}$

(8) $\qquad \left(\dfrac{\partial V}{\partial n_B}\right)_{T,p,n_A} = V_B^{\text{app}} + \left(\dfrac{\partial V_B^{\text{app}}}{\partial \ln n_B}\right)_{T,p,n_A} \tag{9}$

This method offers a more accurate way of treating experimental data than the direct method described in Section 6.2.1.

6.2.3. *From mean molar properties*

The mean molar volume (for example) $V_m(T, p, x_B)$ of a mixture of A and B is

(1), (3) $$V_m = (1 - x_B)V_A + x_B V_B \tag{10}$$

V_A, V_B are the partial molar volumes of A and B respectively. Differentiating with respect to x_B at constant T, p gives

(10) $$\left(\frac{\partial V_m}{\partial x_B}\right) = V_B - V_A + (1 - x_B)\left(\frac{\partial V_A}{\partial x_B}\right) + x_B\left(\frac{\partial V_B}{\partial x_B}\right) \tag{11}$$

(11), (5) $$\left(\frac{\partial V_m}{\partial x_B}\right) = V_B - V_A \tag{12}$$

(10), (12) $$V_m = (1 - x_B)V_A + x_B\left\{V_A + \left(\frac{\partial V_m}{\partial x_B}\right)\right\} \tag{13}$$

(13), (10) $$V_B = V_m + (1 - x_B)\left(\frac{\partial V_m}{\partial x_B}\right) \tag{14}$$

(13), (10) $$V_A = V_m - x_B\left(\frac{\partial V_m}{\partial x_B}\right) \tag{15}$$

The geometrical interpretation of Eqs. (14) and (15) is shown in Figure 6.1, which is the plot of V_m against x_B, over the range of x_B, which is $0 \rightarrow 1$.

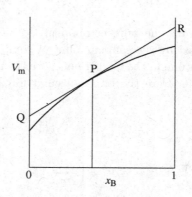

Figure 6.1. Dependence of the mean molar volume V_m on x_B, the mole fraction of component B. The intercepts of the gradient of the curve at P with the ordinates at 0 and 1, are the partial molar volumes V_A and V_B respectively at that mole fraction (see Eqs. (6.2.14) and (6.2.15)).

When x_B is zero, the mean molar volume is the molar volume of pure A and when it is one, the mean molar volume is the molar volume of pure B. For the mixture of mole fraction x_B corresponding to the point P, the partial molar volumes V_A and V_B are given by the intercepts of the gradient of the curve at P with the ordinates at zero and one respectively.

6.3. Mixtures of Gases

6.3.1. *Ideal mixtures*

The partial pressure p_B of a component B in a mixture of gases is defined by

$$p_B = y_B p \tag{1}$$

where y_B is the mole fraction of B. A gas mixture is ideal (a) if the individual components behave ideally and (b) if the chemical potential of each component B is given by

$$\mu_B = \mu_B^*(p) + RT \ln y_B \tag{2}$$

where $\mu_B^*(p)$ is the chemical potential of the pure gas at a pressure p equal to the total pressure of the mixture and y_B is the *mole fraction of B in the gas phase*. But if the individual gases are ideal

$$(5.4.10) \qquad \mu_B^*(p) = \mu_B^*(p^\circ) + RT \ln(p/p^\circ) \tag{3}$$

$$(1),\ (2),\ (3) \qquad \mu_B = \mu_B^*(p^\circ) + RT \ln(p_B/p^\circ) \tag{4}$$

Equation of state

$$V_B = \frac{\partial}{\partial n_B}\left(\frac{\partial G}{\partial p}\right) = \frac{\partial}{\partial p}\left(\frac{\partial G}{\partial n_B}\right) = \frac{\partial \mu_B}{\partial p} = \frac{RT}{p} \tag{5}$$

Thus the partial molar volume of a component B in an ideal gas mixture is equal to the molar volume of the pure gas B at a pressure p equal to that of the mixture

$$(5) \qquad V = \sum_B n_B V_B = \sum_B n_B \frac{RT}{p} = n\frac{RT}{p} \tag{6}$$

On multiplying each side of (6) by y_B

$$(6) \qquad p_B V = n_B RT \tag{7}$$

$$(6) \qquad pV = nRT \tag{8}$$

Thus the definition of an ideal mixture in (a) and (b) ensures that the ideal gas equation of state applies to each component and to the whole mixture.

Gibbs energy of mixing

The Gibbs energy of the mixture is

$$G = \sum_B n_B \mu_B \tag{9}$$

The Gibbs energy of mixing is, therefore

$$\Delta G_{\text{mix}} = \sum_B n_B \mu_B - \sum_B n_B \mu_B^*$$

$$(6) \qquad\qquad = \sum_B n_B RT \ln y_B \tag{10}$$

Entropy of mixing

The molar entropy of mixing is

$$\Delta S_{\text{mix,m}} = -\frac{\partial \Delta G_{\text{mix,m}}}{\partial T} = -\sum_B y_B R \ln y_B \tag{11}$$

Enthalpy of mixing

$$(10),\ (11) \qquad\qquad \Delta H_{\text{mix,m}} = \Delta G_{\text{mix,m}} + \Delta(TS_{\text{mix,m}}) = 0 \tag{12}$$

because no change in temperature occurs if each gas behaves ideally.

Volume change on mixing

$$(10) \qquad\qquad \Delta V_{\text{mix,m}} = -\frac{\partial \Delta G_{\text{mix,m}}}{\partial p} = 0 \tag{13}$$

6.3.2. *Non-ideal mixtures*

We consider only binary mixtures, labelling the two components by subscripts A and B. In a pure gas, the second virial coefficient B accounts for interactions between pairs of molecules. At the same level of approximation, in a binary mixture of A and B, we must account for three types of interaction, AA, AB and BB. The number of pair interactions is proportional to the number of molecules of each type involved in the interaction. The equation of state for a single gas is

$$V_m = \frac{RT}{p} + B \tag{14}$$

For a binary mixture this becomes

$$\frac{V}{n_A + n_B} = \frac{RT}{p} + \frac{n_A^2 B_{AA} + 2n_A n_B B_{AB} + n_B^2 B_{BB}}{(n_A + n_B)^2} \tag{15}$$

Multiply by $n_A + n_B$ and differentiate with respect to n_B

$$(15) \quad V_B = \frac{RT}{p} + \frac{n_B^2 B_{BB} + 2n_A n_B B_{BB} + 2n_A^2 B_{AB} - n_A^2 B_{AA}}{(n_A + n_B)^2} \tag{16}$$

The general result for an r-component mixture is given in Berry, Rice and Ross (2002).

The chemical potential of B is

$$(5) \quad \mu_B = \mu_B^*(p^\circ) + RT\ln(y_B p/p^\circ) + \int_0^p \left(V_B - \frac{RT}{p'}\right) dp' \tag{17}$$

The first two terms on the RS of (17) are the ideal gas part. The third term is the contribution from non-ideality and so the ideal gas molar volume (RT/p') must be subtracted from the partial molar volume of B in the integrand

$$(17), (16) \quad \mu_B = \mu_B^*(p^\circ) + RT\ln(y_B p/p^\circ)$$
$$+ p\{B_{BB} - y_A^2(B_{AA} - 2B_{AB} + B_{BB})\} \tag{18}$$

The general formula for an r-component mixture is given by Berry, Rice and Ross (2002).

Alternatively, the formal simplicity if the ideal gas formula can be maintained by replacing the partial pressure of B by the partial fugacity of B, $\tilde{p}_B$

$$(18) \quad \mu_B = \mu_B^*(p^\circ) + RT\ln(\tilde{p}_B/p^\circ) \tag{19}$$

If the fugacity of B in the mixture is not known, it can be estimated from the *Lewis and Randall rule* that the fugacity of B in the mixture

$$\tilde{p}_B(p) \approx y_B \tilde{p}_B^*(p) \tag{20}$$

where $\tilde{p}_B^*(p)$ is the fugacity of the pure gas B at the same pressure p as the total pressure of the mixture.

6.4. Liquid Mixtures

Liquid mixtures are mixtures of two or more liquids that are homogeneous over at least a substantial concentration range such that a dilute solution of any one of the components may exist. Examples are water + ethyl alcohol, or acetone + carbon disulphide. Concentrations are expressed as mole fractions.

Example 6.4-1

The partial molar volume of ethanol in an aqueous mixture containing 0.600 mole fraction of ethanol is $57.5\,\mathrm{cm^3\,mol^{-1}}$. The density of the solution is $0.8494\,\mathrm{g\,cm^{-3}}$. Calculate the partial molar volume of water in this mixture

$$V_{\mathrm{m}} = \frac{\text{molar mass}}{\text{density}} = \frac{x_A M_A + x_B M_B}{\rho}$$

$$= \frac{(0.400 \times 18.01) + (0.600 \times 46.05)}{0.8494}\,\mathrm{cm^3\,mol^{-1}}$$

(6.2.1)
$$= (0.400 \times V_A) + (0.600 \times 57.5)\,\mathrm{cm^3\,mol^{-1}}$$

$$V_A = 16.2\,\mathrm{cm^3\,mol^{-1}}$$

6.4.1. *Activity*

The *activity* a_B of a component B in a liquid mixture is defined by

$$\mu_B = \mu_B^\ominus + RT \ln a_B \tag{1}$$

μ_B is the chemical potential of B in the liquid mixture and $\mu_B^\ominus$ is the chemical potential of B in the *standard state*, which for a pure liquid or solid, a *liquid* or solid *mixture*, or a solvent, is the pure phase B at the standard pressure $p^\ominus$

$$\mu_B^\ominus(T) = \mu_B^*(T, p^\ominus) \tag{2}$$

(1), (2)
$$\mu_B = \mu_B^*(T, p^\ominus) + RT \ln a_B \tag{3}$$

The mixture is ideal if

$$a_B = x_B, \quad 0 < x_B < 1 \quad \text{(ideal mixture)} \tag{4}$$

More commonly, $a_B \neq x_B$ and

$$(3) \qquad\qquad \mu_B = \mu_B^*(T, p^{\circ}) + RT\ln f_B x_B \qquad\qquad (5)$$

where the *activity coefficient* f_B is a measure of departures from ideality

$$(5), (4) \qquad\qquad f_B \to 1 \quad \text{as} \quad x_B \to 1 \qquad\qquad (6)$$

6.4.2. *Vapour pressure of ideal liquid mixtures*

The condition for equilibrium between a liquid mixture and its vapour is the equality of the chemical potential of each component in the two phases, liquid and vapour. For the vapour over the mixture

$$\mu_B = \mu_B^*(p^{\circ}) + RT\ln(p_B/p^{\circ}) \qquad\qquad (6.3.4)$$

and similarly, for the vapour over the pure component B

$$(6.3.4) \qquad\qquad \mu_B^* = \mu_B^*(p^{\circ}) + RT\ln(p_B^*/p^{\circ}) \qquad\qquad (7)$$

Subtracting (7) from (6.3.4)

$$\mu_B = \mu_B^* + RT\ln(p_B/p_B^*) \qquad\qquad (8)$$

Strictly, $\tilde{p}_B/\tilde{p}_B^*$ should replace p_B/p_B^* but this is not usually necessary because of the low vapour pressures involved. Equating the chemical potential of component B in the liquid mixture and in the vapour phase over the mixture, gives

$$(8), (4) \qquad\qquad a_B = f_B x_B = p_B/p_B^* \qquad\qquad (9)$$

If the mixture is ideal

$$(9) \qquad\qquad p_B = x_B p_B^* \quad \text{(ideal liquid mixture)} \qquad\qquad (10)$$

Equation (10) states that the vapour pressure of each component over a liquid mixture is proportional to the mole fraction of that component in the liquid mixture, which is *Raoult's law* (Figure 6.2). Examples of ideal mixtures are ethylene bromide + propylene bromide and mixtures of the monomethyl and monoethyl ethers of ethylene glycol.

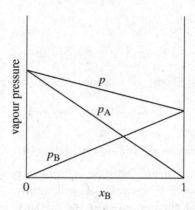

Figure 6.2. Vapour pressures of A and B over an ideal binary liquid mixture. Both components obey Raoult's law over the whole concentration range. p is the total vapour pressure.

From Dalton's law of partial pressures

$$p_B = y_B p \quad \text{(vapour phase)} \tag{11}$$

(10), (11)
$$y_B = x_B \frac{p_B^*}{p} \tag{12}$$

The total vapour pressure of a binary ideal liquid mixture is

(10)
$$p = (1 - x_B)p_A^* + x_B p_B^* = p_A^* + x_B(p_B^* - p_A^*) \tag{13}$$

(Figure 6.2).

Exercise 6.4-1

Show that the composition of the vapour is given by

$$y_B = \frac{x_B p_B^*}{p_A^* + (p_B^* - p_A^*)x_B} \tag{14}$$

(**Hint:** Use Eqs. (12) and (13).)

Because the pure components have different vapour pressures, the vapour and liquid phases have different compositions. The *volatility ratio* is defined by

(12)
$$\alpha_{BA} = \frac{y_B/y_A}{x_B/x_A} = \frac{y_B(1 - x_B)}{x_B(1 - y_B)} = \frac{p_B^*}{p_A^*} \quad \text{(ideal mixture)} \tag{15}$$

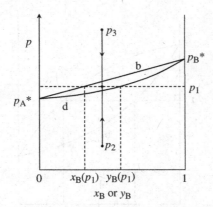

Figure 6.3. Dependence of total vapour pressure on liquid composition (the bubble-point curve, *b*) and on vapour composition (the dew-point curve, *d*). Liquid and vapour coexist between *b* and *d*.

which is independent of composition for ideal mixtures. The total pressure as a function of vapour composition is given by

$$(13), (14), (15) \qquad p = \frac{p_A^* p_B^*}{p_B^* - y_B(p_B^* - p_A^*)} = \frac{p_B^*}{\alpha_{BA} - y_B(\alpha_{BA} - 1)} \qquad (16)$$

which is plotted as the concave-upwards line in Figure 6.3.

Exercise 6.4-2

Provide an alternative derivation for the equation for the total pressure.

The straight line $p(x_B)$, which is called the *bubble-point curve*, shows the dependence of total pressure on liquid composition. The concave-upwards curve, $p(y_B)$ is called the *dew-point curve*. It shows the dependence of total pressure on vapour composition. The horizontal line at p_1, which is the common pressure of liquid and vapour, gives the composition of liquid (x_B) and vapour (y_B) at this pressure. Below the dew-point curve, only vapour is present. If such a system were to be compressed isothermally from an initial pressure p_2 then liquid would start to condense when the vertical line crosses the dew-point curve. Between this point and that where the vertical line crosses the bubble-point curve, liquid and vapour coexist. Above the bubble-point curve only liquid is stable. If the pressure on the liquid at an initial pressure p_3 were to be reduced isothermally, vapour would start to form at the point where the vertical line crosses the bubble-point curve (hence the name "bubble point"). Liquid and vapour would then coexist

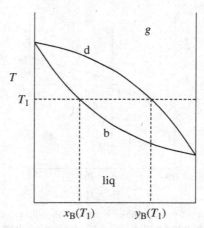

Figure 6.4.   Temperature-composition diagram for the same system as in Figure 6.3, explaining the principles of distillation. d is the dew-point curve, b is the bubble-point curve.

until the point is reached where the vertical line crosses the dew-point curve, below which only vapour is stable.

Figure 6.4 shows a temperature-composition diagram $T(x_B)$ at constant pressure, for the same system. If liquid mixture of composition $x_B(T_1)$ is heated, it boils at T_1. If the vapour of initial composition $y_B(T_1)$ were to be condensed, the condensate would have a composition richer in B than the original liquid mixture and if this were vaporized and the vapour condensed, it would have composition richer still in B. Thus repeated evaporation and condensation leaves behind a liquid richer in the less volatile constituent (A) and a condensate richer in the more volatile constituent (in this case, B). Successive evaporations and condensations are performed automatically in a vertical *fractionating column,* which consists of a number of horizontal plates inside a double-walled condenser. The cooling liquid circulates between the walls, entering from the top, with the result that the top of the condenser is colder than the bottom, so that the less volatile constituent condenses in the bottom part of the condenser and is returned to the evaporating liquid.

6.4.3. *Non-ideal liquid mixtures*

The majority of liquid mixtures display either positive or negative deviations from Raoult's law (Figures 6.5, 6.6) (von Zawidski 1900). These curves illustrate two important features:

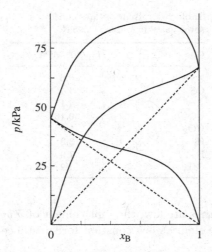

Figure 6.5. Partial pressures and the total vapour pressure above the liquid mixture acetone (A) + carbon disulphide (B) at 308.4 K. This mixture is an example of one displaying positive deviations from Raoult's law. The dashed lines show that Raoult's law is obeyed in the limit of very dilute mixtures.

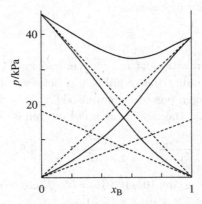

Figure 6.6. Negative deviations from Raoult's law in the mixture acetone (A) + chloroform (B) at 308.4 K. Note the regions of validity of Raoult's law and Henry's law.

(i) as $x_B \to 1$, $p_B \to p_B^* x_B$, $f_B \to 1$ (Raoult's law);

(ii) as $x_B \to 0$, $p_B \to k_{H,B} x_B$, $\gamma_{x,B} \to 1$ (Henry's law).

$\gamma_{x,B}$ is the IUPAC-recommended notation for the activity coefficient of substance B, referred to as Henry's law, when concentrations are measured in mole fractions. Strictly, the pressures in (i) and (ii) should be fugacities

Table 6.1.　Henry's law constants for some gases in water at 298 K.

Gas	$k_H/10^3$ MPa
N_2	8.6
O_2	4.4
Ar	4.1
CO	0.16
CO_2	5.8
He	13.3
H_2	7.9
CH_4	4.2

but when the pressures are low, the replacement of $\tilde{p}$ by p is an acceptable approximation. The Henry's law constant for substance B, is defined by

$$k_{H,B} = \left(\frac{\partial \tilde{p}_B}{\partial x_B}\right)_{x_B=0} \approx \left(\frac{\partial p_B}{\partial x_B}\right)_{x_B=0} \tag{17}$$

Some Henry's law constants are given in Table 6.1. They are often used to calculate the solubility of gases.

Example 6.4-1

Calculate the concentration of oxygen (in $mol\,kg^{-1}$) in the ocean at a depth of 100 m, assuming that equilibrium between the dissolved oxygen and atmospheric oxygen has been established.

From Henry's law, the mole fraction of oxygen is

$$x_{O_2} = \frac{p_{O_2}}{k_{H,O_2}} = \frac{21\,\text{kPa}}{4.4 \times 10^3\,\text{MPa}} = 4.8 \times 10^{-6}$$

Since 1 kg of water contains $10^3/18.016 = 55.51$ mol of H_2O, the amount of O_2 in 1 kg of water is $4.8 \times 10^{-6} \times 55.51\,\text{mol kg}^{-1} = 2.7 \times 10^{-4}\,\text{mol kg}^{-1}$.

Example 6.4-2 Raoult's law activity coefficients from vapour pressure measurements

Table 6.2 gives the activities a (calculated from p/p^*) of chloroform (A) and acetone (B) at the stated mole fractions of chloroform. Plot the activities a_A and a_B as functions of x_B and note the regions of validity of Raoult's law and Henry's law and the negative deviations from Raoult's law. Table 6.2 also gives the activity coefficients of A and B calculated from these data.

Table 6.2. Mole fraction x_B of chloroform, activities a_B and a_A of chloroform and acetone and their corresponding activity coefficients f_A and f_B, in liquid mixtures of acetone + chloroform at 35.2°C, calculated from $a_i = \frac{p_i}{p_i^{\ast}} = f_i x_i, i = $ A,B.

x_B	a_B	a_A	f_B	f_A
0.000	0	1	—	1
0.059	0.031	0.939	0.532	0.997
0.123	0.070	0.869	0.566	0.991
0.185	0.109	0.799	0.588	0.981
0.297	0.189	0.699	0.636	0.951
0.423	0.303	0.506	0.717	0.877
0.514	0.403	0.392	0.782	0.806
0.663	0.581	0.227	0.878	0.680
0.800	0.767	0.109	0.957	0.544
0.9175	0.913	0.038	0.993	0.457
1.000	1	0	1	—

The reader is advised to check the values of at least some of these activity coefficients.

6.4.4. *Application of the Gibbs–Duhem equation*

At constant temperature, the Gibbs–Duhem equation is

$$-V\mathrm{d}p + \sum_B n_B \mu_B = 0 \qquad (18)$$

Altering the composition of the mixture will alter the total vapour pressure p, but we can arrange to keep the total pressure constant by the addition of an inert gas. Dividing (18) by $\sum_B n_B$ and setting $\mathrm{d}p = 0$

(18)
$$\sum_B x_B \mathrm{d}\mu_B = 0 \qquad (19)$$

Since $\sum_B x_B = 1$, μ is a function of T, p and $c - 1$ independent mole fractions. For a two-component system, at constant T and p, therefore

$$\mathrm{d}\mu_B = \left(\frac{\partial \mu_B}{\partial x_B}\right)_{x_A = 0} \mathrm{d}x_B \qquad (20)$$

(20), (19)
$$x_A \left(\frac{\partial \mu_A}{\partial x_A}\right)_{x_B = 0} \mathrm{d}x_A + x_B \left(\frac{\partial \mu_B}{\partial x_B}\right)_{x_A = 0} \mathrm{d}x_B = 0 \qquad (21)$$

But $x_A + x_B = 1$, and, therefore, $dx_A = -dx_B$. Consequently

$$(21) \qquad x_A \left(\frac{\partial \mu_A}{\partial x_A} \right)_{x_B=0} = x_B \left(\frac{\partial \mu_B}{\partial x_B} \right)_{x_A=0} \qquad \text{(constant } T, p) \qquad (22)$$

At equilibrium

$$\mu_B(\text{mixture}) = \mu_B(\text{vapour})$$

$$(6.3.19) \qquad\qquad = \mu_B^*(p^\ominus) + RT\ln(\tilde{p}_B/p^\ominus) \qquad\qquad (23)$$

$$= \mu_B^*(p^\ominus) + RT\ln(p_B/p^\ominus) \quad \text{(low pressures)} \tag{24}$$

$$(22),\ (24) \quad x_A \left(\frac{\partial \ln p_A}{\partial x_A} \right) = x_B \left(\frac{\partial \ln p_B}{\partial x_B} \right) \quad (T,\ \text{total } p \text{ constant}) \qquad (25)$$

which is the *Duhem–Margules equation*. Since T and p are constant and x_A is not an independent variable ($x_A + x_B = 1$)

$$(25) \qquad \frac{-x_A}{p_A} \left(\frac{dp_A}{dx_B} \right) = \frac{x_B}{p_B} \left(\frac{dp_B}{dx_B} \right) \qquad\qquad (26)$$

$$(26) \qquad \left(\frac{d(p_A + p_B)}{dx_B} \right) = \left(\frac{dp_A}{dx_B} \right) + \left(\frac{dp_B}{dx_B} \right) = \left(\frac{dp_B}{dx_B} \right) \left(1 - \frac{x_B p_A}{x_A p_B} \right) \qquad (27)$$

(dp_B/dx_B) is always positive and non-zero; therefore if

$$(27) \qquad \left(\frac{d(p_A + p_B)}{dx_B} \right) > 0, \quad \frac{p_A}{p_B} < \frac{x_A}{x_B}, \quad \frac{y_A}{y_B}\text{(vapour)} < \frac{x_A}{x_B}\text{(liquid)} \qquad (28)$$

and the vapour is relatively richer in component B than is the liquid. Similarly, if

$$(27) \qquad \left(\frac{d(p_A + p_B)}{dx_B} \right) < 0, \quad \frac{p_A}{p_B} > \frac{x_A}{x_B}, \quad \frac{y_A}{y_B}\text{(vapour)} >, \frac{x_A}{x_B}\text{(liquid)} \qquad (29)$$

and the vapour is relatively richer in component A than is the liquid. Thus in both these cases, the two liquids may be separated by distillation.

Now suppose that there is a turning point in the plot of $p_A + p_B$ against x_B, as there is, for example, in acetone + chloroform and ethanol + benzene mixtures. When

$$(26) \qquad\qquad x_B < x_B(Z), \quad \left(\frac{d(p_A + p_B)}{dx_B}\right) < 0 \qquad\qquad (30)$$

and component A distills off preferentially, so that x_B increases. At $x_B = x_B(Z)$

$$(24) \qquad\qquad \left(\frac{d(p_A + p_B)}{dx_B}\right) = 0, \quad \frac{p_A}{p_B} = \frac{x_A}{x_B} = \frac{y_A}{y_B} \qquad\qquad (31)$$

The vapour composition is identical with that of the liquid, which therefore distills without further change in composition. Such mixtures are called *azeotropes*.

The corresponding temperature-composition diagrams are shown in Figures 6.7 and 6.8. Minima in the plot of p against composition correspond to maxima in the $T(x_B)$ curve. Solutions rich in acetone (A) give a distillate of pure acetone and a liquid azeotrope of composition $x_B(Z)$ which would distill without further change in composition. On the other hand, a liquid of initial composition $x_B > x_B(Z)$ would yield a distillate of pure chloroform and liquid azeotrope. Similarly, distillation of a minimum boiling point mixture with initial composition $x_B < x_B(Z)$ (Figure 6.8) would give a distillate of pure azeotrope, leaving pure A in the flask.

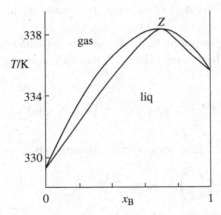

Figure 6.7. Temperature-composition diagram for the system acetone (A) + chloroform (B), which displays a maximum boiling point azeotrope, Z.

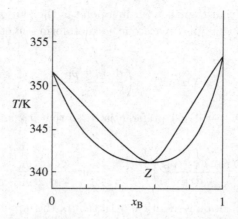

Figure 6.8.   Temperature-composition diagram for the system ethanol (A) + benzene (B), which displays a minimum boiling point azeotrope, Z.

6.4.5. *Regular mixtures*

In ideal mixtures, there is no energy change on mixing and the entropy of mixing is that which would apply if the molecules were assigned randomly to a lattice. For ideal mixtures

$$p_B = x_B p_B^* \tag{10}$$

and they therefore obey Raoult's law and Henry's law over the whole concentration range, $0 < x_B < 1$. But experiment shows that deviations from these laws often occur, as seen, for example, in Section 6.4.3, for mixtures of chloroform + acetone.

Any equation proposed to describe the dependence of p on x must satisfy the Gibbs–Duhem equation (18) and therefore the Duhem–Margules equation

$$x_A \left(\frac{\partial \ln p_A}{\partial x_A} \right) = x_B \left(\frac{\partial \ln p_B}{\partial x_B} \right) \quad (T, \text{ total } p \text{ constant}) \tag{25}$$

A general solution to (25) proposed by Margules is

$$\ln \left(\frac{p_A}{p_A^* x_A} \right) = a_A x_B + \frac{1}{2} b_A x_B^2 + \frac{1}{3} c_A x_B^3 + \cdots \tag{32}$$

$$\ln \left(\frac{p_B}{p_B^* x_B} \right) = a_B x_A + \frac{1}{2} b_B x_A^2 + \frac{1}{3} c_B x_A^3 + \cdots \tag{33}$$

Terminate the expansions at the terms quadratic in x by setting $c_A = c_B = 0$. Then differentiate (32) and (33) and substitute the results in (25), giving

(32), (33), (25) $\qquad a_A = a_B = 0, \quad b_A = b_B = 2w/RT \quad (c_A = c_B = 0)$
$$\tag{34}$$

Equations (32)–(34) define w. The amount by which any thermodynamic function exceeds its ideal value is called the *excess* function. So the *excess chemical potential* of component B in a liquid mixture is

(6), (33), (34) $\qquad \mu_B^E = \mu_B - \mu_B^* - RT\ln x_B = RT\ln f_B \qquad (35)$

$$= w(1 - x_B^2) \tag{36}$$

A mixture for which the excess chemical potential has the form (36) is called a *regular mixture*.

Exercise 6.4-3

What are the dimensions of w? (**Hint:** See Eqs. (33) and (34).)

Exercise 6.4-4

Show how w may be evaluated for a regular mixture in which the deviations from ideality are small. (**Hint:** This implies that exponential terms may be expanded and only the first two terms retained.)

The change in any extensive property E on forming a binary mixture from its pure components is

$$\Delta_{\mathrm{mix}}E = E - (n_A E_A^* + n_B E_B^*). \tag{37}$$

(37) $\qquad\qquad = n_A(E_A - E_A^*) + n_B(E_B - E_B^*) \tag{38}$

Heat of mixing

We first derive an expression for the dependence of activity on temperature

$$\left\{\frac{\partial}{\partial T}(R\ln a_B)\right\}_{p,x_A,x_B\cdots} = \left\{\frac{\partial}{\partial T}\left(\frac{\mu_B}{T}\right)\right\}_{p,x_A,x_B\cdots}$$

$$-\left\{\frac{\partial}{\partial T}\left(\frac{\mu_B^{\ominus}}{T}\right)\right\}_{p,x_A,x_B\cdots} \tag{39}$$

$$\frac{\partial}{\partial T}\left(\frac{\mu_B}{T}\right) = \frac{\partial}{\partial T}\left(\frac{1}{T}\frac{\partial G}{\partial n_B}\right)$$

$$= -\frac{1}{T^2}\frac{\partial G}{\partial n_B} + \frac{1}{T}\frac{\partial}{\partial n_B}\left(\frac{\partial G}{\partial T}\right)$$

$$= -\frac{1}{T^2}\frac{\partial}{\partial n_B}(G + TS) = \frac{-H_B}{T^2}$$

(39)
$$\frac{\partial}{\partial T}(\ln a_B) = -\frac{H_B - H_B^{\ominus}}{RT^2} = -\frac{L_B}{RT^2} \qquad (40)$$

where $L_B = H_B - H_B^{\ominus}$ is the *relative partial molar enthalpy*. For mixtures, the pure liquid is the standard state. Therefore

(40), (38) $$\Delta_{\text{mix}} H = RT^2 \left(n_A \frac{\partial \ln a_A}{\partial T} + n_B \frac{\partial \ln a_B}{\partial T} \right) \qquad (41)$$

Exercise 6.4-5

Write down an expression for the excess heat of mixing. Show that the heat of mixing $\Delta_{\text{mix}} H$ is zero for an ideal mixture.

Volume change on mixing

We need an expression for the dependence of activity on pressure

$$\left\{ \frac{\partial}{\partial p}(R \ln a_B) \right\}_{T, x_A, x_B \dots} = \left\{ \frac{\partial \mu_B}{\partial p} \right\}_{T, x_A, x_B \dots} - \left\{ \frac{\partial \mu_B^{\ominus}}{\partial p} \right\}_{T, x_A, x_B \dots} \qquad (42)$$

$$\frac{\partial \mu_B}{\partial p} = \frac{\partial}{\partial p}\left(\frac{\partial G}{\partial n_B}\right) = \frac{\partial}{\partial n_B}\left(\frac{\partial G}{\partial p}\right) = V_B \qquad (43)$$

(42), (43) $$RT\frac{\partial \ln a_B}{\partial p} = V_B - V_B^{\ominus} \qquad (44)$$

(44), (38) $$\Delta_{\text{mix}} V = RT \left(n_A \frac{\partial \ln a_A}{\partial p} + n_B \frac{\partial \ln a_B}{\partial p} \right) \qquad (45)$$

Exercise 6.4-6

Write down an expression for the excess volume change on mixing two miscible liquids. Show that the volume change on mixing $\Delta_{\mathrm{mix}}V$ is zero for an ideal mixture.

Gibbs energy of mixing

$$\Delta_{\mathrm{mix}}G = n_A(\mu_A - \mu_A^*) + n_B(\mu_B - \mu_B^*) \tag{46}$$

(46) $$\Delta_{\mathrm{mix}}G_m = RT(x_A \ln a_A + x_B \ln a_B) \tag{47}$$

(45) $$\Delta_{\mathrm{mix}}^{\mathrm{id}}G_m = RT(x_A \ln x_A + x_B \ln x_B) \quad \text{(ideal mixture)} \tag{48}$$

(47), (48) $$\Delta_{\mathrm{mix}}G_m^{\mathrm{E}} = RT(x_A \ln f_A + x_B \ln f_B) \tag{49}$$

is the excess Gibbs energy of mixing.

Entropy of mixing

(47) $$\Delta_{\mathrm{mix}}S_m = R(x_A \ln a_A + x_B \ln a_B) \tag{50}$$

Exercise 6.4-7

Write down an expression for the entropy change on forming an ideal binary liquid mixture from its two pure components. What factors do you think might contribute to a non-ideal value for the entropy of mixing?

Example 6.4-3

Calculate the molar excess Gibbs energy of mixing for a mixture of chloroform (A) + acetone (B) containing 0.800 mole fraction of chloroform.

Using the activity coefficient data in Table 6.1

$$\Delta_{\mathrm{mix}}G_m^{\mathrm{E}} = RT(x_A \ln f_A + x_B \ln f_B)$$

$$= 8.3145 \, \mathrm{J \, K^{-1} \, mol^{-1}} \times 308.35 \, \mathrm{K}$$

$$\times (0.800 \ln 0.957 + 0.200 \ln 0.544)$$

$$= -402 \, \mathrm{J \, mol^{-1}}$$

Figures showing the variation of G, H and TS with mole fraction for several liquid mixtures are given by Rowlinson (1969).

Activity coefficients may be obtained from measurements of the dependence of total vapour pressure p upon liquid composition. Eliminating p_A from the Duhem–Margules equation

$$\frac{-x_A}{p_A}\left(\frac{dp_A}{dx_B}\right) = \frac{x_B}{p_B}\left(\frac{dp_B}{dx_B}\right) \quad \text{(constant } T) \tag{26}$$

using

$$p = p_A + p_B \tag{51}$$

yields

$$\left(\frac{\partial p_A}{\partial x_B}\right)_T = \frac{x_B p_A}{p_A - (1 - x_B)p}\left(\frac{\partial p}{\partial x_B}\right)_T \tag{52}$$

from which p_A may be calculated from the dependence of the total vapour pressure p on liquid composition. p_B may then be found from (51) and f_A and f_B then calculated from (9) and the corresponding equation for component A.

Exercise 6.4-8

Complete the derivation of (52) from (26).

6.4.6. Bubble-point and dew-point curves

We need the concentration dependence of the equation governing the isothermal liquid-vapour equilibrium. The chemical potential of B in the vapour over the mixture is

(6.3.4) $$\mu_B = \mu_B^*(p^\ominus) + RT\ln(\phi_B y_B p/p^\ominus) \tag{53}$$

And for component B in the liquid mixture, it is

$$\mu_B = \mu_B^*(p\ominus) + RT\ln(f_B x_B) \tag{54}$$

Since the pressures are low, we approximate (53) by assuming that the vapour behaves as an ideal gas and therefore set the fugacity coefficient $\phi_B = 1$. Equating chemical potentials in the vapour and liquid phases then gives

$$\ln\left(\frac{y_B}{f_B x_B}\right) = \ln\left(\frac{p_B^*}{p}\right) \tag{55}$$

and similarly

$$\ln\left(\frac{y_A}{f_A x_A}\right) = \ln\left(\frac{p_A^*}{p}\right) \tag{56}$$

(55), (56)

$$x_B = \frac{f_A p_A^* - p}{f_A p_A^* - f_B p_B^*} \tag{57}$$

$$y_B = \frac{f_A\, p_A^*\, f_B\, p_B^* - f_B\, p_B^* p}{(f_A\, p_A^* - f_B\, p_B^*)p} \tag{58}$$

Equation (57) is the equation to the *bubble-point curve* and (58) that to the *dew-point curve*.

Exercise 6.4-9

Provide an alternative derivation of Eq. (57). (**Hint:** Write down an expression for the total pressure above a binary liquid mixture.)

(39)

$$\frac{\partial}{\partial T}(\ln a_B) = -\frac{H_B - H_B^\ominus}{RT^2} = -\frac{L_B}{RT^2} \tag{40}$$

(40)

$$\ln\left(\frac{\phi_B y_B}{f_B x_B}\right) = \int_{T_{B,b}}^{T} \frac{H_B^\ominus - H_B^*}{RT'^2}\,dT' \tag{59}$$

with a similar equation for component A. $T_{B,b}$ is the *standard boiling point* of B, that is the temperature at which the vapour pressure of the pure substance B is equal to the standard pressure. The numerator under the integral sign is the enthalpy change $\Delta H_{B,vap}^\ominus$ when one mole of substance B is transformed from the state of pure liquid B into vapour in its standard state, that is, pure B in the gaseous state at the standard pressure. (This quantity used to be called the latent heat of vaporization.) Making the approximation that these standard enthalpies of vaporization are independent of temperature

(59)

$$\ln\left(\frac{y_B}{f_B x_B}\right) = \frac{\Delta H_{vap,B}^\ominus}{R}\left[\frac{1}{T_{B,b}} - \frac{1}{T}\right] = \Lambda_{B,b} \tag{60}$$

(60)

$$\ln\left(\frac{y_A}{f_A x_A}\right) = \frac{\Delta H_{vap,A}^\ominus}{R}\left[\frac{1}{T_{A,b}} - \frac{1}{T}\right] = \Lambda_{A,b} \tag{61}$$

(60), (61) $$x_B = \frac{\lambda_A - f_A}{\lambda_A \lambda_B f_B - f_A} \qquad (62)$$

(60), (61) $$y_B = \frac{(\lambda_A - f_A)\lambda_B f_B}{\lambda_A \lambda_B f_B - f_A} \qquad (63)$$

where

$$\lambda_A = \exp(\Lambda_{A,b}) \qquad (64)$$

and $\Lambda_{A,b}$, $\Lambda_{B,b}$ are defined by the second equality in (60) and (61).

6.4.7. *Partial miscibility*

If a liquid mixture shows positive deviations from Raoult's law then, as T is lowered, these deviations become more pronounced until the *critical solution temperature* $T = T_c$ is reached, below which the mixture separates into two separate phases. (For mixtures of three or more components and many details on solubility, see Gamsjäger *et al.* 2008.) Above T_c, c $= 2$, $\phi = 2$, v $= 2$ and the temperature and composition of the liquid phase (which is described by one mole fraction) are independent variables. Below T_c, $\phi = 3$ and the system is univariant. Therefore, at any fixed $T < T_c$, the composition of both liquid phases is fixed and the partial pressures of both components have fixed values. An example of a partially miscible system is provided in the problems.

Exercise 6.4-10

What can be said about the *amounts* of the two components in this system?

In the extreme case of total immiscibility, the total vapour pressure

$$p = p_A^* + p_B^* \qquad (65)$$

and consequently the mixture boils at a temperature lower than the boiling point of either A or B. The composition of the vapour phase is given by

$$\frac{p_A^*}{p_B^*} = \frac{y_A}{y_B} = \frac{m_A/M_A}{m_B/M_B} \qquad (66)$$

where m_A is the mass of A and M_A the relative molar mass of A and similarly for component B. Since v $= 1$, isobaric distillation yields a condensate of constant composition. The relative amounts of the two components in the distillate may be calculated from (66). An application of (66) is *steam distillation*, which may be used to distill a liquid with a high boiling point at a temperature below 373 K.

6.4.8. *Critical mixing in regular mixtures*

For a regular mixture

(Section 6.4.5) $$\ln f_A = \frac{w}{RT}x^2, \quad \ln f_B = \frac{w}{RT}(1-x)^2 \qquad (67)$$

whence

$$p_A = p_A^*(1-x)\exp(x^2 w/RT) \qquad (68)$$

$$p_B = p_B^* x \exp\{(1-x)^2 w/RT\} \qquad (69)$$

Plots of p/p^* against x (the mole fraction of component B) for $w/RT = 1$, 2 and 2.75 are shown in Figure 6.9. For Figure 6.9a, $w/RT < 2$, so that $T > T_c$, and a single liquid phase is stable over the whole concentration range $0 < x < 1$. For $w/RT = 2$, (Figure 6.9b)

(65), (67) $$\frac{w}{RT}2x = \frac{1}{1-x}, \quad \frac{w}{RT}2 = \frac{1}{(1-x)^2} \qquad (70)$$

(70) $$x = \frac{1}{2}, \quad \frac{w}{RT} = 2 \qquad (71)$$

Equation (71) is the condition for *critical mixing*. When

$$w/RT > 2, \quad T < T_c \qquad (72)$$

one liquid phase is unstable and the system has two liquid phases, as shown in Figure 6.9c for the case $w/RT = 2.75$.

Answers to Exercises 6.4

Exercise 6.4-1

(13), (12) $$y_B = x_B \frac{p_B^*}{p} = \frac{x_B p_B^*}{p_A^* + (p_B^* - p_A^*)x_B} \qquad (14)$$

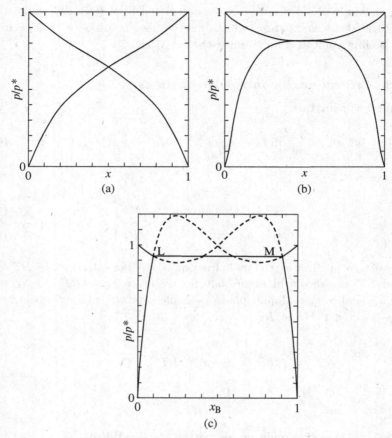

Figure 6.9.   Partial vapour pressures of the two components A and B in a binary regular mixture. The values of w/RT are (a) $w/RT = 1$, $T > T_c$; (b) $w/RT = 2$, $T = T_c$; (c) $w/RT = 2.75$, $T < T_c$. States corresponding to the dashed lines near the centre of the diagram are unstable and will not occur. States on the dashed lines close to L and M may be found but are metastable.

Exercise 6.4-2

(14)
$$x_B = \frac{y_B p_A^*}{p_B^* - y_B(p_B^* - p_A^*)}$$

(15), (17)
$$p = p_A^* + \frac{(p_B^* - p_A^*)y_B p_A^*}{p_B^* + (p_B^* - p_A^*)y_B}$$

$$= \frac{p_A^* p_B^*}{p_B^* - y_B(p_B^* - p_A^*)} \tag{16}$$

Exercise 6.4-3

$$(41) \qquad \Delta_{\text{mix}}H^{\text{E}} = RT^2 \left(n_{\text{A}} \frac{\partial \ln f_{\text{A}}}{\partial T} + n_{\text{B}} \frac{\partial \ln f_{\text{B}}}{\partial T} \right)$$

For an ideal mixture, $f_{\text{A}} = f_{\text{B}} = 1$ so that $\Delta_{\text{mix}}H = 0$

Exercise 6.4-4

$$p = p_{\text{A}} + p_{\text{B}} = f_{\text{A}}x_{\text{A}}p_{\text{A}}^* + f_{\text{B}}x_{\text{B}}p_{\text{B}}^*$$

$$= x_{\text{A}}p_{\text{A}}^* \exp\left(\frac{wx_{\text{B}}^2}{RT}\right) + x_{\text{B}}p_{\text{B}}^* \exp\left(\frac{wx_{\text{A}}^2}{RT}\right)$$

Expand the exponentials and retain only the first two terms, giving

$$p = x_{\text{A}}p_{\text{A}}^* \left(1 + \frac{wx_{\text{B}}^2}{RT}\right) + x_{\text{B}}p_{\text{B}}^* \left(1 + \frac{wx_{\text{A}}^2}{RT}\right)$$

$$\frac{w}{RT} = \left(\frac{p - (x_{\text{A}}p_{\text{A}}^* + x_{\text{B}}p_{\text{B}}^*)}{x_{\text{A}}x_{\text{B}}[p_{\text{A}}^* + x_{\text{A}}(p_{\text{B}}^* - p_{\text{A}}^*)]}\right)$$

Exercise 6.4-5

$$(41) \qquad \Delta_{\text{mix}}H^{\text{E}} = RT^2 \left(n_{\text{A}} \frac{\partial \ln f_{\text{A}}}{\partial T} + n_{\text{B}} \frac{\partial \ln f_{\text{B}}}{\partial T} \right)$$

Exercise 6.4-6

$$(45) \qquad \Delta_{\text{mix}}V^{\text{E}} = RT \left(n_{\text{A}} \frac{\partial \ln f_{\text{A}}}{\partial p} + n_{\text{B}} \frac{\partial \ln f_{\text{B}}}{\partial p} \right)$$

In an ideal mixture, the activity coefficients are unity and therefore the excess volume change on mixing is zero.

Exercise 6.4-7

For an ideal mixture

$$(50) \qquad \Delta_{\text{mix}}S_{\text{m}} = R(x_{\text{A}} \ln x_{\text{A}} + x_{\text{B}} \ln x_{\text{B}})$$

The entropy change in forming an ideal mixture is that which would result if the molecules of type A and B were arranged randomly on a lattice.

This is only possible if the molecules A and B are of similar size and shape. A non-zero configurational entropy in (50) implies that the molecules A and B are of different size and/or shape.

Exercise 6.4-8

$$\left(\frac{\partial p_A}{\partial x_B}\right)_T = \frac{x_B p_A}{p_A - (1 - x_B)p}\left(\frac{\partial p}{\partial x_B}\right)_T \tag{52}$$

Exercise 6.4-9

$$p = f_A x_A p_A^* + f_B x_B p_B^*$$

Substitute $x_A = 1 - x_B$ and solve for x_B to yield

$$x_B = \frac{f_A p_A^* - p}{f_A p_A^* - f_B p_B^*} \tag{57}$$

Exercise 6.4-10

For T above T_c, temperature and the mole fraction of one component are independent variables, but below T_c the presence of another phase reduces the variance to one. At fixed temperature, the composition is fixed and independent of the amount of one component, say A. Once that is fixed, so is the amount of B. The amounts of each component are irrelevant, only their relative amounts are fixed at each temperature.

6.5. Liquid–Solid Equilibrium

6.5.1. *Both phases miscible*

The phase diagram of the Bi + Cd system is shown in Figure 6.10. The vapour pressures of Bi and Cd are sufficiently low that their variations can be ignored in comparison with the much larger constant pressure of inert gas. Consequently, in calculating the variance

$$v = c + 2 - \phi \tag{1}$$

the vapour phase can be neglected and the pressure considered to be constant. Consequently, the variance is

$$\text{(1)} \qquad\qquad v = c + 1 - \phi' \tag{2}$$

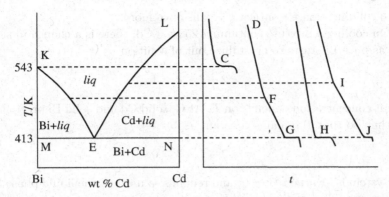

Figure 6.10. (a) Phase diagram of the system Bi + Cd. (b) Cooling curves for four different compositions of the system Bi + Cd.

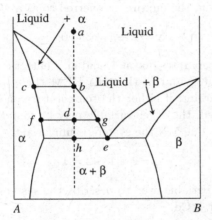

Figure 6.11. Phase diagram of a system A + B displaying partial miscibility in the solid state.

with ϕ' the number of condensed phases and the single variable 1 being the temperature.

If a pure liquid component, say Bi, is cooled, the plot of $T(t)$ (see Figure 6.11) exhibits an almost linear decrease until the fusion temperature is reached, when solid Bi starts to crystallize out. The cooling curve then becomes horizontal at C, as long as two phases are present because then

(2) $$v = 1 + 1 - 2 = 0$$ (3)

When solidification is complete, T falls once more.

On cooling a mixture containing 25 wt% Cd, there is a change in slope at F as pure Bi starts to crystallize out of solution

$$(2) \qquad\qquad v = 2 + 1 - 2 = 1 \qquad\qquad (4)$$

so that concentration depends on T. At G, solid Cd and solid Bi precipitate together, so that T remains constant

$$(2) \qquad\qquad v = 2 + 1 - 3 = 0 \qquad\qquad (5)$$

The system is invariant ($v = 0$) and remains so until all the liquid phase has disappeared. The cooling curve H shows what happens when a mixture of composition x_E is cooled. At H, Bi + Cd precipitate together *without change in composition* x_E. When the liquid and solid phases are both mixtures with the same composition, the mixture is referred to as a *eutectic*. At H

$$(2) \qquad\qquad v = 2 + 1 - 3 = 0 \qquad\qquad (6)$$

Similarly, pure Cd separates out at I, and at J eutectic precipitates.

Along the line KE and in the area KEM there are two phases and the system is univariant: if either the temperature or the composition of a mixture is specified, the other variable may be found from the line KE. Similarly, below the line MN the system is univariant

$$(2) \qquad\qquad v = 2 + 1 - 2 = 1 \qquad\qquad (7)$$

Here, T is the only variable used to describe the system, which consists of two solid phases, Bi + Cd.

6.5.2. *Partial miscibility*

In Figure 6.11, there are two regions representing solid solutions of B in A (phase α) and of A in B (phase β).

Exercise 6.5-1

A liquid mixture A + B (Figure 6.11) which contains 0.3 mole fraction of B is cooled from the fusion temperature of pure A. Describe what happens and state the composition of the final mixture. (***Hint:*** The phase diagram is displayed in Figure 6.11. The points marked by small filled circles are special points that deserve mention.)

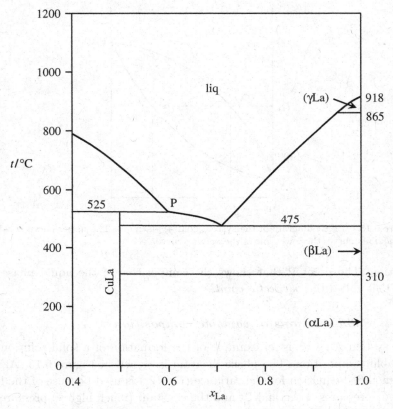

Figure 6.12.   Phase diagram of the system Cu + La, showing two eutectic points and a peritectic point P.

6.5.3. *Incongruent melting points*

Cu and La form four compounds with stoichiometry LaCu, $LaCu_2$, $LaCu_4$ and $LaCu_6$. The phase diagram of the system Cu + La is shown in Figure 6.12. Reading from left (pure Cu) to right (pure La), the first eutectic at 840°C results from the mixture $Cu + LaCu_6$. At higher concentrations of La, this compound decomposes into $LaCu_2 + LaCu_4$, which form a eutectic at 725°C. At still higher concentrations of La (more than 0.5 mole fraction of La) the stable compound is LaCu, the system consisting of a solid mixture of La + LaCu below the eutectic temperature of 468°C. There is no eutectic between $LaCu_2$ and LaCu because LaCu shows an *incongruent melting point*, that is, it decomposes at its melting point to give $LaCu_2$ and a liquid

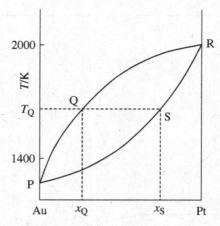

Figure 6.13.   Phase diagram of the system gold + platinum. The upper curve is called the *liquidus* curve, the lower one is the *solidus* curve.

solution. The point P that shows the composition of the liquid phase at 551°C is called the *peritectic point*.

6.5.4. *Solid solutions of variable composition*

The system $Au + Pt$ is an example of the formation of a solid solution of variable composition. The phase diagram is shown in Figure 6.13. Again the modified equation for the variance (Eq. (2)) is used because of the low vapour pressures of Au and Pt and the constant (much higher) pressure of inert gas. Above PQR

$$(2) \qquad\qquad\qquad v = 2 + 1 - 1 = 2 \qquad\qquad\qquad (8)$$

Between PQR and PSR, there are two condensed phases, solid and liquid. Therefore

$$(2) \qquad\qquad\qquad v = 2 + 1 - 2 = 1 \qquad\qquad\qquad (9)$$

At any T, say T_Q, the liquid and solid phases have the compositions x_Q and x_S, respectively. The solid phase is richer in Pt because Pt has a higher fusion temperature than gold. Below PSR, the system is divariant.

Exercise 6.5-2

Justify this last statement by calculating the variance in a manner analogous to Eq. (93).

Exercise 6.5-3

A system at room temperature consisting of a mixture of Au and Pt of composition x_Q is heated to T_Q. What happens?

6.5.5. *Systems of three components*

For systems of three components the phase diagram is a plot of the mole fraction of each component at constant temperature and pressure, using a triangular coordinate system. Consider the point P within the equilateral triangle ABC in Figure 6.14a, with perpendicular height AT of unit length. The length of the line PQ perpendicular to the side BC is the mole fraction

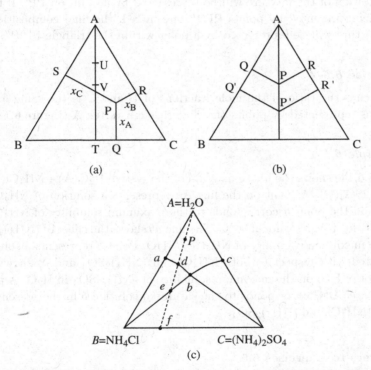

Figure 6.14. (a) Explanation of the construction of the phase diagram of a three-component system A + B + C. The lengths of a, b, c have been made equal to the lengths of x_A, x_B, x_C, respectively, which are the mole fractions of A, B and C. The scale of the drawing is such that AT is of unit length. (b) This figure shows that the ratio of the mole fractions of B and C is the same for all systems represented by points on a line through vertex A. (c) Phase diagram of the three-component system $H_2O + NH_4Cl + (NH_4)_2SO_4$.

of A, 0.3 in this case. Similarly, the length of the line PR perpendicular to AC is the mole fraction of B, $x_B = 0.2$, and the length of the line PS perpendicular to AB is the mole fraction of C, $x_C = 0.5$. It is a property of the equilateral triangle ABC that the sum of the three straight lines through any point within the triangle and perpendicular to the three sides, is equal to the perpendicular height of the triangle AT, so that the condition

$$x_A + x_B + x_C = 1 \tag{10}$$

is satisfied. If two systems of composition P and P′ are mixed, the composition of the mixture will be represented by a point on PP′. If three systems represented by points PP′P″ are mixed, the final composition of the mixture will be represented by a point within the triangle PP′P″.

Exercise 6.5-4

Show that the ratio of the mole fractions of B and C is the same for all systems represented by points on a line through vertex A (Figure 6.14b).

Example 6.5-1

Figure 6.14c shows the phase diagram of the system $H_2O(A) + NH_4Cl(B) + (NH_4)2SO_4(C)$. A point on the line Aa represents a solution of NH_4Cl in H_2O and the point a corresponds to the maximum solubility of NH_4Cl in H_2O at this temperature of 30°C. The line ab shows the effect of $(NH_4)_2SO_4$ on the maximum solubility of NH_4Cl in H_2O. Point b represents a solution saturated with respect to both NH_4Cl and $(NH_4)_2SO_4$ and cb shows the effect of NH_4Cl on the maximum solubility of $(NH_4)_2SO_4$ in H_2O. A point in the area BbC corresponds to the saturated solution b in the presence of excess NH_4Cl and $(NH_4)_2SO_4$.

Answers to Exercises 6.5

Exercise 6.5-1

The system is a bivariant liquid mixture. On cooling it to a, crystals of α with the composition b form. The system is now univariant and on further cooling, the concentration of α shifts along bc and that of the liquid along ad. At f, the liquid has the eutectic composition e and the system is invariant, consisting of solid phases α and β and liquid eutectic. Once this liquid is transformed totally into α and β, the system is once again univariant.

Exercise 6.5-2

There is only one condensed phase, so the variance is

$$v = 2 + 1 - 1 = 2 \tag{2}$$

Exercise 6.5-3

When T rises to the point where the ordinate at x_Q meets the curve PS, the mixture melts, giving liquid and solid mixtures of variable composition. Their compositions at any T correspond to the points where the abscissa at T cuts the curves PQ and PS. When the ordinate at x_Q crosses the curve PQ, the solid has melted completely and only a liquid mixture of composition x_Q is left.

Exercise 6.5-4

Please refer to Figure 6.14b. P, P′ are *any* two points on the line AR. PQ, P′Q′ are normal to AB and PR, P′R′ are normal to AC. The triangles APQ, AP′Q′ are similar and so are APR, AP′R′. Therefore

$$\frac{PQ}{P'Q'} = \frac{AP}{AP'} = \frac{PR}{P'R'} \tag{11}$$

Hence, the ratio of the mole fractions of B and C is the same for all systems represented by points on a line through vertex A.

6.6. Solutions

Solutions are mixtures of two or more components in which one component, the *solvent*, is a liquid in large excess. The minor components — *the solutes* — may be solids, liquids or gases that dissolve in the solvent to yield a homogeneous *solution*. The component in excess may be a solid, but then it is not usual to speak of it as the "solvent".

Consider a system consisting of solid B in contact with a solution of B in A. For equilibrium between the solid phase and solution

$$\mu_B^*(s) = \mu_B^\ominus + RT \ln a_B \tag{1}$$

A lower case (s) in parentheses, or a right superscript ^{s}, indicates the solid phase (of B in this case)

(1)
$$\ln a_B = \frac{\mu_B^*(s) - \mu_B^\ominus}{RT} \tag{2}$$

(2)
$$\left(\frac{\partial \ln a_B}{\partial T}\right)_p = -\frac{h_B^s - h_B^\ominus}{RT^2} = \frac{\Delta h_B^{sol}}{RT^2} \tag{3}$$

$\Delta h_B^{sol} = h_B^\ominus - h_B^s$ is the enthalpy change on transferring one mole of solute B from the solid state to its standard state in solution, that is, the infinitely dilute solution in the case of a solution.

6.6.1. *Elevation of boiling point*

Solid solutes generally have a negligibly small vapour pressure. The vapour pressure of the solution is then just that of the solvent

$$p_A = (p_A^*/f_A)(1 - x_B) \tag{4}$$

where x_B is the mole fraction of the (only) solute. Since a liquid boils when its vapour pressure becomes equal to the external pressure, the boiling point of a solution (T) is higher than that of the pure solvent (T_0), under the same external pressure, which we take to be the standard pressure, $p^\ominus$. Because solution and vapour are in equilibrium

$$\mu_B^\ominus + RT \ln a_B = \mu_B^g(p^\ominus) + RT \ln(\tilde{p}_B/\tilde{p}_B^\ominus) \tag{5}$$

Because the elevation of boiling point is so small, the fugacity of the vapour at the boiling point may be equated to that at the standard pressure, so that

(5)
$$R\left(\frac{\partial \ln a_A}{\partial T}\right)_p = -\frac{\partial}{\partial T}\left(\frac{\mu_A^\ominus}{T}\right)_p + \frac{\partial}{\partial T}\left(\frac{\mu_A^g(p^\ominus)}{T}\right)_p$$

$$= \frac{h_A^\ominus - h_A^g(p^\ominus)}{RT^2} = -\frac{\Delta h_A^{vap}}{RT^2} \tag{6}$$

Δh_A^{vap} is the heat absorbed at constant pressure during the conversion of one mole of solvent in its standard state (the pure liquid) to vapour at the standard pressure; it is called *the molar enthalpy of vaporization*.

Integrating (6) between T_0 and T yields

$$(6) \qquad \int_0^{\ln a_A} d\ln a_A = \frac{1}{R} \int_{T_0}^{T} -\frac{\Delta h_A^{vap}}{T^2} dT \qquad (7)$$

$$(7) \qquad \ln a_A = -\frac{\Delta h_A^{vap}}{R} \left[\frac{1}{T} - \frac{1}{T_0} \right] \qquad (8)$$

We have assumed the solution to be dilute so that Δh_A^{vap} could be taken as independent of T over this small temperature range. We also assume the solution to be ideal (which is justified for dilute solutions). Then

$$T \approx T_0, \quad \ln a_A = \ln x_A = \ln(1 - x_B) \approx -x_B \quad \text{(ideal dilute)} \qquad (9)$$

$$x_B = \frac{n_B}{n_A + n_B} = \frac{m}{(1/M_A) + (n_B/\text{mass of A})} \approx m M_A \qquad (10)$$

where m is the *molality* of the solution. (The molality of a solution is the number of moles of solute B in 1 kg of solvent A.)

$$(9) \qquad \left[\frac{1}{T} - \frac{1}{T_0} \right] = \frac{T_0 - T}{TT_0} \approx -\frac{\Delta T}{T_0^2} \qquad (11)$$

$$(8), (11) \qquad \Delta T = \frac{RT_0^2 M_A m}{\Delta h_A^{vap}} = K_{b,A} m \qquad (12)$$

$$(12) \qquad K_{b,A} = \frac{RT_0^2 M_A}{\Delta h_A^{vap}} \qquad (13)$$

is the *boiling point constant* of A. In principle, Eq. (8) could be used to measure the activity coefficient of the solvent A but it would be difficult to measure T_0 and T with sufficient accuracy to obtain useful results.

6.6.2. *Osmotic pressure of a solution*

If a solution is separated from pure solvent by a semi-permeable membrane (that is permeable to solvent molecules but not solute particles) then diffusion of solvent will occur into the solution (that is, from the phase in which its chemical potential is higher into the phase in which the chemical potential of solvent is lower). This flow of solvent can be opposed by increasing the pressure over the solution. The pressure at which the solution is in equilibrium with pure solvent, when they are separated by a semi-permeable membrane, is called the *osmotic pressure* Π of the solution. Designating the pure solvent as phase α and the solution as phase β, then we have, at equilibrium

$$\Pi = p^\beta - p^\alpha \tag{14}$$

The equilibrium condition for the solvent A is

$$\mu_A^\alpha(p^\alpha) = \mu_A^\beta(p^\beta) \tag{15}$$

(15) $$\mu_A^*(p^\alpha) = \mu_A^*(p^\beta) + RT \ln a_A^\beta \tag{16}$$

(16) $$-RT \ln a_A^\beta = \int_{p^\alpha}^{p^\beta} V_{m,A}^* \, dp \tag{17}$$

Despite the fact that the osmotic pressure might be large, the molar volume of pure A, $V_{m,A}^*$, may be taken to be independent of pressure, because of the low compressibility of liquids in general, so that

(17), (14) $$-RT \ln a_A^\beta = \Pi V_{m,A}^* \tag{18}$$

(9) $$RT \ln a_A = -RT x_B \approx -RT n_B / n_A \quad \text{(ideal dilute)} \tag{19}$$

(18), (19) $$\Pi = \frac{n_B RT}{n_A V_{m,A}^*} = cRT \tag{20}$$

where c is the concentration of A in mol m^{-3}. Corrections for non-ideality may be made by the inclusion of the *osmotic coefficient* ϕ_x, giving instead of (20)

(20) $$\Pi = c\phi_x RT \tag{21}$$

ϕ_x is the osmotic coefficient defined on the mole fraction basis. On the molality basis the osmotic coefficient is defined by

$$\phi_m = -\ln a_A / M_A m \tag{22}$$

To make these formulae on colligative properties (that is, properties that depend on the number of particles of solute) appear more straightforward, I am assuming that only a single solute is present in the solution. For generalizations to more than one solute see Cohen *et al.* (2007).

6.6.3. *Depression of freezing point*

The freezing point of all liquids is lowered by their dissolving a solute. The condition for equilibrium between pure solid solvent and solution is

$$\mu_A^*(s) = \mu_A^\ominus + RT \ln a_A \tag{23}$$

$$(23) \qquad \ln a_A = \frac{\mu_A^*(s) - \mu_A^\ominus}{RT} \tag{24}$$

$$(24) \qquad \left(\frac{\partial \ln a_A}{\partial T}\right)_p = -\frac{h_A^*(s) - h_A^\ominus}{RT^2} = \frac{\Delta h_{fus}}{RT^2} \tag{25}$$

Δh_{fus} is the enthalpy change (heat absorbed at constant pressure) as one mole of pure solvent is converted from the pure solid into the pure liquid (which is the standard state for the solvent), that is, the molar enthalpy of fusion of solvent A.

The depression of freezing point can be measured with sufficient accuracy $(1\,\text{mK})$ to justify allowing for the temperature dependence of Δh_{fus}.

$$(25) \qquad \frac{\partial \Delta h_{fus}}{\partial T} = \frac{\partial H_A^\ominus}{\partial T} - \frac{\partial H_A^*(s)}{\partial T} = C_{p,A}^s - C_{p,A}^{liq} = \Delta C_p^{fus} \tag{26}$$

Since the temperature range $T_0 - T$ is small, we may assume ΔC_p to be independent of T. Consequently

$$(26) \qquad \Delta h_{fus}(T) = \Delta h_{fus}(T_0) + \Delta C_p(T - T_0) \tag{27}$$

$$(25),\ (27) \qquad \ln a_A = \int_{T_0}^{T} \left(\frac{\Delta h_{fus}(T_0) - \Delta C_p(T_0 - T)}{RT^2} + \frac{\Delta C_p}{RT}\right) dT \tag{28}$$

$$= \left(\frac{\Delta h_{fus}(T_0) - \Delta C_p T_0}{R}\right)\left(\frac{1}{T_0} - \frac{1}{T}\right) + \frac{\Delta C_p}{R} \ln \frac{T}{T_0} \tag{29}$$

Exercise 6.6-1

Why is there no lower limit in the integral on the LS of (28)?

Define the *freezing point depression* ΔT (which is a small negative number of K) by

$$T = T_0 + \Delta T \tag{30}$$

Substitute for T from (30) in (29) and expand $\ln [1 + (\Delta T/T_0)]$ and $[1 + (\Delta T/T_0)]^{-1}$ using Taylor's theorem as far as terms in $(\Delta T/T_0)^2$, giving

$$\ln a_A = \frac{\Delta h_{\text{fus}}(T_0)\Delta T}{RT_0^2} + \frac{(\Delta T)^2}{RT_0^2}\left(-\frac{\Delta h_{\text{fus}}(T_0)}{T_0} + \frac{\Delta C_p}{2}\right) \tag{31}$$

If the solution is ideal dilute so that ΔT is small

$$(31),\ (9) \quad \Delta T = -\frac{RT_0^2}{\Delta h_{\text{fus}}(T_0)}x_B = -\left(\frac{RT_0^2 M_A}{\Delta h_{\text{fus}}(T_0)}\right)m = -K_{f,A}m \tag{32}$$

where the freezing point constant of the solvent

$$K_{f,A} = \frac{RT_0^2 M_A}{\Delta h_{\text{fus}}(T_0)} \tag{33}$$

Example 6.6-1

The freezing point depression ΔT of an aqueous solution of molality 0.1 mole kg^{-1} is -0.345 K. Calculate the activity of water and the osmotic coefficient of water in this solution. The freezing point T_0 of pure water is 273.15 K, the enthalpy of fusion of pure water at T_0 is 6017 J mol^{-1} and ΔC_p is 37.7 J K^{-1} mol

$$(31) \quad \ln a_A = \frac{6017 \times (-0.345)}{8.3145 \times 273.15^2} + \frac{(-0.345)^2}{8.3145 \times 273.15^2}\left\{-\frac{6017}{273.15} + \frac{37.7}{2}\right\}$$

$$= -3.3463 \times 10^{-3} - 1.92 \times 10^{-7}$$

$$= -3.3465 \times 10^{-3}$$

$$a_A = 0.99666$$

$$(22) \quad \phi_m = 3.3465 \times 10^{-3}/(18.016 \times 0.345)$$

$$= 0.5384$$

6.6.4. *Solubility in two immiscible liquids*

If a solute is soluble in two immiscible liquids, it will distribute between them, forming two immiscible solutions and therefore two liquid phases α and β. At equilibrium

$$\mu_B^\alpha = \mu_B^\beta \tag{33}$$

(33)
$$\mu_B^{\ominus,\alpha} + RT \ln a_B^\alpha = \mu_B^{\ominus,\beta} + RT \ln a_B^\beta \tag{34}$$

(34)
$$\frac{a_B^\alpha}{a_B^\beta} = \exp\left\{ \frac{\mu_B^{\ominus,\beta} - \mu_B^{\ominus,\alpha}}{RT} \right\} = K_D \tag{35}$$

K_D is called the *partition* or *distribution* coefficient. If the solutions are both ideal dilute

(35)
$$K_D = \frac{x_B^\alpha}{x_B^\beta} \quad \text{(ideal dilute)} \tag{36}$$

The dependence of K_D on T is given by

(35)
$$\left(\frac{\partial \ln K_D}{\partial T} \right)_p = \frac{1}{R} \frac{\partial}{\partial T} \left(\frac{\mu_B^{\ominus,\beta}}{T} \right) - \frac{\partial}{\partial T} \left(\frac{\mu_B^{\ominus,\alpha}}{T} \right)$$

$$= -\left(\frac{h_B^{\ominus,\beta} - h_B^{\ominus,\alpha}}{RT^2} \right) = -\frac{\Delta h_D}{RT^2} \tag{37}$$

Δh_D is the enthalpy change on transferring one mole of solute from its standard state in phase α to its standard state in phase β. Important practical applications are *liquid phase extraction* of a solute and *partition chromatography*, in which a liquid β flows through a column in which another liquid phase α is supported on a finely divided solid. Solutes dissolved in the fixed liquid phase α are eluted by the flow of solvent that forms the second liquid phase β. Separation of the solutes in the fixed liquid phase is usually achieved because many successive partitions occur as the liquid β flows through the column containing the finely divided solid.

In vapour phase chromatography the flowing phase β is a gas, usually helium, and the solutes are separated because of their differing volatilities (as shown by their Henry's law constants, e.g. Table 6.1).

Answer to Exercise 6.6-1

T_0 is the temperature of fusion of the pure solvent. On integrating d $\ln a_A$ between T_0 and T, the lower limit vanishes because the activity of pure A is unity.

Problems 6

6.1 Write down the result for the chemical potential of component B in a two-component gas mixture. Hence obtain the Gibbs function, entropy and enthalpy of such a mixture. (***Hint***: See Eq. (6.2.18).)

6.2 Estimate the fugacity of CO in a mixture of CO, O_2 and CO_2 at 673 K and 35 MPa containing 0.25 mole fraction of CO. For CO, $T_c = 134.4$ K and $p_c = 34.3$ MPa.

6.3 Derive the equation to the dew-point curve, Eq. (6.4.16). (***Hint***: Use Eqs. (6.4.14) and (6.4.15).)

6.4 Use the data in Table 6.2 to test if acetone + chloroform form a regular mixture. (***Hint***: See Eqs. (6.4.35) and (6.4.36).)

6.5 Verify by substitution that the solution given in (6.4.62)–(6.4.64) satisfies (6.4.60).

6.6 Table P6.3 gives values of the partial pressure of Zn at 1027 K over some alloys of zinc + lead, as a function of the mole fraction of Zn.

Table P6.3.   Partial pressure of Zn over Zn + Pb alloys at 1027 K.

x_{Zn}	1.0	0.939	0.881	0.576	0.291	0.143
p_{Zn}/kPa	17.7	17.3	17.3	17.3	17.3	11.4

What conclusions can be drawn about this system? Estimate a value for the activity coefficient of Zn at infinite dilution, taking pure Zn as the standard state.

Calculate the change in chemical potential that accompanies the dissolution of pure zinc in a zinc + lead alloy of 0.143 mole fraction Zn.

6.7 The heat evolved per mole of mixture, when small quantities of sulphuric acid and water are added to a mixture I containing 0.5 mole fraction of each component are 7.91 kJ and 20.92 kJ, respectively. The corresponding quantities for a mixture II containing 0.75 mole fraction of water are 31.46 kJ and 6.06 kJ. Find:

(a) the relative partial molar enthalpies of water and sulphuric acid in I;

(b) the heat evolved when one mole of water and one mole of sulphuric acid are mixed;

(c) the heat evolved when 2 moles of water are added to this mixture.

6.8 For a series of ethanol + chloroform mixtures, the total vapour pressure p and the mole fraction of ethanol (component A) in the liquid (x_A) and vapour (y_A) phases are given in Table P6.4. Calculate both the Raoult's law and Henry's law activity coefficients of ethanol in these liquid mixtures. State the ideal reference state and the standard state used for each calculation. It may be assumed that Henry's law holds up to a concentration of ethanol of at least mole fraction 0.05.

Table P6.4. Mole fractions of ethanol in the liquid (x_A) and vapour (y_A) phases of ethanol + chloroform mixtures of total vapour pressure p.

x_A	0	0.05	0.2	0.4	0.6	0.8	1.0
y_A	0	0.0392	0.1382	0.1864	0.2254	0.4246	1.0
p/kPa	39.34	40.00	40.83	38.69	34.29	25.90	13.70

6.9 Explain the method of *reverse osmosis* which is used to obtain drinking water from sea water.

6.10 Show that Eq. (6.4.79) reduces to Henry's law when $x \ll 1$ and find $\gamma_{x,B}$.

6.11 Prove Viviani's theorem that, in Figure 6.14a, PQ + PR + PS = AT.

6.12 A solution represented by the point P in Figure 6.14c is evaporated isothermally. Describe what happens.

7

Surfaces and Interfaces

7.1. Plane Interface

Consider a liquid α in contact with its own vapour and a neutral gas (air, perhaps). In the interior of α, each molecule is surrounded by z (the average coordination number) of like (or unlike) molecules and so is subject to short-range attractive forces which have (on average) spherical symmetry. In contrast, molecules in the surface layers experience a net inward attraction. This inward acting force is called the *surface tension* γ. The work done (Section 1.4) in expanding the area of the surface reversibly by dA is $\gamma \, dA$, so that this term must appear as part of the change in Helmholtz energy (see Eq. (2.4.4)). The equilibrium state is one in which F is a minimum. Thus droplets of liquid and gas bubbles in a liquid both tend to be spherical in shape, though this shape is often distorted by the presence of other forces (e.g. gravity) or constraints. Anticipating the observed order of magnitude, we note that the units of γ are mJ m^{-2} (or mN m^{-1}). The surface tension is temperature dependent and so the temperature of measurement should be quoted. As one might expect, the temperature dependence of γ is an example of the principle of corresponding states. The following empirical relation has been tested by Guggenheim (1945)

$$\frac{\gamma}{\gamma_0} = \left(1 - \frac{T}{T_c}\right)^n \tag{1}$$

where $n = 11/9$ for Ne, Ar, Kr and Xe, as well as for many organic liquids but seems to be closer to 1 for metals (Grosse 1962). At the interface between two immiscible liquids α and β, there is an inequality in the attractive forces acting on molecules which are in the interface. This asymmetry in the short-range forces normal to the surface leads to an *interfacial tension*. Values of the surface tension for some representative liquids, as well as a few values for the interfacial tension, are given in Table 7.1.

Table 7.1. Values of the surface tension γ and of interfacial tensions (between α and β) for some representative liquids (formerly used names are in brackets). A more extensive tabulation may be found in Adamson and Gast (1997).

$t/°C$	$\gamma/mJ\ m^{-2}$	Liquid
(a) *liquid–vapour interface*		
(for nomenclature, see webbook.nist.gov/chemistry/names-ser.html)		
−183	11.86	argon
20	28.88	benzene
25	26.43	carbon tetrachloride
20	22.39	ethanol
30	21.55	ethanol
25	20.14	ethoxyethane (ethyl ether)
25	47.3	1,2-ethanediol (ethylene glycol)
20	486.5	mercury
25	485.5	mercury
−198	9.41	nitrogen
20	21.62	octane
130	198	sodium
1073	115	sodium chloride
332	543.8	tin
20	72.94	water
25	72.13	water
−110	18.6	xenon

$t/°C$	$\gamma/mJ\ m^{-2}$	α	β
(b) *liquid–liquid interface*			
20	415	mercury	water
20	1.8	water	1-butanol
20	6.8	water	ethyl acetate
20	35.0	water	benzene

Exercise 7.1-1

The values of the surface tension in Table 7.1 cover a large range. Try to rationalize this qualitatively in terms of the forces between the molecules in the various liquids.

Answer to Exercise 7.1-1

Since the origin of the macroscopic property surface tension lies in the forces acting on a surface molecule, we might expect the magnitude of γ to correlate with the attractive forces between the molecules in the liquid. The rare gases argon and xenon have closed shell configurations and low cohesive energies in the crystalline and liquid states because the attractive intermolecular forces are relatively weak, being due entirely to the induced dipole–induced dipole interaction (London dispersion force). Since surface energies result from the absence of other molecules on one side of surface molecules, we anticipate low surface energies, as is indeed the case. Xenon has more electrons than argon and therefore a higher polarizability and cohesive energy. We therefore expect it to have a higher surface energy (Table 7.1). The high cohesive energies of metals result in high surface energies (Na, Hg and Sn in Table 7.1). Again, the high cohesive energy of NaCl (due to ionic interactions) is paralleled by a high surface energy. Organic compounds mostly have a surface energy of around 20 to 30 mJ m^{-2}, but that of water and 1,2-ethanediol (ethylene glycol) are higher because of hydrogen bonding.

7.2. Curved Interface

The pressure in the interior of a soap bubble $p + \delta p$ is greater than the pressure p outside the bubble. (In fact δp can be quite surprisingly high for a small bubble.) To see this, consider the *reversible* expansion of a bubble with radius r to $r + \mathrm{d}r$, with consequent change in bubble volume $\mathrm{d}V$. The change in Helmholtz energy is

$$\mathrm{d}F = -(p + \delta p)\mathrm{d}V + \gamma \mathrm{d}A + p\,\mathrm{d}V = 0 \tag{1}$$

The first two terms on the RS of (1) refer to the bubble, the third term to the gas phase outside the bubble

$$(1) \qquad\qquad (\delta p)4\pi r^2 \mathrm{d}r = 8\pi\gamma r\mathrm{d}r$$

$$\delta p = \frac{2\gamma}{r} \tag{2}$$

Equation (2) is the equilibrium condition. It shows that $\delta p > 0$ as long as r is finite. δp decreases as r increases and when $r \to \infty$, corresponding to a plane surface, $\delta p \to 0$. In general, a non-spherical surface has two principal radii of curvature (Figure 7.1). The dashed lines in Figure 7.1 are arcs of circles that lie in two spheres centred on C_1 and C_2, of radius r_1 and r_2, which touch S at P. Hence they are tangent circles. The solid lines labelled

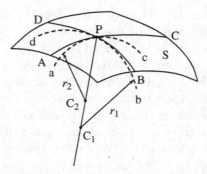

Figure 7.1. The two principal radii of curvature r_1 and r_2 of a non-spherical surface S. r_1 is the radius of a circle on a sphere that forms a tangent to the curve (the tangent circle) at the point P. When the surface is non-spherical there is a second radius of curvature r_2. In this illustration r_1 and r_2 are both positive. The dashed lines ac and bd are arcs of these tangent circles. APC and BPD are principal normal sections (of S at P) and are perpendicular to one another. (They are called 'principal' normal sections because the perpendicular line at P normal to S contains C_1 and C_2.)

AB and CD lie in S; they are *principal normal sections* of S. In this case

(2)
$$\delta p = \gamma \left(\frac{1}{r_1} + \frac{1}{r_2} \right)$$
(3)

which is the Young–Laplace equation. This result (3) is a generalization of (2); a derivation is given by Adamson and Gast (1997). The difference in pressure across a curved surface is the reason for capillary rise or depression, which can be used to measure γ. Generally, the liquid wets the glass walls of the capillary, so that the liquid surface inside the capillary is curved. For the present, we assume that it is hemi-spherical in shape (Figure 7.2a). The pressure p just above the surface (inside the sphere) is greater than that outside the sphere, which is p', by $2\gamma/r$ where r is the radius of the capillary, equal to the radius of curvature of the surface because of the assumption of a hemi-spherical surface inside the capillary. But the pressure at P is $p' + (\rho - \rho_0)gh$ (see Figure 7.2a) where ρ is the density of the liquid, ρ_0 is the density of the gas and g is the acceleration of free fall, which we take to be equal to the standard value

$$g_0 = 9.80665 \, \text{ms}^{-2}$$
(4)

(2)
$$\gamma = \frac{(\rho - \rho_0)g_0 h r}{2}$$
(5)

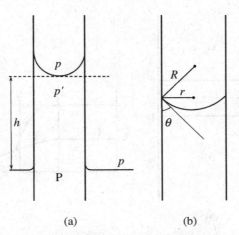

Figure 7.2. (a) For a hemi-spherical surface inside the capillary, the contact angle θ is zero. (b) If $0 < \theta < \pi/2$, the surface is a hemi-spherical cap of radius R.

If the contact angle θ is not zero, then the surface is a spherical cap of radius R. (The contact angle is the angle between the tangent to the curved interface and the vertical glass wall, as shown in Figure 7.2b.) Then it is evident from the figure that $r = R\cos\theta$ and so Eq. (5) becomes

$$(5) \qquad\qquad \gamma = \frac{(\rho - \rho_0)g_0 h r}{2\cos\theta} \qquad\qquad (6)$$

However, it is difficult to measure θ accurately and, therefore, preferable to choose a tube of sufficiently small radius to make θ zero. Liquids — such as mercury — that do not wet the glass have contact angles greater than 90° so that $\cos\theta$ is negative and therefore h is negative. Equation (6) should be corrected for the mass of liquid above the bottom of the meniscus. Assuming a hemi-spherical meniscus, the volume of this liquid is

$$\pi r^3 - \frac{2}{3}\pi r^3 = \frac{1}{3}\pi r^3 \qquad\qquad (7)$$

$$(6), (7) \qquad\qquad \gamma = \frac{(\rho - \rho_0)g_0 r}{2\cos\theta}\left(h + \frac{r}{3}\right) \qquad\qquad (8)$$

7.3. Thermodynamics of Interfaces

The interface between two immiscible liquids is not absolutely sharp. Instead, there is an intermediate region of variable composition called the

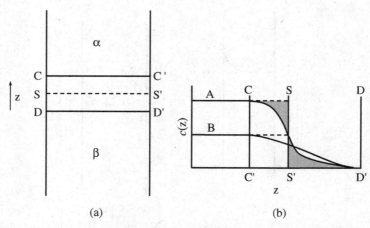

Figure 7.3.   (a) The volume between CC′ and DD′ is of variable composition and is the surface phase σ. (b) The location of SS′ is fixed by the condition that the amount of substance A to the left of SS′ is equal to the amount of A to the right of SS′ (Gibbs convention). The excess amount of B to the right of SS′ is called the surface excess concentration or the adsorption.

surface phase σ in Figure 7.3(a). The composition of this surface phase is described by quantities like n_B^σ, which is the amount of component B between CC′ and DD′ less the amount of B that would be between CC′ and DD′ if the phases α and β extended uniformly with unchanged bulk properties up to SS′. (The location of SS′ will be described shortly.) Assuming that all chemical species A, B, ... are electrically neutral, or that the electric potential is uniform, we have, for the surface phase, which extends from CC′ to DD′

$$(2.4.7), (1.4.9)^{\cdot} \qquad dG = -S\,dT + V\,dp + \sum_B \mu_B\,dn_B + \gamma\,dA \qquad (1)$$

For the imaginary system from CC′ to SS′, which has properties that are uniform and identical with those of phase α

$$dG^\alpha = -S^\alpha dT + V^\alpha dp + \sum_B \mu_B dn_B^\alpha \qquad (2)$$

For the imaginary system from SS′ to DD′, which has properties that are uniform and identical with those of phase β

$$dG^\beta = -S^\beta\,dT + V^\beta dp + \sum_B \mu_B\,dn_B^\beta \qquad (3)$$

Subtract (2) and (3) from (1), which gives

$$(2), (3), (1) \qquad dG^\sigma = -S^\sigma dT + V^\sigma dp + \sum_B \mu_B \, dn_B^\sigma + \gamma \, dA \qquad (4)$$

where

$$\left. \begin{array}{l} G^\sigma = G - (G^\alpha + G^\beta) \\[4pt] S^\sigma = S - (S^\alpha + S^\beta) \\[4pt] V^\sigma = V - (V^\alpha + V^\beta) \\[4pt] n_B^\sigma = n_B - (n_B^\alpha + n_B^\beta) \end{array} \right\} \qquad (5)$$

G, S, V, A are all homogeneous functions of degree 1 in the n_B (provided the change in the n_B is performed in such a way that V and A change in the same ratio). Therefore, by Euler's theorem (see Appendix A1)

$$(4) \qquad G^\sigma = -TS^\sigma + pV^\sigma + \sum_B \mu_B \, n_B^\sigma + \gamma A \qquad (6)$$

On differentiating (6) and substituting for dG^σ from (4), we get

$$S^\sigma dT - V^\sigma dp + A d\gamma + \sum_B n_B^\sigma \, d\mu_B = 0 \qquad (7)$$

which is the Gibbs–Duhem equation for the phase σ. On dividing (7) by A, we get

$$S_a^\sigma \, dT - \tau dp + d\gamma + \sum_B \Gamma_B \, d\mu_B = 0 \qquad (8)$$

$$(6), (1) \qquad 0 = SdT - Vdp - Ad\gamma + \sum_B n_B \, d\mu_B \qquad (7)$$

In (7.3.8), S_a^σ is the excess entropy of the surface phase σ divided by the area of σ and Γ_B is the excess amount of B in σ divided by the area of σ

$$\Gamma_B = \frac{n_B^\sigma}{A} \qquad (9)$$

which is the *surface excess concentration* or the *adsorption*. At constant T and p,

$$(8) \qquad d\gamma = -\sum_B \Gamma_B \, d\mu_B \qquad (10)$$

which is the *Gibbs adsorption equation*. Suppose that the solvent is labelled by subscript A and that there is just one solute, which for brevity we leave unlabelled. Then

(10)
$$d\gamma = -\Gamma_A d\mu_A - \Gamma d\mu$$
(11)

μ and μ_A cannot be varied independently and so (11) cannot be used to determine Γ and Γ_A independently. Gibbs suggested that the surface SS′ be located so that

$$\Gamma_A = 0 \ (\text{Gibbs convention})$$
(12)

This means that SS′ should be chosen such that the shaded area to the right of SS′ in Figure 7.3(b) is equal to the shaded area on the left of SS′. Figure 7.3(b) shows the operation of this convention for two condensed phases, α and β. From Figure 7.3(b), the amount of B in σ, less the amount of B that would have been present had the concentration B remained constant up to SS′ and then dropped to zero, that is n_B^σ, is positive in this example. n_B^σ/A is the surface excess concentration Γ_B.

Exercise 7.3-1

Suggest a procedure by which V, A, n_B may all be altered in the same ratio.
   Using the Gibbs convention (12)

(11)
$$d\gamma = -\Gamma d\mu = -\Gamma RT d\ln a$$
(13)

(12)
$$\Gamma = -\frac{1}{RT}\frac{d\gamma}{d\ln a}$$
(14)

Answer to Exercise 7.3-1

Imagine a system at equilibrium in the shape of a unit cube and effect a reduction in volume by removing a thin slice of thickness $d\xi$ normal to the surface. Then n_B, V, A and all extensive properties are reduced in the ratio $1 - d\xi$, while all intensive properties such as T, p and μ_B remain unchanged. The former are homogeneous functions of degree one and the latter homogeneous functions of degree zero. Remember that Euler's theorem is a theorem about homogeneous functions (Appendix A1).

7.4. Some Typical Results

Materials like short-chain fatty acids and alcohols are soluble both in water and in hydrocarbon solvents. At an air–water interface or an oil–water interface, such molecules tend to locate with the hydrophilic head-groups $-COOH$ or $-OH$ in the water and the hydrocarbon chains in the vapour or oil phase. The preferential adsorption of such molecules at the interface is energetically favourable compared with solution in either phase. Strong adsorption at an interface is called *surface activity* and the solute molecules are consequentially called *surfactants*. The tendency of surfactant molecules to pack into the surface favours an expansion of the surface and so lowers the surface tension.

Exercise 7.4-1

The surface tension of aqueous solutions of 1-butanol decreases with concentration but that of sodium chloride solutions increases slightly with increasing concentration of solute. Explain.

For dilute aqueous solutions of many organic materials such as short-chain fatty acids, alcohols, phenols etc. with OH or COOH end groups

$$\gamma - \gamma_0 = -b \ln(1 + am) \tag{1}$$

where γ is the surface tension of the solution, γ_0 that of pure water, a and b are positive numbers and m is the molality

$$(7.3.14), (1) \quad \Gamma = -\frac{m}{RT}\frac{d\gamma}{dm} = -\frac{m}{RT}\left(\frac{-ab}{1+am}\right) = \frac{b}{RT}\left(\frac{am}{1+am}\right) \tag{2}$$

At high values of m; $am \gg 1$ and $\Gamma = b/RT$. This equation (2) resembles that for a Langmuir isotherm (see Section 7.6) and therefore suggests that Γ at high values of m corresponds to monolayer coverage of the surface.

Example 7.4-1

For butyric acid at 291 K, $a = 19.64 \, \text{kg mol}^{-1}$, $b = 1.29 \times 10^{-4} \, \text{N m}^{-1}$. For $m > 0.5 \, \text{mol kg}^{-1}$, $am > {\sim}10$ and

$$\Gamma \approx \frac{1.29 \times 10^{-4} \, \text{N m}^{-1}}{8.31 \, \text{J K}^{-1} \, \text{mol}^{-1} \times 291 \, \text{K}} = 5.33 \times 10^{-6} \, \text{mol m}^{-2} \tag{3}$$

If this were to correspond to a close-packed monolayer of solute, then we would conclude that the surface area occupied by one molecule is

$$1/N\Gamma = \frac{1}{6.02 \times 10^{23}\,\mathrm{mol}^{-1} \times 5.33 \times 10^{-6}\,\mathrm{mol\,m}^{-2}} = 0.31\,\mathrm{nm}^2 \quad \mathrm{or} \quad 31\,\mathrm{\mathring{A}}^2 \tag{4}$$

However, the surface area occupied by one molecule in a close-packed monolayer (a Gibbs monolayer) of solute molecules at the surface is about the same ($\sim 0.2\,\mathrm{nm}^2$) for a number of fatty acids, supporting the model which has them standing up on their polar groups. For NaCl and other electrolytes, Γ is negative, which means that there is less NaCl in the surface than in the bulk of the solution. In an aqueous NaCl solution of molality $m = 1\,\mathrm{mol/kg}$, the value of Γ is equivalent to a surface layer of water one molecule thick.

In dilute solutions

$$(1) \qquad\qquad\qquad\qquad \gamma = \gamma_0 - ab\,m \tag{5}$$

so that γ varies linearly with the molality of solute

$$(5),(2) \qquad\qquad\qquad\qquad \Gamma = \frac{\gamma_0 - \gamma}{RT} \tag{6}$$

Define the surface pressure π by

$$\pi = \gamma_0 - \gamma \tag{7}$$

For a Gibbs monolayer, $\Gamma = 1/A_\mathrm{m}$, where A_m is the molar surface area of solute, so that

$$(6),(7) \qquad\qquad\qquad\qquad \pi A_\mathrm{m} = RT \tag{8}$$

which is the equation of state for an *ideal 2-D layer*.

There are other methods of measuring the excess concentration in the surface. In McBain's method, the surface layer is skimmed off the solution using a fast-moving microtome, and the excess amount in the surface determined by analysis. Then

$$\Gamma = \frac{w_0 \Delta c}{MA} \tag{9}$$

where Δc is the excess mass ratio (mass solute/mass solvent) in the surface, w_0 is the mass of solvent in the slice, M is the molar mass of the solute and A is the area of the surface.

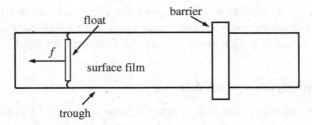

Figure 7.4. Diagrammatic representation of a Langmuir trough.

A *Langmuir trough* (Figure 7.4) is a shallow trough filled with water and containing a fixed barrier and a moveable float, which consists of a Teflon strip suspended from a torsion wire. A drop of surfactant, dissolved in a volatile solvent which is immiscible with water — for example, octadecanoic acid (stearic acid) dissolved in benzene — is put on the surface, and the solvent (benzene) evaporates, leaving a film of solute (octadecanoic acid) on the surface. The Teflon strip is moved towards the fixed barrier in steps and the force acting on the mica float measured at each step by measuring the force required to restore the torsion balance to equilibrium. The force per length acting on the Teflon strip is the surface pressure. A plot of π against the area per molecule for 2-hydroxytetracosanoic acid (cerebronic acid) is shown in a paper by Ries (1984). The steep rise in π as A decreases is indicative of the formation of a close-packed monolayer, and extrapolation of the linear portion of the curve gives a value for the size of the molecule. The curved portion preceding the linear rise corresponds to the compression of a 2-D layer.

If a glass slide is dipped into a solution of a solute with a polar end-group, the solute adheres to the glass with the polar end-group attached to the glass surface. If the dipping is repeated, a second layer consisting of solute from the water surface adheres to the glass. Repeated dips result in a layer sufficiently thick for its thickness to be determined by interference colours. Dividing the thickness of the film by the number of times the slide was dipped in the solution gives the length of the molecule. This simple method, due to Langmuir and Blodgett, was one of the first used to determine the size of molecules.

Answer to Exercise 7.4-1

1-butanol is a surfactant and the increased concentration of 1-butanol molecules in the surface layer reduces the surface tension of the solvent.

In contrast, in electrolyte solutions generally (and NaCl in particular) the strong solvent–solute interactions cause the solute to migrate away from the surface into the bulk solvent, increasing the surface tension.

7.5. Thermodynamics of Wetting

Consider two phases α and β in contact (Figure 7.5a and 7.5b) with a drop of liquid χ in contact with the interface between α and β. Depending on the Gibbs energy change, this drop may spread so as to cover the whole interface between phases α and β with a thin layer of phase χ (Figure 7.5b and 7.5d) or it may remain as a droplet, though perhaps of a different shape. The Gibbs energy change in this system when χ is added, is

$$\Delta G = (\gamma^{\alpha\chi} + \gamma^{\beta\chi}) - \gamma^{\alpha\beta}dA = -W\,dA \qquad (1)$$

Equation (1) defines the *wetting coefficient* (or spreading coefficient) W. If $W \geq 0$, ($\Delta G \leq 0$) the droplet of χ spreads spontaneously, wetting the interface between α and β. But if $W < 0$, χ remains as a droplet and does not wet the interface. No matter how carefully machined or polished, a solid surface is always rough, so that, when two solids are brought together, they are only in contact at certain points with gaps between them, allowing a liquid to spread between the surfaces if $W \geq 0$. This is shown

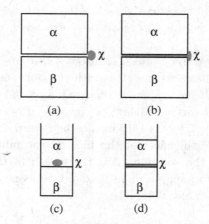

(a)   (b)

(c)   (d)

Figure 7.5.   (a), (b) Wetting of the interface between two solid phases by a liquid χ. As explained in the text, this diagram is schematic in that the meeting of two rough surfaces is shown as a narrow separation of α and β. (c), (d) Wetting of the interface between two liquid phases by a third liquid χ.

diagrammatically in Figure 7.5 by leaving a small gap between the two solid surfaces.

Exercise 7.5-1

Can the three phases α, β and χ attain equilibrium if $W < 0$?

Answer to Exercise 7.5-1

No. If $W < 0$, $\Delta G > 0$ and the three phases cannot attain the equilibrium state.

7.6. The Gas–Solid Interface

It is customary to distinguish between *physical* and *chemical* adsorption, the latter being generally called *chemisorption*. Physical adsorption is due to the attractive van der Waals forces between the molecules and the solid. Chemisorption involves the rearrangement of the electron distribution in the adsorbed molecules and the solid, which may be verified by calculations of the electron density in both adsorbent and adsorbate before and after chemisorption occurs (e.g. Zhukovskii *et al.* 2000, Zhukovskii *et al.* 2003). One might therefore expect physical adsorption to be instantaneous but chemisorption to be a slow process, requiring activation energy, but this criterion can be deceptive. For example, physical adsorption on a porous solid (such as a zeolite) may be slow because the gas has to diffuse into the pores of the solid, while the chemisorption of oxygen on a clean metal surface is very rapid. Adsorption is accompanied by the evolution of energy called the *heat of adsorption*, $-\Delta_{ads}H$. For physical adsorption, the magnitude of the heat of adsorption is generally in the range 8 to $25 \, kJ \, mol^{-1}$ and usually not more than two or three times the heat of liquefaction of the *adsorbate*. For chemisorption, $-\Delta_{ads}H$ is usually higher than that, being generally more than $80 \, kJ \, mol^{-1}$. A physically adsorbed gas may often be removed rapidly by warming to a moderate temperature, but for a chemisorbed gas the magnitude of the heat of desorption is often of the same order as that of the heat of adsorption, so that a high temperature may be required to desorb the gas completely. Chemisorption may involve specific chemical interactions, for example the dissociation of O_2 and the absorption of O atoms into a metal surface.

The amount adsorbed depends on the pressure of the gas and the amount adsorbed as a function of the pressure p, at constant temperature, is called the *adsorption isotherm*. Let the number of surface sites on which a gas molecule can be adsorbed be N_s and let θ be the fraction of these that are occupied by an adsorbed molecule. Each adsorbed molecule occupies a site at which there is a potential energy minimum along the z coordinate normal to the surface. It is also a potential energy minimum in the x and y directions parallel to the surface. If the height V_s of the potential barrier between one site and the next one adjacent to it in the x and y directions is large compared with kT, the molecule will remain on that site, except for occasional desorption and adsorption events that maintain equilibrium between the adsorbed layer and the gas phase. On the other hand, if kT is larger than the potential barrier V_s between one site and the next adjacent one, a *mobile* monolayer results.

The rate of collision of gas molecules per area of surface is

$$(3.7.15) \qquad\qquad Z_w = \frac{p}{\sqrt{2\pi mkT}} \qquad\qquad (1)$$

The rate of adsorption is therefore

$$\frac{p}{\sqrt{2\pi mkT}} S_{st}(1 - \theta) \exp(-E_{ads}/kT) \qquad (2)$$

S_{st} is called the *sticking coefficient* and allows for the fact that not every collision with an adsorption site on bare surface might be successful. For example, a molecule may approach the surface with a z-component of velocity sufficiently large that the kinetic energy exceeds E_{ads}, where E_{ads} is the energy necessary to overcome any potential barrier at an adsorption site. This in-coming molecule now meets a repulsive wall so that it is repelled with a z-component of velocity now directed away from the surface. It may overcome the potential barrier of height E_{ads} and be repelled away from the surface or, if it lacks sufficient kinetic energy to do this, it will be trapped in the potential well and become adsorbed. Hence s_{st} may be less than one.

The rate of desorption is

$$\theta \nu_{des} \exp[-(E_{ads} + \Delta U_{ads}/kT) \qquad (3)$$

where ν_{des} is the vibrational frequency of the adsorbed molecule in the z-direction and ΔU_{ads} is the depth of the potential well that binds the adsorbed molecule to the surface

$$(2), (3) \qquad \frac{\theta}{1-\theta} = \frac{pS_{st}}{(2\pi mkT)^{1/2}\nu_{des}\exp[-(\Delta U_{ads}/kT)} = b(T)p \qquad (4)$$

The difficulties about using (4) in a first-principles calculation of $b(T)$ are that not only would one need to calculate both ΔU_{ads} and ν_{des} but also estimate s_{st}. Let the volume of gas adsorbed at a coverage θ be V^{a} and its maximum value (at $\theta = 1$) be $V^{\text{a}}_{\text{max}}$. Then

$$(5) \qquad \frac{V^{\text{a}}}{V^{\text{a}}_{\text{max}}} = \theta = \frac{b(T)p}{1 + b(T)p} \qquad (5)$$

which is the Langmuir adsorption isotherm (*cf.* Eq. (7.4.2) for Γ). Equation (6) suggests that $V^{\text{a}}/V^{\text{a}}_{\text{max}}$ should saturate at high pressures. But for many systems (e.g. nitrogen on silica) the isotherm does not saturate. Instead V^{a} rises steeply at pressures approaching the saturation vapour pressure, p_0. This can be accounted for by assuming *multilayer adsorption*, with $-\Delta_{\text{ads}}H$ for the second and subsequent layers being approximately equal to $\Delta_{\text{vap}}H$. These assumptions lead to the Brunauer, Emmett, Teller (BET) isotherm

$$\frac{p}{V^{\text{a}}(p_0 - p)} = \frac{1}{V^{\text{a}}_{\text{max}}c} + \frac{c-1}{V^{\text{a}}_{\text{max}}c}\left(\frac{p}{p_0}\right) \qquad (6)$$

which may be tested by plotting the LS of (6) against p/p_0. If a straight line results, the intercept is $1/V^{\text{a}}_{\text{max}}c$ and the slope $(c-1)/V^{\text{a}}_{\text{max}}c$. We thus obtain the volume of a monolayer, $V^{\text{a}}_{\text{max}}$. Usually V and $V^{\text{a}}_{\text{max}}$ are expressed as the volume of gas adsorbed at STP. A derivation of the BET isotherm may be found in Atkins (1986 or later editions). The surface area of the adsorbent can be determined from $V^{\text{a}}_{\text{max}}$ and the BET equation is widely used for this purpose. The area for a given mass of adsorbent is given by

$$A = \left(\frac{p^{\ominus}V_{\text{max}}}{RT^{\ominus}}\right) A^{\text{a}}_{\text{m}} \qquad (7)$$

where the term in parentheses is the number of moles adsorbed to give a monolayer and A^{a}_{m} is the area of adsorbent occupied by one mole of adsorbate. Generally, (6) and (7) are used with nitrogen as the adsorbate and a value of $0.162\,\text{nm}^2$ for $A^{\text{a}}_{\text{m}}/N_{\text{A}}$.

It is implicit in the derivation of both the Langmuir and BET isotherms that $-\Delta_{\text{ads}}H$ is assumed to be independent of θ within a monolayer. Experimentally, the heat of adsorption is often found to decrease with increasing θ (e.g. for nitrogen on rutile, Drain and Morrison 1953). If this decrease is proportional to log θ, then $\theta(p)$ obeys the *Freundlich isotherm*

$$\theta = \frac{V}{V_m} = k\,p^{1/n}, \quad n > 1 \tag{8}$$

But if $-\Delta_{\text{ads}}H$ decreases linearly with increasing θ

$$\theta = \frac{V}{V_m} = d^{-1}\ln(ap) \tag{9}$$

in which a and d are constants. Equation (9) is the Temkin isotherm.

7.6.1. *Statistical thermodynamics of adsorption (Fowler and Guggenheim 1939)*

Because the gas molecules are indistinguishable, the partition function (Chapter 3) for the gas is

$$Q^g = \frac{(q^g)^N}{N!} \tag{10}$$

q^g is the partition function per molecule and is given by

$$q^g = q_t^g q_r^g q_v^g \tag{11}$$

where q_t^g is the translational contribution to the partition function, q_r^g the rotational contribution and q_v^g the vibrational contribution. The rotational and vibrational factors are replaced by unity for a monatomic molecule like xenon. In the adsorbed state

$$q^a = q_t^a q_r^a q_v^a \tag{12}$$

The partition function in the adsorbed state depends on which model we use. If the adsorbed molecules are trapped on adsorption sites in shallow potential wells, then this is referred to as a *localized adsorbed monolayer*. Nevertheless, we expect that the adsorbate molecules will still be able to diffuse over the surface by acquiring sufficient energy (from the solid) to overcome the potential barriers of height V_s. In this model, therefore, of a localized adsorbed monolayer, the three translational degrees of freedom in the gas phase are replaced by two translational degrees of freedom on the surface and one vibrational mode normal to the surface. In polyatomic molecules, in the absence of either experimental information or theoretical calculations, some assumption must be made as to whether the rotational and vibrational frequencies are affected by the adsorption process or not. In the adsorbed state, then

$$Q^a = \frac{N_s!(q^a)^N \exp(-\Delta_{ads}U/kT)}{(N_s - N)!N!} \tag{13}$$

The number of available adsorption sites is N_s and of these, N are occupied by adsorbed molecules. These N molecules are mobile, and indistinguishable, and hence the configurational factor $N!$ in (13). The common energy zero for adsorbed and gaseous molecules is that of a molecule at rest at a sufficient distance from the crystal not to be affected by any fields present in the crystal. $\Delta_{ads}U$ is the change in potential energy in taking an adsorbed molecule from the surface to a point at rest well outside the crystal. On using Stirling's approximation, the Helmholtz energy of the adsorbed gas is

$$(13) \qquad F^a = -kT[N_s \ln N_s - (N_s - N)\ln(N_s - N)$$
$$- N \ln N + N \ln(q^a_{ads}\, q^a_{int})] + \Delta_{ads}U \tag{14}$$

q^a_{ads} is the contribution to the molecular partition function from the three degrees of freedom that were translations in the gas phase (and which therefore depends on which model we use for the adsorbed state) and q^a_{int} is the partition function for the internal degrees of freedom (rotations and vibrations) in the adsorbed state. The chemical potential of an adsorbed molecule is, therefore,

$$(14) \qquad \mu^a = kT \ln \frac{N}{(N_s - N)} - kT \ln(q^a_{ads}\, q^a_{int}) - \Delta_{ads}U/kT \tag{15}$$

Equating this to the chemical potential per molecule in the gas phase

$$(10) \qquad \mu^g = -kT \ln q^g \tag{16}$$

gives

$$\frac{\theta}{1 - \theta} = \frac{q^a}{q^g} \exp(-\Delta_{ads}U_m/RT) \tag{17}$$

A gas-phase molecule has three degrees of translational freedom and so

$$(3.6.12) \qquad q^g_{tr} = \left(\frac{2\pi mkT}{h^2}\right)^{3/2} \frac{kT}{p} \tag{18}$$

If we assume that the partition function for the internal degrees of freedom is unaffected by adsorption and that the mobile adsorbed molecule has two degrees of translational freedom, then

$$(15), (17), (18) \qquad \frac{\theta}{1 - \theta} = \frac{(A_m/N_A)p}{\nu_a (2\pi m k T)^{1/2}} \exp(-\Delta_{ads} U_m/RT) \qquad (19)$$

A_m/N_A is the surface area per molecule and we have assumed that $h\nu_a \ll kT$, which is expected to be so for a loosely-bound adsorbed molecule. The factor multiplying p is the constant $b(T)$ in the Langmuir equation (1). If the rotational and vibrational degrees of freedom are perturbed in the adsorbed state, then this factor $b(T)$ in (15) needs to be multiplied by Q^a_{int}/Q^g_{int}.

Exercise 7.6-1

Check the dimensions of the RS of Eq. (15).

Answer to Exercise 7.6-1

The LS of (15) is dimensionless. The RS is

$$(15) \qquad \frac{(A_m/N_A)p}{\nu_a (2\pi m k T)^{1/2}} \exp(-\Delta_{ads} U/RT)$$

which has units of $m^2 N\,m^{-2}/s^{-1}(kg\,J)^{1/2} = 1$, and so also is dimensionless.

7.7. Thermodynamics of Adsorption

We define a surface phase σ as the surface layer together with that part of the solid below the surface layer which interacts with both the surface layer and any molecules adsorbed on the surface. One cannot say precisely how many layers of the adsorbent are contained in σ, since this depends on the range of the forces involved. The thermodynamic description of the surface phase σ is effected by the inclusion of the necessary work term $\pi\,dA^\sigma$ in the fundamental equations for the open surface phase σ. π is the spreading pressure; its dimensions are force per unit length or energy per unit area

$$(7.4.7) \qquad \qquad \pi = \gamma_0 - \gamma \qquad \qquad (1)$$

where γ_0 is the surface energy per unit area of the clean surface and γ that of the surface partly covered by molecules of adsorbate. The internal energy of σ is

$$U^\sigma(S^\sigma, V^\sigma, n_B^\sigma, A^\sigma) \tag{2}$$

and its complete differential is therefore

$$
dU^\sigma = \left(\frac{\partial U^\sigma}{\partial S^\sigma}\right)_{V^\sigma, n_B^\sigma, A^\sigma} dS^\sigma + \left(\frac{\partial U^\sigma}{\partial V^\sigma}\right)_{S^\sigma, n_B^\sigma, A^\sigma} dV^\sigma
$$
$$
+ \sum_B \left(\frac{\partial U^\sigma}{\partial n_B^\sigma}\right)_{S^\sigma, V^\sigma, n_{A \neq B}^\sigma, A^\sigma} dn_B^\sigma + \left(\frac{\partial U^\sigma}{\partial A^\sigma}\right)_{S^\sigma, V^\sigma, n_B^\sigma} dA^\sigma \tag{3}
$$

(***Warning!*** Please do not confuse the area A^σ of the surface phase σ with the substance A, which is usually designated as the adsorbent.)

(3) $\qquad dU^\sigma = T dS^\sigma - p dV^\sigma + \sum_B \mu_B^\sigma dn_B^\sigma + \pi dA^\sigma \qquad$ (4)

Exercise 7.7-1

Obtain from (4) the fundamental equation for dS^σ.

Integrating (4) at constant intensive variables yields

(4) $\qquad U^\sigma = TS^\sigma - pV^\sigma + \sum_B \mu_B^\sigma n_B^\sigma + \pi A^\sigma \qquad$ (5)

As for a bulk phase, the enthalpy H^σ, Helmholtz energy F^σ and Gibbs energy G^σ of the surface phase σ are defined by

$$H^\sigma = U^\sigma + pV^\sigma, \quad F^\sigma = U^\sigma - TS^\sigma, \quad G^\sigma = U^\sigma + pV^\sigma - TS^\sigma \tag{6}$$

Differentiating H^σ, F^σ, and G^σ and substituting for dU^σ from (4), gives

(6), (4) $\qquad dH^\sigma = T dS^\sigma + V^\sigma dp + \sum_B \mu_B^\sigma dn_B^\sigma + \pi dA^\sigma \qquad$ (7)[a]

(6), (4) $\qquad dF^\sigma = -S^\sigma dT - p dV^\sigma + \sum_B \mu_B^\sigma dn_B^\sigma + \pi dA^\sigma \qquad$ (8)

(6), (4) $\qquad dG^\sigma = -S^\sigma dT + V^\sigma dp + \sum_B \mu_B^\sigma dn_B^\sigma + \pi dA^\sigma \qquad$ (9)

which are the fundamental equations for the open surface phase σ.

---

[a]The difference in the sign before the work term πdA^σ in (7) and Eq. (4.1) in Everett (1950a) is due to the fact that Everett used the old convention that work done by the system is positive. Everett also used the symbol ϕ for the spreading pressure. The

Exercise 7.7-2

Express the chemical potential of substance B in phase σ as a partial derivative of U^σ, H^σ, F^σ and G^σ and explain how it is possible for μ_B^σ to be the partial derivative of four different functions.

Exercise 7.7-3

Write down four alternative expressions for the spreading pressure π. Be sure to state which variables are being held constant during partial differentiation.

Heat of Adsorption

To keep the notation as simple as possible, consider a system containing only a fixed mass of adsorbent A in equilibrium with a gas B. If an amount dn_B of B is adsorbed reversibly at constant volume, temperature and adsorbent area then the heat absorbed during this process is

$$(4) \qquad \frac{dq}{dn_B^\sigma} = \left(\frac{\partial U^\sigma}{\partial n_B^\sigma} \right)_{T,V^\sigma,A^\sigma} \qquad (10)$$

The negative of this quantity (a positive number) is called the *differential heat of adsorption*. If the pressure of B is maintained constant instead of the volume, then

$$(7) \qquad \frac{dq}{dn_B^\sigma} = \left(\frac{\partial H^\sigma}{\partial n_B^\sigma} \right)_{T,p,A^\sigma} \qquad (11)$$

The negative of this quantity is called the *differential isosteric heat of adsorption* (that is, the heat of adsorption at constant coverage). More precisely, Eq. (10) and (11) define the differential energy and enthalpy of adsorption. Their variation with coverage is given by equations like

$$H_B^\sigma(\theta) = H_{m,B}^\sigma(0) + \theta \left(\frac{\partial H_{m,B}^\sigma}{\partial \theta} \right)_{T,p,A^\sigma} \qquad (12)$$

Equation (12) states that the partial molar enthalpy of adsorbate B is equal to the molar value of the enthalpy of B at zero coverage (that is, the limiting value as $\theta \to 0$) plus the derivative of the molar enthalpy of B

---

sign conventions and symbols used in the book agree as closely as possible with the recommendations of Cohen *et al.* (2007).

with respect to the fraction of surface covered, multiplied by the fraction of surface covered, θ

$$(7.6.15) \qquad \mu^{\mathrm{a}} = kT \ln \frac{N}{(N_{\mathrm{s}} - N)} - kT \ln(q^{\mathrm{a}}_{\mathrm{ads}} \, q^{\mathrm{a}}_{\mathrm{int}}) - \Delta_{\mathrm{ads}} U / kT \qquad (13)$$

Equating, this to the chemical potential per molecule in the gas phase

$$(10) \qquad \mu^{\mathrm{g}} = -kT \ln q^{\mathrm{g}} \qquad (14)$$

gives

$$\frac{\theta}{1 - \theta} = \frac{q^{\mathrm{a}}}{q^{\mathrm{g}}} \exp(-\Delta_{\mathrm{ads}} U_m / RT) \qquad (15)$$

A gas-phase molecule has three degrees of translational freedom and so

$$(3.6.12) \qquad q^{\mathrm{g}}_{\mathrm{tr}} = \left(\frac{2\pi mkT}{h^2} \right)^{3/2} \frac{kT}{p} \qquad (16)$$

If we assume that the partition function for the internal degrees of freedom is unaffected by adsorption and that the mobile adsorbed molecule has two degrees of translational freedom, then

$$(15), (16) \qquad \frac{\theta}{1 - \theta} = \frac{(A_{\mathrm{m}}/N_{\mathrm{A}})p}{\nu_{\mathrm{a}}(2\pi mkT)^{1/2}} \exp(-\Delta_{\mathrm{ads}} U_m / RT) \qquad (17)$$

$A_{\mathrm{m}}/N_{\mathrm{A}}$ is the surface area per molecule and we have assumed that $h\nu_a \ll kT$.

The dependence of each of the heats of adsorption on temperature is given by an equation like that of Clausius and Clapeyron for the heat of vaporization and may be derived in a similar fashion. For example

$$\left(\frac{\partial \ln p_{\mathrm{B}}}{\partial T} \right)_{p,\theta} = -\frac{\Delta H_{\mathrm{B}}}{n_{\mathrm{B}} RT^2} \qquad (18)$$

Constant p in Eq. (18) refers to constant total pressure of inert gas and B.

The localized monolayer model was introduced in Section 7.6. We reiterate that localized means that gas molecules are adsorbed preferentially at particular sites on the surface of the solid, but does not necessarily imply that the adsorbed molecules are immobile. It has also been assumed that the adsorption energy is the same on all sites, that the adsorbed molecules do not interact significantly with one another and that there are no changes in the frequencies of rotation and vibration in the adsorbed state. All these assumptions must be regarded as approximations. Since we usually have insufficient knowledge of the surface and of the adsorbed state to construct an exact model, these effects are generally combined in an imprecisely defined non-configurational entropy of adsorption ΔS^a. The adsorption isotherm then becomes

$$(7.6.15) \qquad \frac{\theta}{1-\theta} = p \exp(\Delta S^a_m / R) \exp(-\Delta U^a_m / RT) \qquad (19)$$

And of course we must always bear in mind the possible onset of multilayer adsorption. This thermodynamic treatment can be extended to more than one adsorbate.

We now consider some experimental results. Generally the adsorbent is enclosed in an evacuated, constant-volume system. The requisite amount of vapour is admitted and the heat evolved is measured. The heat evolved per mole of adsorbate is the *integral heat of adsorption*; it is equal to the change in the mean molar energy of the adsorbate B

$$-q(n^\sigma_B) = -\Delta U^a_{m,B} \qquad (20)$$

Repetition of the above procedure establishes the functional dependence of $q(n^\sigma_B)$ on n^σ_B. Differentiation of $q(n^\sigma_B)$ with respect to n^σ_B gives the differential heat of adsorption. Alternatively, $\delta q(n^\sigma_B)$ can be measured as small increments δn^σ_B are added to the adsorbed phase, giving the differential heat of adsorption directly.

Since the molar Gibbs energy change accompanying adsorption from the gas phase is

$$\Delta G^a_{m,B} = G^a_{m,B} - G^{\ominus}_{m,B} = RT \ln p \qquad (21)$$

The corresponding entropy change on adsorption is

$$\Delta S^a_{m,B} = \frac{\Delta H^a_{m,B} - \Delta G^a_{m,B}}{T} \qquad (22)$$

where $\Delta H^a_{m,B}$ is given by

$$(13) \qquad \frac{\partial(\Delta G^a_{m,B}/T)}{\partial T} = R \frac{\partial \ln p}{\partial T} = -\frac{\Delta H^a_{m,B}}{T^2} \qquad (23)$$

Thus from the adsorption isotherm and its variation with temperature, both the enthalpy (usually called the 'heat') and the entropy of adsorption can be calculated.

Mostly, experimental results do not agree with the rather simple model that assumes (or implies) that the heat of adsorption is independent of coverage. The differential heat of adsorption generally decreases with increasing θ, for example for argon, oxygen and nitrogen on rutile (Drain and Morrison 1953) and for nitrogen on carbon black. Possible contributory factors are the interaction between adsorbate molecules and the heterogeneity of crystal surfaces on an atomic scale. The probability of one of the N_s sites being occupied is $\theta = N/N_s$. Adsorption of a neutral molecule, whether by physical adsorption or chemisorption, results in an induced dipole normal to the surface (i.e. parallel to the z-axis). The energy of interaction between two (not necessarily identical) adsorbed molecules separated by a distance R, with dipole moments μ_1, μ_2 and polarizabilities α_1, α_2, is (e.g. Berry *et al.* 2002)

$$V(R) = \frac{\mu_1 \mu_2}{4\pi\varepsilon_0 R^3}[-2\cos\theta_1\cos\theta_2 + \sin\theta_1\sin\theta_2\cos(\phi_2 - \phi_1)]$$
$$-\frac{\alpha_2^2\mu_1^2(3\cos^2\theta_1 + 1)}{2(4\pi\varepsilon_0)^2 R^6} - \frac{\alpha_1^2\mu_2^2(3\cos^2\theta_2 + 1)}{2(4\pi\varepsilon_0)^2 R^6} \tag{24}$$

The first term on the RS is the interaction energy between the two dipoles and the next two are the dipole-induced dipole terms. The magnitude of each of the latter is about 2% of that of the dipole–dipole term at a separation of 0.5 nm.

Exercise 7.7-4

Make a units check on Eq. (19).

The two dipoles are both normal to the surface (parallel to the z axis) $\theta_1 = \theta_2 = 0$ and the dipole–dipole term is zero in this configuration. Therefore, there is an attraction between the adsorbed molecules, but because of the R^{-6} dependence of $V(R)$ on R, this would only be significant between nn. Because of the overlap repulsion between molecules that are close together, the net potential between two adsorbed molecules is of the Lennard-Jones form. If R is less than the distance of closest approach in the gas phase, the net interaction is repulsive.

The probability of occupation of a particular site is θ and the probability that one of the z nn sites is occupied is $z\theta$. If the interaction energy between a pair of adsorbate molecules on nn sites is w, the total interaction energy is

$$V_{\text{int}} = \frac{z\theta^2 w}{2} \tag{25}$$

where the factor 2 in the denominator ensures that the same pair is not counted twice. This interaction energy should be added to the RS of (15) to give the integral energy of adsorption that includes the interaction of adsorbed molecules. It follows that the contribution from pair interactions to the differential heat of adsorption is $zw\theta$ (*cf.* Adamson and Gast 1997, p. 613). The factor $b(T)$ in the Langmuir equation should therefore be replaced by

$$b'(T) = b(T)\exp(zw\theta/RT) \tag{26}$$

Answers to Exercises 7.7

Exercise 7.7-1

Rearrange (4) to give

$$(4) \qquad dS^\sigma = \frac{1}{T}\,dU^\sigma + \frac{p}{T}\,dV^\sigma - \sum_B \frac{\mu_B^\sigma}{T}\,dn_B^\sigma - \frac{\pi}{T}\,dA^\sigma$$

Exercise 7.7-2

$$\mu_B^\sigma = \sum_B \left(\frac{\partial U^\sigma}{\partial n_B^\sigma}\right)_{S^\sigma,V^\sigma,n_{A\neq B}^\sigma,A^\sigma} = \sum_B \left(\frac{\partial H^\sigma}{\partial n_B^\sigma}\right)_{S^\sigma,p,n_{A\neq B}^\sigma,A^\sigma}$$

$$= \sum_B \left(\frac{\partial F^\sigma}{\partial n_B^\sigma}\right)_{V^\sigma,p,n_{A\neq B}^\sigma,A^\sigma} = \sum_B \left(\frac{\partial G^\sigma}{\partial n_B^\sigma}\right)_{T,p,n_{A\neq B}^\sigma,A^\sigma}$$

In calculating the partial derivative of U^σ, H^σ, F^σ and G^σ with respect to n_B^σ, different sets of independent variables are being held constant.

Exercise 7.7-3

$$(11) \qquad \pi = \sum_B \left(\frac{\partial U^\sigma}{\partial A^\sigma}\right)_{S^\sigma,V^\sigma,n_B^\sigma} = \sum_B \left(\frac{\partial H^\sigma}{\partial A^\sigma}\right)_{S^\sigma,p,n_B^\sigma}$$

$$= \sum_B \left(\frac{\partial F^\sigma}{\partial A^\sigma}\right)_{T,V^\sigma,n_B^\sigma} = \sum_B \left(\frac{\partial G^\sigma}{\partial A^\sigma}\right)_{T,p,n_B^\sigma}$$

Exercise 7.7-4

The units of the RS of (19) are

$$\text{first term} = \frac{C^2 m^2}{CV^{-1} m^{-1} m^3} = CV; \quad \text{second term} \frac{CV^{-1} m^2 C^2 m^2}{(CV^{-1} m^{-1})^2 m^6} = CV;$$

third term, same as second term. Therefore, units of LS = units of RS.

7.8. Nucleation

If $\mu_\alpha > \mu_\beta$, then the transformation of α into β is a natural process. But in phase changes, the small first-formed particles of the new phase β have a higher Gibbs energy than the bulk phase β due the surface energy associated with small particles or bubbles. Consider the formation of a small sphere (a *nucleus*) of solid of radius r within a super-cooled liquid at a temperature $T < T_f$. Then the Gibbs energy change accompanying the formation of this nucleus is

$$\Delta_{l \to s} G(r) = 4\pi r^2 \gamma^{sl} - (4\pi r^3 \rho^s / 3M) \Delta_f G_m(T) \tag{1}$$

One is naturally reluctant to use a macroscopic value for the surface energy in an equation in which the size of the nucleus is of the order of nm. However, Benson and Shuttleworth (1951) found that the surface energy of a crystallite containing only 13 atoms was only 15% less than that of a planar surface. If this result is generally valid, then the Gibbs energy of particles containing only a few atoms will not be affected seriously by using the surface energy for a planar surface in Eq. (1).

The radius of a nucleus corresponding to the maximum value of $\Delta G(r)$ is given by

(1)
$$\frac{\partial \Delta G(r)}{\partial r} = 8\pi r \gamma^{sl} - (4\pi r^2 \rho^s / M) \Delta_f G_m(T) = 0 \tag{2}$$

(2)
$$r_{\max} = \frac{2\gamma^{sl} M}{\Delta_f G_m(T) \rho^s} \tag{3}$$

The critical radius $r = r_c$, is obtained by setting $\Delta G(r) = 0$, giving

(1)
$$r_c = \frac{3\gamma^{sl} M}{\Delta_f G_m(T) \rho^s} = \frac{3}{2} r_{\max} \tag{4}$$

$$\Delta_f G_{\mathrm{m}}(T) = \mu^l(T) - \mu^s(T)$$

$$= \mu^l(T) - \mu^s(T_f) - \frac{\partial \mu^s(T_f)}{\partial T}[T - T_f]$$

$$= \mu^l(T) - \mu^l(T_f) + S_{\mathrm{m}}^s[T - T_f]$$

$$= -S_{\mathrm{m}}^l[T - T_f] + S_{\mathrm{m}}^s[T - T_f]$$

$$= -\Delta_f S_{\mathrm{m}}[T - T_f] = -\Delta_f H_{\mathrm{m}} \times \frac{\Delta T}{T_f} \tag{6}$$

where $\Delta T = T - T_f$ and $\Delta T / T_f$ is the *super-cooling*.

Exercise 7.8-1

Comment on the signs of ΔT and $\Delta_f G_{\mathrm{m}}(T)$

$$(3), (6) \qquad\qquad r_{\max} = -\frac{2\gamma^{sl} M T_f}{(\Delta T) \rho^s \Delta_f H_{\mathrm{m}}} \tag{7}$$

$$(1), (6) \qquad \Delta_{l \to s} G(r) = 4\pi r^2 \gamma^{sl} - (4\pi r^3 \rho^s / 3M) \Delta_f H_{\mathrm{m}} \times \frac{-\Delta T}{T_f} \tag{8}$$

Example 7.8-1

Consider the formation of gold nuclei in molten gold, for which $T_f = 1337.6\,\mathrm{K}$, $\Delta H_f = 12675\,\mathrm{J\,mol^{-1}}$ and $\rho^s = 18.88\,\mathrm{g\ cm^{-3}}$ (at $20°\mathrm{C}$). (a) Suppose that $\Delta T = -40\,\mathrm{K}$ and $\gamma^{sl} = 50\,\mathrm{mJ\,mol^{-1}}$

$$(8) \qquad\qquad \Delta G(r) = 4\pi (r/\mathrm{nm})^2 \times 0.050 \times 10^{-18}\,\mathrm{J}$$

$$- \frac{4\pi (r/\mathrm{nm})^3 \times 10^{-27}\mathrm{m}^3 \times 18.88\,\mathrm{g\,cm^{-3}} \times 12675\,\mathrm{J\,mol^{-1}} \times 40\,\mathrm{K})}{196.966 \times 1.6656 \times 10^{-27} \times 6.02214 \times 10^{23}\,\mathrm{kg\,mol^{-1}} \times 1337.6\,\mathrm{K}}$$

$$= 0.62832\,(r/\mathrm{nm})^2\,\mathrm{aJ} - 0.15176\,(r/\mathrm{nm})^3\,\mathrm{aJ} \tag{9}$$

Recall that $1\,\mathrm{aJ} = 10^{-18}\,\mathrm{J}$ (see the list of SI prefixes in Cohen *et al.* 2007). Plots of $\Delta G(r)$ in aJ against r in nm are shown in Figure 7.6 for (a) the above values of the parameters, (b) super-cooling of magnitude 80 K, which doubles the second term in (9), (c) a surface energy of $70\,\mathrm{mJ\,m^{-2}}$, which multiplies the first term in (9) by 1.4. A shortcoming of formula (9) is the use of a room temperature value of the density of gold. To correct this

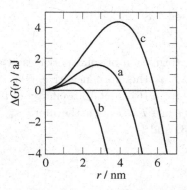

Figure 7.6. Gibbs energy of nucleus formation as a function of the radius of the nucleus plotted from Eq. (7.8.9). The values of the necessary parameters are given in the text.

would require data for the expansivity of gold at temperatures almost up to its fusion temperature.

Exercise 7.8-2

Calculate the critical size of a nucleus r_c under the conditions specified in Example 7.8-1(a).

The assembly of n_c molecules of a super-cooled liquid B into a nucleus of solid B occurs by the process

$$n_c B = B_{n_c} \tag{7}$$

Thus the initiation of a solid phase within a super-cooled liquid requires the coming together, within a short space of time, of several molecules of B, and their formation of a fragment of the crystal lattice of B. The assembling of this crystal fragment is associated with a local loss of both the energy and entropy of fusion and it is this which makes the formation of a nucleus a highly improbable event in the homogeneous phase of liquid B. Consequently, nucleation is more likely to occur on the walls of the containing vessel. Theory (Allnatt and Jacobs 1968) indicates that the number of nuclei formed in time t will be proportional to t^n, thus accounting for the long induction periods followed by a rapid increase in the number of nuclei. Once nuclei of the new phase of B are formed they will grow rapidly and ingest many smaller nuclei and potential nucleus-forming sites.

Small nuclei are inherently unstable because of their surface energy (Figure 7.6) and this is the lesson to be learned from the above thermodynamic analysis.

Answers to Exercises 7.8

Exercise 7.8-1

We are discussing the formation of solid particles in a super-cooled liquid, i.e. one at $T < T_f$. ΔT is therefore negative. The enthalpy of fusion is positive and therefore from (6) the Gibbs energy of fusion at T is positive, as indeed it must be, since $T < T_f$. Furthermore, r_{max} and r_c are positive quantities (see Eqs. (3) and (4)).

Exercise 7.8-2

$$(7) \quad r_c = -\frac{3\gamma^{sl} M T_f}{(\Delta T)\rho^s \Delta_f H_m}$$

$$= \frac{3 \times 0.050\,\text{J m}^{-2} \times 196.966 \times 1.66054 \times 10^{-27} \times 6.02214 \times 10^{23}\,\text{kg mol}^{-1} \times 1337.6\,\text{K}}{40\,\text{K} \times 18.888\,\text{g cm}^{-3} \times 12675\,\text{J mol}^{-1}}$$

$$= 4.1415\,\text{nm}$$

which clearly agrees with curve (a) in Figure 7.6.

Problems 7

Mathcad (or your favourite mathematical package) should be used when answering problems marked by *.

7.1 Write down the fundamental equation for dU for an open phase for which the surface contribution of the energy may not be neglected. Obtain the equation for the Gibbs energy of this phase and hence derive the Gibbs–Duhem equation for this phase. (**Hint:** It is not necessary to label this phase, since it is the only one of interest.)

7.2 Calculate the pressure inside an air bubble of diameter 400 nm in water at 20°C and 101 kPa pressure.

7.3 The contact angle of mercury on glass is 140°. Calculate the capillary depression of the meniscus of mercury in a glass tube of 0.75 mm diameter at 20°C. (**Hint:** The surface tension of mercury at 20°C is given in Table 7.1.)

7.4 A drop of liquid l is placed on the surface of a solid s, in air. The measured contact angle is θ. Write down an equation for the wetting coefficient and hence derive the equilibrium condition.

7.5 2.3 g of an aqueous solution of hydrocinnamic acid (relative molar mass 150 g mol^{-1}) at 298.15 K were skimmed from 310 cm^2 of surface, using a microtome. The difference Δc in the mass ratio (mass solute/

mass solvent) between the surface and the bulk solution was found to be 1.30×10^{-5} g solute per g of water. The concentration of hydrocinnamic acid in the bulk solution was 0.00400 g of the acid per g of water. In separate experiments, measurements of the surface tension of hydrocinnamic acid yielded the result $d\gamma/dc = -4.00$ N m^{-1} (with c in the above units of g acid per g water). Find γ from the skimming data and compare the result with that calculated from the Gibbs adsorption equation. (**Hint:** The solution may be assumed to behave ideally.)

7.6 The vibrational frequency ν_a in Eq. (7.6.19) may be taken to be 5×10^{12} s^{-1}. Use this equation to calculate θ for N_2 at a temperature of 78 K and a pressure of 0.1 MPa on an adsorbent for which the adsorption energy ΔU_a is 1500 J mol^{-1}. The surface area per molecule of adsorbed N_2 is given in the text as 16.2 nm^2. Assume that the internal degrees of freedom are unaffected by the adsorption process.

7.7 Derive a thermodynamic equation of state for a surface, analogous to Eq. (2.3.8). (**Hint:** That is, find $(\partial U/\partial A)_{T,V}$. Remember that $\gamma(T)$.)

7.8 Sometimes it is useful to use the spreading pressure π as an independent variable instead of the area A^σ. Define a new set of thermodynamic potentials, $\mathcal{U}^\sigma, \mathcal{H}^\sigma, \mathcal{F}^\sigma, \mathcal{G}^\sigma$, by

$$\mathcal{U}^\sigma = U^\sigma - \pi A^\sigma, \quad \mathcal{H}^\sigma = H^\sigma - \pi A^\sigma,$$
$$\mathcal{F}^\sigma = F^\sigma - \pi A^\sigma, \quad \mathcal{G}^\sigma = G^\sigma - \pi A^\sigma$$

Hence derive the fundamental equations for $d\mathcal{U}^\sigma, d\mathcal{H}^\sigma, d\mathcal{F}^\sigma, d\mathcal{G}^\sigma$.
Derive four alternative expressions for the chemical potential of B. (**Hint:** Be sure to state which variables are held constant during partial differentiation.)
Derive four alternative expressions for the spreading pressure π.
Define differential heats of adsorption at constant spreading pressure for the thermodynamic potentials $\mathcal{U}^\sigma$ and $\mathcal{H}^\sigma$.

*7.9 Consider the case of super-saturated water vapour at $20°$C, when the partial pressure of water in air is twice times the equilibrium vapour pressure p.

(a) What is the sign of ΔG when n is (i) 1000 molecules and (ii) a million molecules of water, when p is twice p^*?

*7.10 Plot a graph of ΔG against the number of molecules n in a droplet of condensed water.

7.11 Explain briefly the operation of a Wilson cloud chamber.

8

Chemical Equilibrium

8.1. Stoichiometry

The general chemical reaction may be represented quantitatively by

$$0 = \sum_{B} \nu_B B \tag{1}$$

B is the chemical symbol for the substance B involved in the reaction, as for example, NH_3 (ammonia). The ν_B are *stoichiometric numbers*, positive for products and negative for reactants. They tell us the relative amounts (number of moles) of the substances involved in the reaction. For example, the formation of two moles of ammonia from the reaction of one mole of nitrogen and three moles of hydrogen is represented by

$$0 = -N_2 - 3H_2 + 2NH_3 \tag{2}$$

The *extent of reaction* ξ is defined by

$$\xi = \frac{n_B(\xi) - n_B(0)}{\nu_B} \tag{3}$$

$n_B(\xi)$ is the amount of substance B present when the extent of reaction is ξ and $n_B(0)$ is the amount of B present when $\xi = 0$, corresponding to the specified initial conditions. Equation (3) shows that the unit of ξ is the same as that of the amount of substance, namely mole. For example, if 1 mol of N_2 and 3 mol of H_2 are converted completely into 2 mol of NH_3

$$(3) \qquad \xi = \frac{n_{NH_3}(\xi) - n_{NH_3}(0)}{\nu_{NH_3}} = \frac{2\,\text{mol} - 0\,\text{mol}}{2} = 1\,\text{mol} \tag{4}$$

$$(3) \qquad \xi = \frac{n_{N_2}(\xi) - n_{N_2}(0)}{\nu_{N_2}} = \frac{0\,\text{mol} - 1\,\text{mol}}{-1} = 1\,\text{mol} \tag{5a}$$

$$(3) \qquad \xi = \frac{n_{H_2}(\xi) - n_{H_2}(0)}{\nu_{H_2}} = \frac{0\,\text{mol} - 3\,\text{mol}}{-3} = 1\,\text{mol} \tag{5b}$$

Equations (4) and (5) illustrate that ξ is independent of B, whereas Δn_B does depend on B, being negative for reactants and positive for products.

Exercise 8.1-1

For the general chemical reaction (1), write down the relation between the infinitesimal change in the extent of reaction $d\xi$ and the resulting changes in the amounts of the reactants and products, B. Does this result depend on the sign of ν_B?

Answer to Exercise 8.1-1

(3)
$$d\xi = \frac{dn_B}{\nu_B} \tag{6}$$

$d\xi$ is independent of the sign of ν_B since, if B is a reactant, ν_B is negative and so is dn_B.

8.2. Affinity and Reaction Enthalpy

The fundamental equation for $G(T, p, n_B)$ for an open phase is

$$dG = -S\,dT + V\,dp + \sum_B \mu_B dn_B \tag{1}$$

$(1), (8.1.6)$
$$dG = -S\,dT + V\,dp + \sum_B \mu_B \nu_B d\xi \tag{2}$$

The *reaction affinity* is defined by

$$A = -\sum_B \mu_B \nu_B \tag{3}$$

$(2), (3)$
$$dG = -S\,dT + V\,dp - A\,d\xi \tag{4}$$

which gives for the affinity

$$A = -\left(\frac{\partial G}{\partial \xi}\right)_{T,p} \tag{5}$$

If a reaction is occurring spontaneously, $d\xi$ is positive, dG is negative and the affinity is therefore positive. Thus the definition of affinity is in accordance with common usage, in which the statement that substances B and C have an affinity for one another implies that they react.

Exercise 8.2-1

Derive fundamental equations for the thermodynamic potentials U, S, H, F, J and Y, which include the extent of reaction ξ as an independent variable.

At constant T and p

$$(1), (3) \qquad\qquad dG = -A\,d\xi = \sum_B \nu_B \mu_B d\xi \qquad\qquad (6)$$

Since G is function of state, integrating (6) between $\xi = 0\,\text{mol}$ and $\xi = 1\,\text{mol}$ yields

$$A = -\Delta G = -\sum_B \nu_B \mu_B \geq 0 \qquad\qquad (7)$$

which is the affinity for the general chemical reaction (8.1.1). Remember that the ν_B are negative for reactants, positive for products. Equation (7) is the condition for the feasibility of a reaction, but whether or not the reaction actually does take place depends on kinetic factors. If A is negative, it will not occur under the stated conditions.

Exercise 8.2-2

What are the units of each term in Eq. (7)?

Exercise 8.2-3

In Chapter 2 we wrote the expression

$$dS = d_e S + d_i S$$

which separates the total entropy production dS into the sum of two terms, $d_e S$, which is the entropy produced by the interaction of the system with its surroundings, and $d_i S$, which is the entropy produced by irreversible processes going on inside the system (for example chemical reactions). When the only interaction with a system's surroundings is the absorption of heat dq and the only irreversible process going on within the system is a chemical reaction, write explicit expressions for $d_e S$ and $d_i S$.

Consider the complete transformation of stoichiometric amounts of reactants at T, p into products at T, p, so that $\Delta\xi = 1\,\text{mol}$. Because H is a function of the state of the system, the enthalpy change due to the reaction

$$\Delta_r H = q_p \qquad\qquad (8)$$

is equal to the heat absorbed at constant pressure. This quantity was formerly called the heat of reaction at constant pressure, but is now called the *reaction enthalpy* (Cohen *et al.* 2007). Subscript $_r$ tells us that this is the enthalpy change during a chemical reaction. $\Delta_r H$ is the change in an extensive property, but numerical data always refer to $\Delta \xi = 1\,\mathrm{mol}$ so that the units of $\Delta_r H$ are $\mathrm{J\,mol^{-1}}$, or more commonly $\mathrm{kJ\,mol^{-1}}$.

Exercise 8.2-4

In an actual experiment, the reaction was exothermic and the temperature of the system rose to $T' > T$ and its pressure rose to $p' > p$. When reaction was complete, the temperature was restored to T and the pressure to p. Does this make any difference to the measured enthalpy change?

If the same reaction is carried out at constant volume and the final temperature again adjusted to be equal to the initial temperature, the heat absorbed is

$$\Delta_r U = q_V \tag{9}$$

which is the internal energy change at constant volume (formerly, the heat of reaction at constant volume) for $\Delta \xi = 1\,\mathrm{mol}$.

Since H and U are functions of the state of the system, they are independent of the path by which the system is brought from state one to state two and the enthalpy and energy changes in (8) and (9) are therefore additive. This is sometimes known as *Hess' law*. Although the general chemical reaction (8.1.1) is the most convenient way of describing reactions in general, the forms

$$\text{reactants} = \text{products} \tag{10}$$

$$\text{reactants} \rightarrow \text{products} \tag{11}$$

are commonly used to describe a specific chemical reaction, as in Example 8.2-1 below.

Example 8.2-1

The heat of combustion of graphite to CO may not be determined directly since some CO_2 is always formed. Therefore, we make use of Hess' law (that is, the additivity of reaction enthalpies).

$$\Delta_r H^{\ominus}(298.15\,\text{K})/\text{kJ}\,\text{mol}^{-1}$$

(a) $\qquad C(\text{gr}) + O_2(\text{g}, p^{\ominus}) = CO_2(\text{g}, p^{\ominus})$ $\qquad\qquad -393.5$

(b) $\qquad CO(\text{g}, p^{\ominus}) + \frac{1}{2}O_2(\text{g}, p^{\ominus}) = CO_2(\text{g}, p^{\ominus})$ $\qquad\quad -283.0$

(a)–(b) $\qquad C(\text{gr}) + \frac{1}{2}O_2(\text{g}, p^{\ominus}) = CO(\text{g}, p^{\ominus})$ $\qquad\quad -110.5$

It is necessary to specify the physical state of each participant. $C(\text{gr})$ signifies that the carbon is in the form of graphite. The standard pressure $p^{\ominus} = 0.1\,\text{MPa}$ is here stated explicitly, but will in future *be taken to be understood*, unless a different value of p is stated. These data are for the standard temperature, as shown by including the standard temperature $T^{\ominus} = 298.15\,\text{K}$ in parentheses. It is always necessary to specify the physical state of all reactants and products. When all reactants and products are in their *standard states*, the reaction enthalpy is the *standard reaction enthalpy* (unit $\text{J}\,\text{mol}^{-1}$), as signified by the superscript $^{\ominus}$ on $\Delta_r H$

(8.1.1) $$\Delta_r H^{\ominus} = \sum_{\text{B}} \nu_\text{B} H_\text{B}^{\ominus} \qquad (12)$$

Standard States (see Cohen *et al.* 2007, pp. 61–62 for precise definitions of $\mu_\text{B}^{\ominus}$). The standard state of a *gas* B (either the pure gas B or of the component B in a gas mixture) is the hypothetical ideal state of the pure gas B at the standard pressure. Accordingly, the chemical potential of B in a mixture of gases, at temperature T, pressure p and mole fractions y_C, is

$$\mu_\text{B}(T, p, y_\text{C}) = \mu_\text{B}^{\ominus}(T) + RT\ln(y_\text{B}p/p^{\ominus}) + \int_0^p \left[V_\text{B}(T, p, y_\text{C}) - \frac{RT}{p} \right] \mathrm{d}p \qquad (13)$$

$\mu_\text{B}(T, p, y_\text{C})$ and $V_\text{B}(T, p, y_\text{C})$ are the chemical potential and partial molar volume of B in a mixture of gases B, C, ... Notice that the standard chemical potential $\mu_\text{B}^{\ominus}(T)$ of B is a function of T only and independent of the pressure p and of the set of mole fractions y_C that describe the composition of the mixture. (For further details on this point, consult McGlashan (1979).) The first two terms on the RS of (13) are the chemical potential of the hypothetical ideal gas B in a mixture of ideal gases, while the integral describes the correction for non-ideality.

The standard state of a liquid or solid substance, whether pure or in a mixture, or a solvent, is the state of the pure substance B, in its normal physical state at the standard pressure. The standard chemical potential of B is, therefore

$$\mu_B^{\ominus}(T) = \mu_B^*(T, p^{\ominus}) \tag{14}$$

For a solute in a solution, the standard state is the hypothetical ideal solution of standard molality $m^{\ominus}$. The standard chemical potential of B is therefore defined by

$$\mu_B^{\ominus}(T) = \lim_{m_B \to 0} [\mu_B - RT \ln(m_B/m^{\ominus})] \tag{15}$$

Consequently

$$\mu_B(T, m_B) = \mu_B^{\ominus}(T) + RT\ln(m_B \gamma_{B,m}/m^{\ominus})] \tag{16}$$

where $\gamma_{B,m}$, the activity coefficient of B at molality m, accounts for deviations from ideality.

   Although the standard pressure, standard molality and standard concentration could have any specified values, in practice the most common - choices (and the ones used in this book) are:

$$\left. \begin{array}{r} p^{\ominus} = 0.1\,\text{MPa}\ (=1\,\text{bar}) \\ m^{\ominus} = 1\,\text{mol}\,\text{kg}^{-1} \\ c^{\ominus} = 1\,\text{mol}\,\text{dm}^{-3} \end{array} \right\} \tag{17}$$

$H_B^{\ominus}$ in Eq. (13) is the *standard enthalpy* of substance B. If tables of $H_B^{\ominus}$ were available, then the standard reaction enthalpy $\Delta_r H^{\ominus}$ could be calculated. But since absolute values of $H_B^{\ominus}$ may not be determined experimentally, the *convention* is adopted of *setting the enthalpy of formation $\Delta_f H^{\ominus}$ of any pure element in its standard state to zero*. It follows that the standard enthalpy of a compound is its standard enthalpy of formation from its elements in their standard states and therefore that

(12)
$$\Delta_r H^{\ominus} = \sum_B \nu_B \Delta_f H_B^{\ominus} \tag{18}$$

Exercise 8.2-5

What are the standard states of the halogens chlorine, bromine and iodine? What is the standard enthalpy of formation of gaseous bromine? (***Hint:*** The enthalpy of vaporization of Br_2 at 298.15 K is 30.907 kJ mol^{-1}.)

The same notation applies to other thermodynamic functions. For example, the standard reaction entropy is

(13)
$$\Delta_r S^\ominus = \sum_B \nu_B S_B^\ominus \qquad (19)$$

and the standard Gibbs energy of reaction is

(12)
$$\Delta_r G^\ominus = \sum_B \nu_B \mu_B^\ominus \qquad (20)$$

Extensive tables of thermodynamic data may be found in Wagman *et al.* (1982), Chase *et al.* (1985), Cox *et al.* (1989), Knacke *et al.* (1991), Barin (1995), Chase (1998), and Binnewies and Milke (2002). (See also Appendix A3.)

Example 8.2-2

Determine the heat of hydrogenation of ethylene at 298.15 K from the following data.

$$\Delta_r H^\ominus \ (298.15\,,\mathrm{K})/\mathrm{kJ\,mol}^{-1}$$

(a) $C_2H_4(g) + 3O_2(g) = 2CO_2(g) + 2H_2O(l)$ -1411.3

(b) $H_2(g) + \dfrac{1}{2}O_2(g) = H_2O(l)$ -286.2

(c) $C_2H_6(g) + \dfrac{7}{2}O_2(g) = 2CO_2(g) + 3H_2O(l)$ -1559.8

(a) + (b) − (c) $C_2H_4(g) + H_2(g) = C_2H_6(g)$ $-1411.3 - 286.2 + 1559.8$
$$= -137.7$$

Exercise 8.2-6

The reaction enthalpy for the complete hydrolysis of liquid $POBr_3$

$$POBr_3(liq) + 3H_2O(liq) = 3HBr(aq) + H_3PO_4(aq)$$

is $\Delta_r H^\ominus(298.15\,\mathrm{K}) = -333.5\,\mathrm{kJ\,mol}^{-1}$, where aq means a dilute aqueous solution.

The following enthalpies of formation are available from tabulated data:

$$\Delta_f H^{\ominus}(298.15\,\text{K})/\text{kJ mol}^{-1}$$

$H_2O(\text{liq})$	-286.2
$HBr(\text{aq})$	-118.6
$H_3PO_4(\text{aq})$	-1277.4

Find the enthalpy of formation of $POBr_3(\text{liq})$.

Answers to Exercises 8.2

Exercise 8.2-1

Since

$$G = U + pV - TS$$

(4) $\qquad dU = T\,dS - p\,dV - A\,d\xi$ $\qquad\qquad$ (21)

(20) $\qquad dS = (1/T)dU + (p/T)dV + (A/T)\,d\xi$ $\qquad$ (22)

(20) $\qquad dH = T\,dS + V\,dp - A\,d\xi$ $\qquad\qquad$ (23)

(20) $\qquad dF = -S\,dT - p\,dV - A\,d\xi$ $\qquad\qquad$ (24)

(24), (P3.3.3) $\qquad dJ = (U/T^2)dT + (p/T)dV + (A/T)\,d\xi$

(3), (P3.3.5) $\qquad dY = (H/T^2)dT - (V/T)dp + (A/T)\,d\xi$

Exercise 8.2-2

The SI unit of each term in (7) is J, because of the integration of (6) with respect to $d\xi$, the unit of which is mol.

Exercise 8.2-3

The entropy produced due to the absorption of heat dq from the surroundings is

$$d_e S = \frac{dq}{T}$$

The entropy produced by the chemical reaction going on within the system is

$$d_i S = \frac{A\,d\xi}{T}$$

The total entropy production is, therefore,

$$dS = d_e S + d_i S$$

Exercise 8.2-4

No. The enthalpy is a function of the state of the system and its change depends only on the initial and final values of the variables T, p.

Exercise 8.2-5

$$Cl_2(g, p = p^\ominus), \quad Br_2(\text{liq}, p = p^\ominus), \quad I_2(\text{cr}, p = p^\ominus).$$

The element Br_2 is a liquid at 298.15 K and therefore $\Delta_f H^\ominus$ Br_2(liq, 298.15 K) = 0. Consequently, $\Delta_f H^\ominus$ Br_2 (g, 298.15 K) = 30.907 kJ mol^{-1}.

Exercise 8.2-6

$$(18) \qquad \Delta_r H^\ominus = \sum_B \nu_B \Delta_f H_B^\ominus$$

$$-333.5 = -\Delta_f H^\ominus[POBr_3(\text{liq})]/\text{kJ mol}^{-1} - 3(-286.2)$$
$$+ 3(-118.6) + 1(-1277.4)$$

$$\Delta_f H^\ominus[POBr_3(\text{liq}), 298.15 \text{ K}] = 441.1 \text{ kJ mol}^{-1}.$$

8.3. Temperature Dependence of the Reaction Enthalpy

In Section 8.2 it was shown that the reaction enthalpy at 298.15 K could be calculated from standard heats of formation $\Delta_f H^\ominus(298.15 \text{ K})$. To find $\Delta_r H^\ominus(T)$ at any other temperature T, differentiate (8.2.12) at constant pressure

$$(8.2.12) \qquad \left(\frac{\partial \Delta_r H^\ominus}{\partial T}\right)_p = \sum_B \nu_B \left(\frac{\partial H_B^\ominus}{\partial T}\right)_p = \sum_B \nu_B C_p(B) = \Delta_r C_p \qquad (1)$$

which is the *Kirchhoff equation*. In (1), molar values of H_B and $C_p(B)$ are implied because the reaction enthalpy always refers to $\Delta\xi = 1$

$$(1) \qquad \Delta_r H^\ominus(T) - \Delta_r H^\ominus(T_1) = \int_{T_1}^T \Delta_r C_p dT \qquad (2)$$

The temperature dependence of the heat capacity at constant pressure is described by the series

$$C_p(B) = a(B) + b(B)(T/K) + c(B)(K/T)^2 + d(B)(T/K)^2 \qquad (3)$$

All four terms are not necessarily needed for a particular substance

$$(3) \quad \Delta_r C_p = \sum_B \nu_B C_p(B)$$

$$= \sum_B \nu_B[a(B) + b(B)(T/K) + c(B)(K/T)^2 + d(B)(T/K)^2]$$

$$= \Delta_r a(B) + \Delta_r b(B)(T/K) + \Delta_r c(B)(K/T)^2 + \Delta_r d(B)(T/K)^2 \qquad (4)$$

Recall that the Δ_r operator was defined in (8.2.12) as

$$\Delta_r = \sum_B \nu_B \qquad (5)$$

When it is clear that a chemical reaction is involved, the subscript $_r$ is often omitted, particularly in (4). Given the value of $\Delta_r H$ at $T_1 = 298.15\,\mathrm{K}$ (from tables) its value at another temperature T may be calculated from (2), (4) and (5)

$$\Delta_r H(T) = \Delta_r H(T_1) + \Delta a(T - T_1)$$

$$+ \frac{1}{2}\Delta b(T^2 - T_1^2)/K - \Delta c\left(\frac{1}{T} - \frac{1}{T_1}\right)K^2$$

$$+ \frac{1}{3}\Delta d(T^3 - T_1^3)/K^2 \qquad (6)$$

Since Δa is of O(1) (that is, of order 1), Δb of O(10^{-3}), Δc of O(10^6) and Δd of O(10^{-6}), a variation in the notation (see Knacke *et al.* 1991) is useful. Replace $\Delta_f H(B)$ by h_B (for brevity) and set the dimensionless variable[a]

$$y = 10^{-3}\,T/K \qquad (7)$$

---

[a]It is true that the symbol y_B is used for the mole fraction of B in the gas phase, but confusion with y is unlikely to occur.

Then for any substance B, the heat capacity at constant pressure (with subscript p understood) is represented by

$$(3) \qquad\qquad C = a + by + cy^{-2} + dy^2 \qquad\qquad (8)$$

Note that the coefficients a, b, c, d, have two distinct meanings which are easily distinguished by noting whether the independent variable is T or y. When numerical values are substituted for a, b, c, d, in (8), in contrast to Eq. (3), the expression (8) contains no awkward powers of 10, except for one factor 10^3, which arises from the change in the independent variable.

If one mole of B is heated from T_1 to T at constant pressure, its enthalpy becomes

$$H_m(T) = H_m(T_1) + \int_{T_1}^{T} C_p(T)\,dT \qquad\qquad (9)$$

In the new notation

$$(9), (8) \qquad H_m(T) = 10^3 K h(y) = 10^3 K \left[h(y_1) + \int_{y_1}^{y} C(y)dy \right]$$

$$= 10^3 K \left[c' + ay + \frac{1}{2}by^2 - cy^{-1} + \frac{1}{3}dy^3 \right] \qquad (10)$$

c' is a numerical constant for each substance and is tabulated along with the coefficients in the heat capacity (e.g. Knacke *et al.* (1991), who use the symbol h^+ for our c'). Comparing (10) with (6), we see that c' is the sum of all the terms in y_1, the lower limit of the integral. With $T_1 = 298.15\,\text{K}$, the standard enthalpy of formation of each B is needed to evaluate $c'(\text{B})$, but this has already been done in the tables of Knacke *et al.* (1991).

For any chemical reaction

$$(8) \qquad\qquad \Delta C = \Delta a + (\Delta b)y + (\Delta c)y^{-2} + (\Delta d)y^2 \qquad\qquad (11)$$

The Δ operator was defined in (5), it being generally understood that this expression (11) refers to a chemical transformation, so that subscript $_r$ may be omitted. The reaction enthalpy at (some yet to be specified value) y is

$$(10) \qquad \Delta h(y) = \Delta h(y_1) + \int_{y_1}^{y} \Delta C(y)\,dy$$

$$= \Delta c' + (\Delta a)y + \frac{1}{2}(\Delta b)y^2 - (\Delta c)y^{-1} + \frac{1}{3}(\Delta d)y^3 \qquad (12)$$

where

$$\Delta c' = \sum_{B} \nu_B c'(B) \tag{13}$$

Alternatively, one could use the NBS tables (Wagman *et al.* 1982) or the NIST JANAF tables (Chase 1998) and interpolate at the required temperatures.

The thermal contribution to the entropy of a substance is

$$(8) \qquad S - S_0 = s' + 2.3026(3a + a \log y) + by - \frac{c}{2}y^{-2} + \frac{d}{2}y^2 \tag{14}$$

with s' a numerical constant specific to each substance.

Exercise 8.3-1

What are the units of s' and c'?

Example 8.3-1

Use (10) and (14) to calculate the enthalpy and entropy of ammonia at 298.15 K. From the tables in Knacke *et al.* (1991) the relevant information for NH_3 is:

$a/\mathrm{J\,K^{-1}\,mol^{-1}}$	$b/\mathrm{J\,K^{-1}\,mol^{-1}}$	$c/\mathrm{J\,K^{-1}\,mol^{-1}}$	$c'/\mathrm{J\,K^{-1}\,mol^{-1}}$	$s'/\mathrm{J\,K^{-1}\,mol^{-1}}$
37.321	18.661	−0.649	−60.244	−29.402

$$
\begin{aligned}
(10) \qquad \Delta_f H(NH_3, 298.15\,\mathrm{K}) &\equiv 10^3 h(NH_3) \\
&= 10^3[-60.244 + 11.127 + 0.829 \\
&\quad + (0.649 \times 3.354)\,\mathrm{J\,mol^{-1}}] \\
&= -46.111\,\mathrm{kJ\,mol^{-1}}
\end{aligned}
$$

which agrees with the tabulated value (see Table 8.1). Don't be misled into thinking that you have found an absolute value for the enthalpy of ammonia. You have but calculated the enthalpy of formation of NH_3 at 298.15 K, referred to the usual zero, namely, that H is taken to be zero for every element in its standard state. The (thermal) entropy of ammonia at 298.15 K is

$$
\begin{aligned}
(14) \qquad S - S_0 &= -29.402 + 257.806 - 45.165 + 5.650 + 3.650\,\mathrm{J\,K^{-1}\,mol^{-1}} \\
&= 192.45\,\mathrm{J\,K^{-1}\,mol^{-1}}
\end{aligned}
$$

which again agrees with the tabulated value.

Table 8.1. Standard enthalpy of formation of ammonia from various sources.

Source	$\Delta_f H^{\ominus}(298.15\,\text{K})/\text{kJ}\,\text{mol}^{-1}$
Cox *et al.* (1989)	-45.94 ± 0.35
Knacke *et al.* (1991)	-46.111
Barin (1995)	-45.940
Chase (1998)	-45.898 ± 0.4

In this scheme, the molar Gibbs energy is

$$(8),(14) \qquad \mu = 10^3\text{K}\left[c' - s'y - ay[\ln(T/\text{K}) - 1] - \frac{b}{2}y^2 - \frac{c}{2}y^{-1} - \frac{d}{6}y^3\right]$$

$$(15a)$$

$$= 10^3\text{K}\left[c' - s'y + ay - 2.3026ay(3 + \log y)\right.$$
$$\left. - \frac{b}{2}y^2 - \frac{c}{2}y^{-1} - \frac{d}{6}y^3\right]$$

$$(15b)$$

Exercise 8.3-2

Check the units of each term in (15b).

References to six compilations of thermodynamic data were given in Section 8.2. In some of these, data are given to six significant figures. This is done to ensure numerical consistency, but the reader should not be led to believe that such accuracy is always meaningful. For example, the enthalpy of formation of ammonia from several sources is given in Table 8.1; the estimated uncertainty is $\pm 0.4\,\text{kJ}\,\text{mol}^{-1}$.

Answers to Exercises 8.3

Exercise 8.3-1

The units of s' are the same as those of heat capacity, namely $\text{J}\,\text{K}^{-1}\,\text{mol}^{-1}$. The units of c' are also $\text{J}\,\text{K}^{-1}\,\text{mol}^{-1}$. (Notice the K outside the [ ] in (10).)

Exercise 8.3-2

The units of μ are $\text{J}\,\text{mol}^{-1}$. The units of c' and of s' are $\text{J}\,\text{K}^{-1}\,\text{mol}^{-1}$, as are the units of a, b, c and d. The SI unit of y is 1. Hence the units of each term on the RS of (15) are $\text{J}\,\text{mol}^{-1}$.

8.4. The Standard Affinity, Reaction Isotherm and Equilibrium Constant

With all reactants and products in their standard states

$$A^{\ominus} = -\Delta G^{\ominus} = -\sum_B \nu_B \mu_B^{\ominus} \tag{1}$$

is the *standard affinity* for the reaction and $\Delta G^{\ominus}$ is the *standard Gibbs energy change*. If we could set up tables of standard chemical potentials $\mu_B^{\ominus}(298.15\,\text{K}, p^{\ominus})$, then we would be able to calculate the standard molar Gibbs energy change for all the reactions involving the substances in this table. But absolute values of standard chemical potentials cannot be determined experimentally (see Section 3.11 and Section 5.4). Some arbitrary zero in the standard Gibbs energy scale must therefore be selected and the one chosen is that $\mu^{\ominus}(T^{\ominus}, p^{\ominus})$ *for all elements is set equal to zero*. The value of $\mu^{\ominus}(T^{\ominus}, p^{\ominus})$ for a *compound* is then equal to the Gibbs energy change on the formation of one mole of that compound, in a designated physical state, from stoichiometric amounts of its separated elements in their standard states. This quantity is called the *standard Gibbs energy of formation* of that compound and is often written as $\Delta_f G^{\ominus}$ (298.15 K, $p^{\ominus}$). Numerical values of $\Delta_f G^{\ominus}$ are to be found in tables of thermodynamic data. But in such compilations the subscript $_f$ might well be omitted, as well as the standard temperature and pressure. For example, one might find that $\Delta G^{\ominus}[\text{H}_2\text{O(g)}] = -228.597\,\text{kJ}\,\text{mol}^{-1}$, the standard temperature of 298.15 K and the standard pressure $p^{\ominus} = 0.1\,\text{MPa}$ being understood, or perhaps declared in the caption to the table. The above datum means that

$$\mu^{\ominus}[\text{H}_2\text{O(g)}] - \mu^{\ominus}[\text{H}_2\text{(g)}] - \frac{1}{2}\mu^{\ominus}[\text{O}_2\text{(g)}] = -228.597\,\text{kJ}\,\text{mol}^{-1},$$
$$T^{\ominus} = 298.15\,\text{K}, \quad p^{\ominus} = 0.1\,\text{MPa} \tag{2}$$

The chemical potential of B is

$$\mu_B = \mu_B^{\ominus} + RT \ln a_B \tag{3}$$

(3), (8.2.7)
$$A = -\Delta G = -\sum_B \nu_B \mu_B$$

$$= -\sum_B \nu_B \mu_B^{\ominus} - \sum_B \nu_B RT \ln a_B$$

$$= -\Delta G^{\ominus} - RT \ln \prod_B a_B{}^{\nu_B} \tag{4}$$

Equation (4) is the *reaction isotherm*: it enables us to calculate the affinity for any designated set of activities, provided $\Delta G^{\ominus}$ is known at the required temperature. The argument of the ln function in (4) is the *reaction quotient*.

Exercise 8.4-1

Use the following data to decide whether liquid mercury at $600\,\text{K}$ would react with oxygen gas at a pressure of (a) $13.33\,\text{Pa}$ (b) $1.333\,\text{Pa}$.

For the reaction

$$Hg(\text{liq}) + \frac{1}{2}O_2(g) = HgO(s), \quad \Delta_r G^{\ominus} = -85.77 + 0.1004(T/\text{K})\,\text{kJ}\,\text{mol}^{-1}$$

(**Hint**: Since the pressures are low, ideal behaviour may be assumed, so that the activity of O_2 is $p/p^{\ominus}$, where the standard pressure $p^{\ominus} = 0.1\,\text{MPa}$.)

At equilibrium

$$A = -\Delta G = 0 \tag{5}$$

$$(5),(4) \qquad K^{\ominus}(T) = \exp(-\Delta G^{\ominus}/RT) = \prod_B a_B{}^{\nu_B} \tag{6}$$

$K^{\ominus}(T)$ is the *standard equilibrium constant* for the general chemical reaction (8.1.1). The superscript $^{\ominus}$ informs us that all reactants and products are in their standard states, as described in Section 8.2. $K^{\ominus}$ is therefore independent of pressure.

$K^{\ominus}(T)$ is the *standard equilibrium constant* for the general chemical reaction (8.1.1). Other equilibrium constants besides $K^{\ominus}$ have been defined and these are summarized in Table 8.2.

Exercise 8.4-2

What is the SI unit of $K^{\ominus}(T)$?

Example 8.4-1

The pressure of CO_2 in equilibrium with $CaCO_3 + CaO$ at $1000\,\text{K}$ is $2.067\,\text{kPa}$. Find the standard Gibbs energy change for the reaction

$$CaCO_3(s) = CaO(s) + CO_2(g)$$

at $1000\,\text{K}$.

Table 8.2. Definition of equilibrium constants for the general chemical reaction (8.1.1) (see Cohen *et al.* 2007, pp. 44–45).

Name	Symbol	Definition	Unit
Equilibrium constant (K)	$K^{\ominus}$	$\exp(-\Delta G^{\ominus}/RT)$	1
K, fugacity basis	K_f	$\prod_{B} \tilde{p}_B{}^{\nu_B}$	$Pa^{\Sigma\nu_B}$
K, pressure basis	K_p	$\prod_{B} p_B{}^{\nu_B}$	$Pa^{\Sigma\nu_B}$
K, concentration basis	K_c	$\prod_{B} c_B{}^{\nu_B}$	$(\text{mol m}^{-3})^{\Sigma\nu_B}$
K, molality basis	K_m	$\prod_{B} m_B{}^{\nu_B}$	$(\text{mol kg}^{-1})^{\Sigma\nu_B}$
K, mole fraction basis	K_y, K_x	$\prod_{B} y_B{}^{\nu_B}, \prod_{B} x_B{}^{\nu_B}$	1

Since the activity of pure solid phases is 1 and the fugacity of a gas at low pressures may be equated to its pressure

$$K^{\ominus}(1000\,\text{K}) = \frac{a_{\text{CaO}}\, a_{\text{CO}_2}}{a_{\text{CaO}}} = \frac{p(\text{CO}_2)}{p^{\ominus}} = \frac{2.067\,\text{kPa}}{100\,\text{kPa}} = 0.02067$$

$$\Delta_r G^{\ominus}(1000\,\text{K}) = -8.314 \times 1000 \times \ln(0.02067)\,\text{J mol}^{-1}$$

$$= 32.25\,\text{kJ mol}^{-1}$$

Example 8.4-2

For the participants in the reaction

$$N_2O_4(g) = 2NO_2(g) \tag{7}$$

	$\Delta H^{\ominus}(298.15\,\text{K})/\text{kJ mol}^{-1}$	$(S^{\ominus} - S_0)/\text{J K}^{-1}\,\text{mol}^{-1}$
$N_2O_4(g)$	9.1	304.4
$NO_2(g)$	33.1	240.0

For reaction (7), calculate (a) the reaction enthalpy at 298.15 K; (b) the equilibrium constant $K^{\ominus}(298.15\,\text{K})$; (c) the equilibrium constant K_p at 298.15 K; (d) the partial pressure of N_2O_4 in MPa in nitrogen dioxide gas at 298.15 K and a total pressure of 0.1 MPa.

For reaction (7), (i.e. per mole of N_2O_4)

(a) $\qquad \Delta H^\ominus = 2 \times 33.1 - 9.1\,\text{kJ mol}^{-1} = 57.1\,\text{kJ mol}^{-1}$

(b) $\quad \Delta(S^\ominus - S_0) = 2 \times 240.0 - 304.4\,\text{J K}^{-1}\,\text{mol}^{-1} = 175.6\,\text{J K}^{-1}\,\text{mol}^{-1}$

$\qquad\qquad \Delta G^\ominus = 57.1 - (298.15 \times 175.6)\,\text{J mol}^{-1} = 4745\,\text{J mol}^{-1}$

(6) $\qquad\qquad K^\ominus(T) = \exp(-\Delta G^\ominus/RT) = \prod_B a_B{}^{\nu_B}$

(6) $\qquad\qquad K^\ominus = \exp[-4745/(8.3145 \times 298.15)] = 0.1475 = \prod_B (p_B/p^\ominus)^{\nu_B}$

From Table 8.2, since $\sum_B \nu_B = 1$

(c) $\qquad K_p \doteq K^\ominus(p^\ominus)^1 = 0.1475 \times 10^{-1}\,\text{MPa}$

$$p_{NO_2} = y_{NO_2}p = \frac{2\xi}{1+\xi}p, \quad p_{N_2O_4} = y_{N_2O_4}p = \frac{1-\xi}{1+\xi}p$$

(d) $\qquad K_p = \dfrac{(p_{NO_2})^2}{p_{N_2O_4}} = \dfrac{4\xi^2}{(1-\xi)(1+\xi)}p = \dfrac{4\xi^2}{1-\xi^2}p = 0.1475 \times 10^{-1}\,\text{MPa}$

For $p = 0.1\,\text{MPa}$, $\xi = 0.189\,\text{mol}$, $p_{N_2}O_4 = \frac{1-\xi}{1+\xi}p = 0.682 \times 10^{-1}\,\text{MPa}$

In general, $K^\ominus(298.15\,\text{K})$ is found from tables of $\Delta G^\ominus$. To find $K^\ominus$ at any other T, differentiate (6) with respect to T at constant p

(6) $\qquad\qquad \ln K^\ominus = -\dfrac{\Delta G^\ominus}{RT}$ $\hfill$ (8)

(8) $\qquad\qquad \dfrac{d \ln K^\ominus}{dT} = \dfrac{d}{dT}\left(\dfrac{-\Delta G^\ominus}{RT}\right) = \dfrac{\Delta G^\ominus}{RT^2} + \dfrac{T\Delta S^\ominus}{RT^2}$ $\hfill$ (9)

(9), (8.3.7) $\qquad\qquad \dfrac{d \ln K^\ominus}{d(T/K)} = \dfrac{\Delta H^\ominus}{RT^2/K}$ $\hfill$ (10)

which is the *van 't Hoff equation*. $\Delta H^\ominus$ is the standard enthalpy of reaction and its value at 298.15 K can be obtained from tables of standard enthalpies of formation.

For *exothermic* reactions, the RS of (10) is *negative* and so $K^\ominus$ decreases with increasing T. But decreasing T (to increase the yield) slows down the rate of attainment of equilibrium and so some compromise will be necessary. The rate of reaction may also be increased by finding an effective *catalyst*. The effect of pressure will be discussed later in Section 8.5.

Exercise 8.4-3

Does a catalyst shift the position of equilibrium?

In the y-notation, on using (8.3.12), Eq. (10) becomes

$$\frac{d \ln K^{\ominus}}{10^3 dy} = \frac{10^3 \Delta h}{10^6 R y^2} = \frac{\Delta c' + (\Delta a)y + \left(\frac{1}{2}\Delta b\right) y^2 - (\Delta c)y^{-1} + \left(\frac{1}{3}\Delta d\right) y^3}{10^3 R y^2}$$

(11)

Integrating (11) between y_1 and y gives

$$\ln\left(\frac{K^{\ominus}(T)}{K^{\ominus}(T_1)}\right) = \ln\left(\frac{K^{\ominus}(y)}{K^{\ominus}(y_1)}\right)$$

$$= \frac{1}{R}\left[\Delta c'\left(\frac{1}{y_1} - \frac{1}{y}\right) + \Delta a \ln\left(\frac{y}{y_1}\right) + \frac{\Delta b}{2}(y - y_1)\right.$$

$$\left. + \frac{\Delta c}{2}\left(\frac{1}{y^2} - \frac{1}{y_1^2}\right) + \frac{\Delta d}{6}\left(y^2 - y_1^2\right)\right]$$

(12)

The equilibrium constant at $T_1 = 298.15\,\text{K}$ can be obtained from tables of the Gibbs energy of formation and so $K^{\ominus}(T)$ can be found from (12) at any temperature T, provided the requisite heat capacity data are available.

Alternatively, differentiate $(\Delta G^{\ominus}/T)$ partially with respect to T, giving

$$\frac{\partial}{\partial T}\left(\frac{\Delta G^{\ominus}}{T}\right) = -\frac{(\Delta G^{\ominus} + T\Delta S)}{T^2} = \frac{-\Delta H^{\ominus}}{T^2}$$

(13a)

which is *the Gibbs–Helmholtz equation*. Since from (8), $\Delta G^{\ominus} = -RT \ln K^{\ominus}$, Eq. (13) is an equivalent statement to (10).

Exercise 8.4-4

Derive the alternative form of the Gibbs–Helmholtz equation

$$\frac{\partial}{\partial T^{-1}}\left(\frac{\Delta G^{\ominus}}{T}\right) = \Delta H^{\ominus}$$

(13b)

Example 8.4-3

Larson and Dodge (1923) measured K_p for the ammonia synthesis

$$\frac{1}{2}N_2 + \frac{3}{2}H_2 = NH_3$$

(14)

at $450°\text{C}$ as a function of the total pressure p, with the results given in Table 8.3. Examine whether K_f is independent of p.

Table 8.3. Equilibrium constant K_p for the ammonia synthesis as a function of pressure at $450°\,C$, as measured by Larson and Dodge (1923). The fugacity coefficients needed to calculate the fugacity coefficient product in column 3 were obtained using the Lewis and Randall rule, namely that

$$\tilde{p}_B(p) \approx y_B \tilde{p}_B^*(p) \qquad (6.3.20)$$

p/MPa	$(10^2 K_p)/\mathrm{MPa}^{-1}$	$\prod_{B}\phi_B{}^{\nu_B}$	$(10^2 K_f)/\mathrm{MPa}^{-1}$
1.01	6.50	0.988	6.42
3.03	6.58	0.969	6.37
5.05	6.81	0.953	6.49
10.13	7.18	0.905	6.50
30.4	8.64	0.743	6.42
60.8	12.77	0.573	7.32
101.3	22.98	0.443	10.1

The first two columns contain the experimental data for K_p at pressures up to $\sim$100 MPa

$$K_f = K_p \prod_{B} \phi_B{}^{\nu_B} = K_p \frac{\phi_{NH_3}}{(\phi_{N_2})^{\frac{1}{2}}(\phi_{H_2})^{\frac{3}{2}}} \qquad (15)$$

The results in Table 8.3 show that K_f is constant up to about 30 MPa. Deviations at higher pressures are due to approximations inherent in the Lewis and Randall rule.

Example 8.4-4

The standard enthalpy of formation and Gibbs energy of formation of ammonia (Chase 1998) are

T/K	$\Delta_f H^\ominus/\mathrm{kJ\,mol}^{-1}$	$\Delta_f G^\ominus/\mathrm{kJ\,mol}^{-1}$
298.15	-45.898	16.367
0	-38.907	

In the temperature range of interest, the heat capacities of reactants and product may be represented by the expressions

$$C_p(N_2)/\mathrm{J\,K}^{-1}\,\mathrm{mol}^{-1} = 26.96 + 5.912 \times 10^{-3}(T/\mathrm{K})$$
$$- 0.3376 \times 10^{-6}(T/\mathrm{K})^2$$

$$C_p(H_2)/\mathrm{J\,K^{-1}\,mol^{-1}} = 29.07 + 0.8368 \times 10^{-3}(T/\mathrm{K})$$
$$+ 2.011 \times 10^{-6}(T/\mathrm{K})^2$$
$$C_p(NH_3)/\mathrm{J\,K^{-1}\,mol^{-1}} = 25.89 + 32.58 \times 10^{-3}(T/\mathrm{K})$$
$$- 3.046 \times 10^{-6}(T/\mathrm{K})^2$$

Derive a formula to calculate $\ln K^{\ominus}$ at any temperature in the range in which the heat capacity data are valid. Find $K^{\ominus}$ and $\Delta G^{\ominus}$ for the ammonia synthesis at 450°C.

First calculate the heat capacity terms

$$\Delta a/\mathrm{J\,K^{-1}\,mol^{-1}} = 25.89 - \frac{1}{2}(26.96) - \frac{3}{2}(29.07) = -31.195$$

$$10^3 \times \Delta b/\mathrm{J\,K^{-1}\,mol^{-1}} = 32.58 - \frac{1}{2}(5.912) + \frac{3}{2}(0.8368) = 30.879$$

$$10^6 \times \Delta d/\mathrm{J\,K^{-1}\,mol^{-1}} = -3.046 + \frac{1}{2}(0.3376) - \frac{3}{2}(2.011) = -5.894$$

Because c' has not been determined with this heat capacity data, we must integrate (10) to yield

(10) $\ln K^{\ominus}(T) - \ln K^{\ominus}(T_1)$

$$= \frac{-\Delta H_0}{R}\left(\frac{1}{T} - \frac{1}{T_1}\right) + \frac{\Delta a}{R}\ln\left(\frac{T}{T_1}\right) + \frac{\Delta b \times 10^{-3}}{2R\,\mathrm{K}}(T - T_1)$$
$$+ \frac{\Delta c \times 10^{6}\mathrm{K}^2}{2R}\left(\frac{1}{T^2} - \frac{1}{T_1^2}\right) + \frac{\Delta d \times 10^{-6}}{6R\,\mathrm{K}^2}(T^2 - T_1^2) \quad (16)$$

Note that the fourth term on the RS is absent in these data for NH_3 but that this is not always so. Substitute $R = 8.3145\,\mathrm{J\,K^{-1}\,mol^{-1}}$, $T_1 = 298.15\,\mathrm{K}$, the numerical values for $\Delta H_0^{\ominus}$ and the heat capacity coefficients and $\Delta G^{\ominus} = 16\,367\,\mathrm{J\,mol^{-1}}$ giving

$$\ln K^{\ominus}(T) = \frac{16\,367}{8.3145 \times 298.15} + \frac{4\,679}{T/\mathrm{K}} - 15.693 - 3.752\ln(T/298.15)$$

$$+ \frac{15.44 \times 10^{-3}}{8.3145}\{(T/\mathrm{K}) - 298.15\} - 0.11817\{10^{-6}(T/\mathrm{K})^2 - 0.0889\}\} \quad (17)$$

At $T = 723.15\,\mathrm{K}$,

$$\ln K^{\ominus}(T) = 6.602 + 6.347 - 15.693 - 3.324 + 0.312 - 0.021 = -4.986$$
$$(18)$$

$$K^{\ominus} = 6.83 \times 10^{-3}, \quad \Delta_r G^{\ominus} = -29.98\,\mathrm{kJ\,mol^{-1}} \quad (19)$$

I chose this example to illustrate the fact that it may be more accurate to calculate the equilibrium constant than it is to measure it. Chase (1998) has considered the data for NH_3 very carefully and concluded that the thermodynamic functions calculated from statistical thermodynamics, using spectroscopic data (Haar 1968), are the most accurate available. These data include values of log $K^{\ominus}$ at 100 K intervals. Figure 8.1 shows the plot of log $K^{\ominus}$ against $10^3/T$.

Sometimes only the first (constant) term in the heat capacity is available and then both (12) and (16) reduce to an equation of the form

$$\log K^{\ominus}(T) = \frac{A}{T/K} + B \log(T/K) + C \tag{20}$$

where A, B and C are constants. The data for NH_3 in this temperature range have been fitted to an equation of this form (Barin 1995):

$$\log K^{\ominus} = \frac{2300\,K}{T} - 1.32 \log(T/K) - 1.56 = F(T) \tag{21}$$

which yields the value for $K^{\ominus}$ at 723.15 K, of

$$K^{\ominus} = 7.02 \times 10^{-3} \tag{22}$$

and which may be compared with (19). In Figure 8.1, the single point marked by a filled circle was calculated from Eq. (17). We must conclude

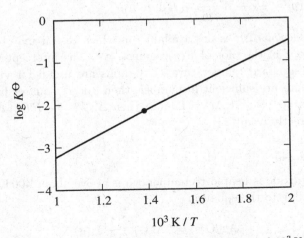

Figure 8.1.   log $K^{\ominus}$ for the ammonia synthesis, as a function of $10^3\,K/T$. The continuous line shows results from data in Chase (1998). The single point was calculated from Eq. (17).

that the pioneering measurements of Larson and Dodge (1923) were in error by about 7%. Considering the experimental difficulties of measuring K_p at high temperatures and high pressures, this is perhaps not too surprising. The result in (19), which is based on the standard enthalpy of formation, and the standard Gibbs energy of formation, as well as experimental measurements of C_p at high temperatures, differs from that in (22) by 2.7%. An uncertainty of 0.3% in ΔH and ΔS would result in an uncertainty of 3% in K.

Since modern compilations of thermodynamic data such as the NIST JANAF tables (Chase 1998) include tables of $\Delta_f G^\ominus(T)$, the reader may perhaps question whether it is worthwhile learning how to do the kind of calculation that we have illustrated by the ammonia synthesis. Indeed it is, not only because of the likelihood of needing data that are not in the tables but because such calculations enhance one's understanding of, and confidence in, thermodynamics.

For reactions for which $\Delta_r G^\ominus$ (298.15 K) is not available from tables of $\Delta_f G^\ominus$ (298.15 K), $K^\ominus(T)$ may be determined at a number of temperatures and the results plotted graphically, using a rearranged form of (12)

(12) $R \ln K^\ominus(y)$

$$- \left[\Delta a \ln \left(\frac{y}{y_1} \right) + \frac{\Delta b}{2}(y - y_1) + \frac{\Delta c}{2} \left(\frac{1}{y^2} - \frac{1}{y_1^2} \right) + \frac{\Delta d}{6} \left(y^2 - y_1^2 \right) \right]$$

$$= F(y) = -\frac{\Delta c'}{y} + \frac{\Delta c'}{y_1} + R \ln K^\ominus(y_1) \tag{23}$$

For some reactions $\Delta c'$ is not available, in which case it may be obtained from the slope of the plot of $F(y)$ against y^{-1}. The intercept then yields ΔG at the standard temperature T_1. Graphs are useful for visualization but if the data are sufficient in number, then a least squares fit may give greater accuracy. But if $\Delta c'$ is known, then $\Delta_r G^\ominus = -RT \ln K^\ominus$ may be obtained directly from (23).

Example 8.4-5

When silver oxide is heated to temperatures above about 400 K it decomposes according to the equation

$$Ag_2O(s) = 2Ag(s) + \frac{1}{2}O_2(g) \tag{24}$$

Table 8.4. Calculation of $\Delta_f G^{\ominus}(298.15\,\mathrm{K})$, from the dissociation pressure of Ag_2O (Benton and Drake 1932). The last column contains values $\Delta_r G^{\ominus}(T_1)$, where T_1 is 298.15 K, calculated from data obtained at the values of $t/^{\circ}C$ given in the first column. The reaction is the decomposition of Ag_2O, so for $\Delta_f G^{\ominus}(298.15\,\mathrm{K})$, reverse the sign.

$t/^{\circ}C$	$y^{-1} = 10^3 \mathrm{K}/T$	$p/10^4\,\mathrm{Pa}$	$\ln K^{\ominus}$	$\dfrac{F(y)}{\mathrm{J\,K^{-1}\,mol^{-1}}}$	$\dfrac{\Delta c' \left(\frac{1}{y} - \frac{1}{y_1} \right)}{\mathrm{J\,K^{-1}\,mol^{-1}}}$	$\dfrac{\Delta_r G^{\ominus}}{\mathrm{J\,mol^{-1}}}$
173.0	2.242	5.625	−0.2877	−0.4479	−36.045	−10 880
178.0	2.217	6.786	−0.1939	0.3878	−36.856	−10 873
183.1	2.192	8.066	−0.1075	1.1630	−36.503	−10 537
188.2	2.168	9.559	−0.0226	2.0318	−36.412	−10 250
191.2	2.154	10.532	0.0608	2.6500	−36.248	−10 178

Values of the dissociation pressure of silver oxide are given in Table 8.4. In this temperature range the heat capacities of reactant and products are:

$$C_p(Ag_2O(s))/\mathrm{J\,K^{-1}\,mol^{-1}} = 59.33 + 40.79 \times 10^{-3}(T/\mathrm{K})$$
$$- 0.46 \times 10^6 (\mathrm{K}/T)^2$$

$$C_p(Ag(s))/\mathrm{J\,K^{-1}\,mol^{-1}} = 24.22 + 2.74 \times 10^{-3}(T/\mathrm{K})$$
$$+ 2.84 \times 10^{-6}(T/\mathrm{K})^2$$

$$C_p(O_2(g))/\mathrm{J\,K^{-1}\,mol^{-1}} = 29.15 + 6.48 \times 10^{-3}(T/\mathrm{K})$$
$$- 0.18 \times 10^6 (\mathrm{K}/T)^2 - 1.02 \times 10^{-6}(T/\mathrm{K})^2$$

$\Delta c\prime = 32.415\,\mathrm{J\,K^{-1}\,mol^{-1}}$ (Knacke *et al.* 1991). Find $\Delta_r H^{\ominus}(298.15\,\mathrm{K})$ and $\Delta_r G^{\ominus}(298.15\,\mathrm{K})$.

From the given heat capacity data

$$\Delta a/\mathrm{J\,K^{-1}\,mol^{-1}} = 3.685, \quad 10^3 \times \Delta b/\mathrm{J\,K^{-1}\,mol^{-1}} = -32.07$$

$$10^{-6} \times \Delta c/\mathrm{J\,K^{-1}\,mol^{-1}} = 0.37, \quad 10^6 \times \Delta d/\mathrm{J\,K^{-1}\,mol^{-1}} = 5.17$$

$$K^{\ominus} = \prod_B a_B{}^{\nu_B} = \prod_B (p_B/p^{\ominus})^{\nu_B} = (p_{O_2}/10^5\,\mathrm{Pa})^{1/2}$$

$R \ln K^{\ominus}$ was calculated from the given pressure data and the values are given in Table 8.4. $F(y)$ was then calculated at the given temperature. Since $\Delta c'$ is known

(12), (16) $\qquad \Delta_r H^{\ominus}(298.15\,\mathrm{K}) = 10^3 \Delta c' = 32\,415\,\mathrm{J\,K^{-1}}$

Answers to Exercises 8.4

Exercise 8.4-1

At $600\,\mathrm{K}$, $\Delta G^{\ominus} = -85.77 + 0.1004 \times 600\,\mathrm{kJ} = -25.530\,\mathrm{kJ}$

The activity of the two condensed phases is unity. Therefore

(4) $\qquad A = -\Delta G^{\ominus} - RT\ln\prod_{B} a_{B}{}^{\nu_{B}} = -\Delta G^{\ominus} - RT\ln\{(p/p^{\ominus})^{-1/2}]$

(a) $\qquad A = 25.530 - 8.3145 \times 600 \times 2.3026$

$$\times\left(-\frac{1}{2}\right)\log(13.33 \times 10^{-5})\,\mathrm{kJ\,mol}^{-1}$$

$$= 25.530 - (5.7435 \times 3.8808) = 3.24\,\mathrm{kJ\,mol}^{-1}$$

(b) $\qquad A = 25.530 - (5.7435 \times 4.8808) = -2.50\,\mathrm{kJ\,mol}^{-1}$

In case (a), $A > 0$ and the reaction is feasible, but in case (b), $A < 0$ and reaction is impossible.

Exercise 8.4-2

Since $K^{\ominus}(T)$ is dimensionless, the SI unit of $K^{\ominus}(T)$ is 1.

Exercise 8.4-3

For a single reaction, the activities of all participants are determined by the equilibrium constant which depends only on temperature. A catalyst alters the rate at which chemical equilibrium is attained, but has no effect on the equilibrium activities. However, if there are several reactions going on simultaneously, a catalyst may alter the product distribution, as in the Fischer–Tropsch synthesis of hydrocarbons by the reduction of CO by H_2.

Exercise 8.4-4

Since $\mathrm{d}T^{-1}/\mathrm{d}T = -1/T^2$, $\mathrm{d}T/\mathrm{d}T^{-1} = -T^2$

$$\frac{\partial}{\partial T^{-1}}\left(\frac{\Delta G^{\ominus}}{T}\right) = \Delta G^{\ominus} - \left(\frac{\Delta S^{\ominus}}{T}\right)(-T^2) = \Delta H^{\ominus} \qquad (13b)$$

which is the Gibbs–Helmholtz equation.

8.5. Effect of Pressure on Reactions Involving Gases

It follows from the definitions of $K^{\ominus}$ and K_f in Table 8.2, that these equilibrium constants are independent of pressure. However, if the sum of the stoichiometric coefficients of *gaseous participants* is not zero

$$\sum_B \nu_B \neq 0 \quad (\text{B} = \text{a gaseous component}) \tag{1}$$

then their concentrations will depend on the total pressure p. If the sum over gaseous participants in (1) is less than zero, then the system responds to an increase in pressure by producing more products, in accordance with Le Chatelier's Principle (formulated in 1888)

$$(\text{Table 8.2}) \qquad K_f = \prod_B \tilde{p}_B{}^{\nu_B} = K_p \prod_B \phi_B{}^{\nu_B} = K^{\ominus} \prod_B (p^{\ominus})^{\nu_B} \tag{2}$$

where, on using Dalton's law

$$K_p = K_y p^{\Sigma \nu_B} \tag{3}$$

From its definition in (8.4.6), $K^{\ominus}$ is independent of p. From (2) and (3), we see that, while K_f is independent of p, K_y is not, when the sum over gaseous participants $\sum_B \nu_B$ is not zero. The statement is often made that K_p is independent of pressure. This is certainly true at low pressures, but at high pressures one may expect some pressure dependence of K_p.

Example 8.5-1

In Example 8.4-2 we calculated the equilibrium constant K_p of the reaction

$$N_2O_4(g) = 2NO_2(g) \tag{8.4.7}$$

at 298.15 K and a total pressure of 0.1 MPa to be 0.1475×10^{-1} MPa and the partial pressure of N_2O_4 to be 0.682×10^{-1} MPa. Assuming ideal behaviour, calculate the partial pressure of N_2O_4 in nitrogen dioxide gas at 298.15 K and pressures of 0.05, 0.2, and 0.4 MPa.

From Example 8.4-2(c), at the standard pressure $p = 10^{-1}$ MPa

$$(8.4.7) \qquad K_p = \frac{(p_{NO_2})^2}{p_{N_2O_4}} = \frac{4\xi^2}{(1-\xi)(1+\xi)} p = 0.1475 \times 10^{-1} \text{ MPa} \tag{1}$$

Table 8.5. Pressure-dependence of the
extent of reaction (8.4.7) and of the partial pressure of N_2O_4 at 298.15 K.

$p/p^\ominus$	ξ/mol	$p_{N_2O_4}/10^{-1}$ MPa
0.5	0.262	0.585
1	0.189	0.682
2	0.134	0.882
4	0.0956	0.909

The variation of ξ and of $p_{N_2O_4}$ with total pressure p calculated from (1) is shown in Table 8.5. Increasing pressure reduces the concentration of NO_2 and favours the formation of N_2O_4, as expected from Le Chatelier's Principle.

Exercise 8.5-1

Calculate ξ and $p_{N_2O_4}$ when 1 mol of an inert gas at $p^\ominus$ and 298.15 K is added to 1 mol of nitrogen dioxide gas at $p^\ominus$ and 298.15 K, at constant pressure.

Answer to Exercise 8.5-1

The total amount of nitrogen dioxide plus inert gas is $(2 + \xi)$ mol and therefore

$$(1) \qquad K_p = \frac{(p_{NO_2})^2}{p_{N_2O_4}} = \frac{[(2\xi/(2+\xi)]^2}{(1-\xi)/(2+\xi)}p$$

$$= \frac{4\xi^2}{(1-\xi)(2+\xi)}p = 0.1475 \times 10^{-1}\,\text{MPa}$$

For $p/p^\ominus = 1$, $\xi = 0.2495$ mol, $p_{N_2O_4} = 0.334 \times 10^{-1}$ MPa.

8.6. The Giaque Function

The method of calculating Gibbs energies $\Delta_r G$, and hence equilibrium constants, which was described in Section 8.4, is straightforward but can be lengthy. It is not practicable to tabulate $\Delta_f G$ since Gibbs energies vary too rapidly with temperature. However, the *Giaque function* Φ (it is sometimes known by the less satisfactory name "free energy function") from which $\Delta_r G$ may be calculated, varies less rapidly with T and has therefore been included in the NIST JANAF tables (Chase 1998). The definition of the Giaque function is

$$\Phi = \frac{G_m^\ominus - H_m^\ominus(T_1)}{T} \quad (T_1 = 298.15\,\text{K}) \tag{1}$$

G_m and H_m are the molar values of the Gibbs energy and enthalpy. T_1 is used for the standard temperature in (1) to save writing and because it might be convenient to use a different reference temperature on some occasions. H_m, and therefore G_m, both contain a term U_0 for the energy zero to which thermodynamic internal energies, and therefore enthalpies and Gibbs energies, are referred. U_0 (see Section 3.11) is the energy of one mole of the substance with its molecules at rest in their ground nuclear, electronic, rotational and vibrational states. It contains the zero-point energy which is inevitably always present even in the ground vibrational state. However, since the same term U_0 appears in both H_m and G_m it vanishes in Φ. For any particular reaction r

$$(1) \qquad \frac{\Delta_r G_m^\ominus}{T} = \Delta_r \Phi + \frac{\Delta_r H_m^\ominus(T_1)}{T} \qquad (T_1 = 298.15\,\text{K}) \tag{2}$$

The LS of (2) is $-R \ln K^\ominus$ for the reaction r. $\Delta_r H_m^\ominus(T_1)$ is readily calculated from tables of standard heats of formation so that (2) yields the reaction Gibbs energy and equilibrium constant at any desired temperature. Table 8.6 shows such a calculation for the ammonia synthesis in the temperature range of practical interest. The values of the Giaque function for ammonia, nitrogen and hydrogen are from Chase (1998) and the evaluation of $\log K^\ominus$ from Eq. (2) is then a matter of simple arithmetic. A plot of $\Delta_r \Phi$ against $10^3 \text{K}/T$ is in Figure 8.2. Interpolation would be possible.

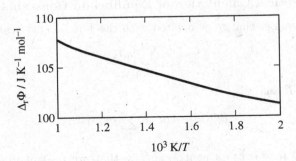

Figure 8.2. Plot of $\Delta \Phi$ against $10^3 \, \text{K}/T$ for the ammonia synthesis.

Table 8.6. Values of the Giaque function Φ (see Eq. (1)) from Chase (1998) for ammonia, nitrogen and hydrogen and of $\Delta_r \Phi$ and $\log K^{\ominus}$ for r the ammonia synthesis.

T/K	Ammonia $-\Phi$ $\mathrm{J\,K^{-1}\,mol^{-1}}$	Hydrogen $-\Phi$ $\mathrm{J\,K^{-1}\,mol^{-1}}$	Nitrogen $-\Phi$ $\mathrm{J\,K^{-1}\,mol^{-1}}$	$\Delta_r \Phi$ $\mathrm{J\,K^{-1}\,mol^{-1}}$	$\log K^{\ominus}$
298.15	192.774	130.680	191.609	99.051	2.867
400	194.209	131.817	192.753	(a)	0.776
500	197.201	133.973	194.917	101.217	−0.501
600	200.302	136.392	197.353	102.963	−1.382
700	203.727	138.822	199.813	104.412	−2.029
800	207.160	141.171	202.209	105.701	−2.524
900	210.543	143.411	204.510	106.710	−2.916
1000	213.849	145.536	206.708	107.809	−3.234

(a) See Exercise 8.6-1

Exercise 8.6-1

Evaluate $\log K^{\ominus}$ for the ammonia synthesis at 400 K (see Table 8.6).

Answer to Exercise 8.6-1

$$\Delta_r \Phi = -194.209 + [(0.5 \times 192.573) + (1.5 \times 131.817)]$$

$$= 99.893 \, \mathrm{J\,K^{-1}\,mol^{-1}}$$

$$\Delta_r \Phi + \frac{\Delta_r H_m^{\ominus}(T_1)}{T} = 99.893 - (45898/400) = -14.857 \, \mathrm{J\,K^{-1}\,mol^{-1}}$$

Divide by $-2.3026\,R = -19.145\,\mathrm{J\,K^{-1}\,mol^{-1}}$, giving 0.776, which is $\log K^{\ominus}$ at 400 K.

8.7. Theoretical Calculation of Equilibrium Constants

The Gibbs energy change associated with the general chemical reaction

$$0 = \sum_B \nu_B B \tag{1}$$

at constant T and p is

$$dG = -A\,d\xi = \sum_B \nu_B \mu_B \, d\xi \tag{8.2.7}$$

The sum (1) may contain neutral species that are present in the reaction mixture but do not participate actively in the chemical reaction, but that in (8.2.7) contains only those substances which are reactants or products. The equilibrium condition is

$$\Delta G = \sum_{\text{B}} \nu_{\text{B}} \mu_{\text{B}} = 0 \tag{2}$$

Equation (2) is generally applicable, but because the activity of a *pure* condensed phase is unity, usually only species present in the gas phase occur in (2). At low pressures (because of the assumption of ideality) the partition function of the ideal gaseous mixture in (2), which contains N_{B} molecules of B and so forth, is

$$(3.8.11) \qquad Q(N_{\text{A}}, N_{\text{B}}, \dots V, T) = \prod_{\text{B}} \frac{[q_{\text{B}}(V, T)]^{N_{\text{B}}}}{N_{\text{B}}!} \tag{3}$$

Since

$$F = -kT \ln Q \tag{3.9.27}$$

the chemical potential of B is given by

$$\mu_{\text{B}} = \left(\frac{\partial F}{\partial N_{\text{B}}} \right)_{T, V, N_{\text{A} \neq \text{B}}} = -kT \ln \left(\frac{q_{\text{B}}(T, V)}{N_{\text{B}}} \right) \tag{4}$$

where each q_{B} is proportional to V

$$(2), (4) \qquad \prod_{\text{B}} \left(\frac{q_{\text{B}}}{N_{\text{B}}} \right)^{\nu_{\text{B}}} = 1 \tag{5}$$

$$(5) \qquad \prod_{\text{B}} \left(\frac{kT q_{\text{B}}}{p^{\ominus} V} \right)^{\nu_{\text{B}}} = \prod_{\text{B}} \left(\frac{N_{\text{B}} kT}{p^{\ominus} V} \right)^{\nu_{\text{B}}} = \prod_{\text{B}} \left(\frac{p_{\text{B}}}{p^{\ominus}} \right)^{\nu_{\text{B}}} = K^{\ominus} \tag{6}$$

which expresses the equilibrium constant $K^{\ominus}$ in terms of the partition function and thus allows a theoretical calculation of $K^{\ominus}$. Alternatively

$$(5) \qquad \prod_{\text{B}} \left(\frac{kT q_{\text{B}}}{V} \right)^{\nu_{\text{B}}} = \prod_{\text{B}} \left(\frac{N_{\text{B}} kT}{V} \right)^{\nu_{\text{B}}} = \prod_{\text{B}} p_{\text{B}}^{\nu_{\text{B}}} = K_p \tag{7}$$

The restriction to ideal gases may be removed by multiplying (7) by the fugacity product, which gives

$$\prod_{\text{B}} \left(\frac{kT \phi_{\text{B}} q_{\text{B}}}{V} \right)^{\nu_{\text{B}}} = K_f \tag{8}$$

but quite likely the experimental information required to evaluate (8) will not be available so that one would need to use some approximation like the Lewis and Randall rule. The theoretical evaluation of (8) is beyond the level aimed at in this book.

Example 8.7.1

Calculate the equilibrium constant $K^\ominus$ for the isotope exchange reaction

$$H_2 + D_2 = 2HD \tag{9}$$

at the temperatures $400\,K$, $500\,K$, $600\,K$ and $700\,K$.

In the Born–Oppenheimer approximation (Atkins 1983) the motion of the electrons is assumed to occur within a fixed nuclear framework. For a diatomic molecule, this means a fixed internuclear separation r. However, the nuclei are vibrating, so the calculation of the electronic energy must be repeated at various values of r to establish the dependence on r of the ground state electronic energy $u(r)$. Since the only consequence of the presence of the two nuclei is their two charges, the electronic energy is the same function of r for all three molecular species H_2, D_2, and HD. Because this energy $u(r)$ is the potential energy for the vibrational motion, all the factors in the partition function quotient cancel except for those involving the atomic masses and the symmetry numbers. The surviving terms are:

translational

$$(3.3.12) \qquad \frac{[q_t(HD)]^2}{q_t(H_2)q_t(D_2)} = \frac{(m_H + m_D)^3}{(4m_H m_D)^{3/2}} \tag{10}$$

rotational

$$(8.10.26) \qquad \frac{[q_r(HD)]^2}{q_r(H_2)q_r(D_2)} = \frac{16(m_H m_D)}{(m_H + m_D)^2} \tag{11}$$

vibrational (v = 0, only)

$$\frac{[q_v(HD)]^2}{q_v(H_2)q_v(D_2)} = \exp\left[-\frac{\Theta_v(HD)}{T}\left(1 - \frac{m_H^{1/2} + m_D^{1/2}}{2^{1/2}(m_H + m_D)^{1/2}}\right)\right] \tag{12}$$

since $q_v = \exp(-\Theta_v/2T) = \exp(-h\nu_v/2kT)$. The vibrational frequency ν_v is proportional to $1/m^{1/2}$, where m is the reduced mass of the diatomic molecule

$$(12) \quad K^{\ominus} = \frac{2(m_H + m_D)}{(m_H m_D)^{1/2}} \exp\left[-\frac{\Theta_v(HD)}{T}\left(1 - \frac{m_H^{1/2} + m_D^{1/2}}{2^{1/2}(m_H + m_D)^{1/2}}\right)\right]$$

$$(13)$$

The equilibrium constant was evaluated from (13) by Hill (1962), using $\nu_v(HD) = 3770\,cm^{-1}$. The calculated values of $K^{\ominus}$ agree with the experimental data (which are scattered).

8.8. Reaction Rate Theory

We will now discuss a specific example of the general chemical reaction (8.1.1). Consider the reaction

$$A + BC = (ABC)^{\ddagger} = AB + C \tag{1}$$

It is reasonable to suppose that before a molecule of A can react with one of BC, the two species must collide and form some kind of transient, metastable complex called the *activated state*, which then either breaks up to yield a molecule of AB plus one of C, or returns to the initial state A + BC. The reason why a discussion of chemical reaction rates appears in a book about thermodynamics is that one method of calculating the rate of reaction (1) is to assume that reactants and activated complex $(ABC)^{\ddagger}$ are in equilibrium, and to emphasize this we shall rewrite (1) in the form

$$A + BC = (ABC)^{\ddagger} \to AB + C \tag{2}$$

This assumption of equilibrium between reactants and the activated state is hard to justify rigorously, but seems reasonable intuitively. It is supported by calculations of potential energy surfaces (e.g. Fowler and Guggenheim 1939, Hill 1962). As an atom of A approaches close to a molecule of BC the potential energy V rises, due to repulsive (overlap) intermolecular forces. It has been verified for the case $H + H_2$ that the angle of approach in which the three H atoms are in a straight line is the most favourable one, requiring the least energy. Let r_{AB}, r_{BC} be the distances between atoms A and B, and between B and C. We want to display the potential energy V as a function of r_{AB} and r_{BC}. This is commonly done by using a 3-D coordinate system in which the z-axis is used for V and the y-axis for r_{AB}. If a potential surface in 3-D is plotted, remember that despite the fact that r_{AB} and r_{BC} are collinear, they will appear to be normal to one another in this kind of plot (e.g. Hill 1962). If the approaching A atom has sufficient kinetic energy, it

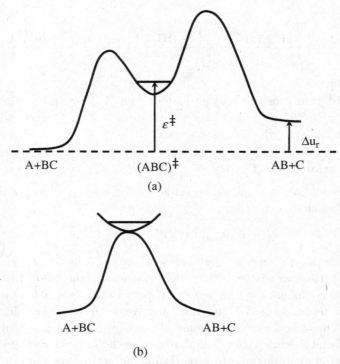

Figure 8.3. (a) Potential energy along a typical reaction path. The activation energy $\varepsilon^{\ddagger}$ includes the zero-point energy in the activated state $(ABC)^{\ddagger}$. Either one or both the potential maxima on either side may be absent. In the latter case, as in (b), it is usual to superimpose the potential energy curve representing the vibrational mode in the activated state which was, and will become, a translational mode. In (b) the energies of the initial and final states happen to be very similar.

will overcome the potential barrier and locate in a position such that the potential energy of the A-B-C complex lies in a shallow minimum. When one of the vibrational modes of this complex acquires sufficient energy, to overcome the bond energy in this loosely-bound complex, it will decompose into either A + BC or AB + C. The potential energy along a typical reaction path is like that shown in Figure 8.3. On forming the reaction complex, three translational degrees of freedom are converted into vibrational modes. Two of these are not of direct interest: our concern is with the vibrational mode along the reaction coordinate (Figure 8.3a and b) which, when sufficiently vigorous, becomes a translational mode that causes the disruption of the activated complex. The concentration $C^{\ddagger}$ (number per volume) of activated complexes $(ABC)^{\ddagger}$ is

(8.7.5)
$$C^{\ddagger} = C_A C_{BC} \left(\frac{q'_{(ABC)^{\ddagger}}}{q'_A q'_{BC}} \right) \exp(-\varepsilon^{\ddagger}/kT) \tag{3}$$

where q' is the partition function per volume and $\varepsilon^{\ddagger}$ is the potential energy difference between the activated complex $(ABC)^{\ddagger}$ and the separated molecules $A + BC$ (see Figure 8.3). Note that $\varepsilon^{\ddagger}$ includes the zero-point energy $\frac{1}{2}h\nu^{\ddagger}$ of the vibration along the reaction path. The rate of reaction (2) is

$$-\frac{dC_A}{dt} = k_2 C_A C_{BC} = \nu^{\ddagger} C^{\ddagger} \tag{4}$$

where $\nu^{\ddagger}$ is the frequency of the crucial vibration, along the reaction path, in the activated state which will become a translational degree of freedom as the complex flies apart. In the first equality k_2 is the *bimolecular rate constant* for the reaction. The equality simply presumes that before A and BC can react, they must collide and that the frequency of these collisions is proportional to the concentrations of the reactants, A and BC. Since the complex is loosely bound, at least with respect to this one bond that is about to break, this is a low frequency vibration ($h\nu^{\ddagger}/kT \ll 1$) and the partition function of this vibrational degree of freedom is

$$q_v^{\ddagger} = [1 - \exp(-h\nu^{\ddagger}/kT)]^{-1} \approx kT/h\nu^{\ddagger} \tag{5}$$

(3), (4), (5)
$$k_2 = \frac{kT}{h} \left(\frac{q''_{(ABC)^{\ddagger}}}{q'_A q'_{BC}} \right) \exp(-\varepsilon^{\ddagger}/kT) \tag{6}$$

The double prime on the partition function for the activated complex indicates that both the volume and one degree of vibrational freedom are to be omitted, since they are already accounted for. The RS of (6) is $(kT/h)K^{\ddagger}$ where $K^{\ddagger}$ is the equilibrium constant for the formation of activated complexes (with the above implications about degrees of freedom)

$$A + BC = (ABC)^{\ddagger}$$

Using $-RT \ln K^{\ddagger} = \Delta G^{\ddagger} = \Delta H^{\ddagger} - T\Delta S^{\ddagger}$,

(6) $\quad k_2 = (kT/h)\exp(-\Delta G^{\ddagger}/kT) = (kT/h)\exp(\Delta S^{\ddagger}/k)\exp(-\Delta H^{\ddagger}/kT)$
$$\tag{7}$$

$\Delta G^{\ddagger}$, $\Delta H^{\ddagger}$ and $\Delta S^{\ddagger}$ are, respectively, the Gibbs energy of activation, the enthalpy of activation and the entropy of activation.

Problems 8

8.1 Derive a relation between the reaction enthalpy at constant pressure q_p and the reaction energy at constant volume, q_V. (**Hint:** Ideal behaviour may be assumed.)

8.2 The standard enthalpy of formation of liquid water at 298.15 K is $-285.830\,\text{kJ}\,\text{mol}^{-1}$ and that of crystalline ferric oxide (Fe_2O_3, cr) is $-824.2\,\text{kJ}\,\text{mol}^{-1}$. Calculate the standard enthalpy of reaction at 298.15 K for the reduction of ferric oxide by hydrogen gas.

8.3 Find $\Delta H^{\ominus}(T)$ for the reaction

$$CO_2(g) + H_2(g) = CO(g) + H_2O(g)$$

	$\Delta H^{\ominus}$ (298.15 K)/kJ mol^{-1}	C_p/J K^{-1} mol^{-1}
CO(g)	-110.52	29.14
H$_2$O(g)	-241.83	33.58
CO$_2$(g)	-393.51	37.13
H$_2$(g)		20.79

8.4 The mole fraction of NH_3 in the equilibrium mixture of N_2, H_2 and NH_3 formed from the stoichiometric mixture of N_2 and H_2 at 723 K and 30.3 MPa is 0.355. Find the extent of reaction and hence calculate K_p and K_f. (**Hint:** The chemical equation for the ammonia synthesis is (8.4.7).) The fugacity coefficients of the *pure gases* at this T and p are:

$$N_2 \quad H_2 \quad NH_3$$
$$1.14 \quad 1.014 \quad 0.908$$

8.5 From the following thermodynamic data, calculate the enthalpy of formation of HI from its gaseous elements at 500 K. The constants in the heat capacity Eq. (8.3.6) are

B	a(B)	b(B)	c(B)
H$_2$(g)	26.88	3.59	0.11
HI(g)	26.36	3.83	0.17
I$_2$(g)	37.25	0.78	-0.05

$$\Delta_f H^\circ (298.15\,\text{K})/\text{kJ}\,\text{mol}^{-1}$$

(a) $\quad\quad \frac{1}{2} H_2(g) + I(s) = HI(g) \quad\quad\quad 26.4 \pm 0.2$

(b) $\quad\quad\quad I(s) = \frac{1}{2} I_2(g) \quad\quad\quad\quad 53.4$

(a) – (b) $\quad \frac{1}{2} H_2(g) + \frac{1}{2} I_2(g) = HI(g) \quad\quad -27.0$

Reaction (b) is the sublimation of iodine.

8.6 It is proposed to reduce the partial pressure of O_2 in commercially available N_2 by passing the nitrogen over copper turnings at 600°C. Is this, in fact, an effective method? For the reaction, $2Cu + \frac{1}{2}O_2 = Cu_2O$

$$\Delta G/\text{kJ}\,\text{mol}^{-1} = -195.4 - 0.0164(T/\text{K})\log(T/\text{K}) + 0.1427\,T/\text{K}$$

8.7 The extent of reaction ξ for the dissociation of $PCl_5(g)$ according to the equation

$$PCl_5(g) = PCl_3(g) + Cl_2(g)$$

at 0.1 MPa pressure is 0.80 at 520 K and 0.92 at 600 K. Find ΔG°, ΔH° and ΔS° at 520 K. What is the effect of the total pressure on this equilibrium?

8.8 A mixture of two moles of NO and one mole of O_2 is heated to 800 K and equilibrium is established at the standard pressure. The partial pressure of O_2 is 7.1×10^4 Pa. Find the standard Gibbs energy change for this reaction at 800 K.

9

Electrolytes

For most of this book so far the systems considered have consisted of neutral particles; electric fields have been absent, except for the derivation of the work done on a system in an electric field. We now consider systems containing ions. The emphasis is on systems consisting of a gas like HCl or an ionic solid like KCl dissolved in an ionizing solvent of high relative permittivity such as water, to form an *electrolyte solution*. The thermodynamics of crystalline solids — *metals, ionic crystals* and *semiconductors* — is deferred until Chapter 11.

9.1. Electrolysis

Solutions of ionic crystals like NaCl or $CuSO_4$ in water conduct an electric current and are therefore presumed to consist mostly of ions dispersed in the solvent. Ionic crystals have high cohesive energies, as illustrated by their high fusion temperatures. However, when dissolved in a solvent of high permittivity the attractive forces between oppositely charged ions are so weakened that the crystal dissolves to form an ionic solution (see the solution to Problem 9.11). Each ion in a solution will tend to have more near neighbours of the opposite charge but the maintenance of the regular crystalline structure is opposed by the thermal motion of the ions. A more detailed description of the structure of an ionic solution will be given in Section 9.7.

If an electric potential difference is applied between two platinum electrodes immersed in an aqueous solution of $CuSO_4$, metallic Cu is deposited on the negative electrode (the cathode). This is an example of *electrolysis* or the decomposition of a substance by the passage of an electric current. The laws of electrolysis were formulated by Michael Faraday in 1834. Consider the general chemical reaction

$$0 = \sum_B \nu_B B \tag{1}$$

in which one of species B is the electron e^- with charge $-e$ where e is the elementary charge, that is, the charge on the proton. An example of (1) is

$$Ag^+ + e^- = Ag \tag{2}$$

in which $\nu_{Ag} = +1$, $\nu_{e^-} = -1$, and $\nu_{Ag^+} = -1$. The stoichiometry of reaction (1) informs us that

$$\Delta\xi = \frac{\Delta n_B}{\nu_B} = \frac{\Delta n_{e^-}}{\nu_{e^-}} \tag{3}$$

where Δn_{e^-} is the amount of electrons involved in the process. The change in the number of electrons due to this reaction is

$$\Delta N_{e^-} = N_A \Delta n_{e^-} = \frac{\nu_{e^-}}{|\nu_{e^-}|}\left(\frac{Q}{e}\right) \tag{4}$$

The first factor on the RS of (4) is the sign of ΔN_{e^-} and the second factor is its magnitude. Q is the charge passed through the solution in Coulombs and e is the elementary charge (the charge[a] on the proton, 1.6022×10^{-19} C[a])

$$(3),\ (4) \qquad\qquad \Delta n_B = \frac{\nu_B}{|\nu_{e^-}|}\left(\frac{Q}{F}\right) \tag{5}$$

where F is the Faraday constant, $96\,485$ C mol^{-1}. The change in the mass of B is

$$(5) \qquad\qquad \Delta m_B = M_B \frac{\nu_B}{|\nu_{e^-}|}\left(\frac{Q}{F}\right) \tag{6}$$

Example 9.1-1

A current I of 200.0 mA is passed through a solution of $CuSO_4$ for 600.0 s. Calculate the mass of Cu deposited on the negative Pt electrode.

$$(6) \qquad \Delta m_{Cu} = 63.546 \text{ g mol}^{-1} \times \frac{1}{2} \times \frac{0.200 \text{ A} \times 600 \text{ s}}{96\,485 \text{ C mol}^{-1}} = 0.03952 \text{ g}$$

---

[a] As noted earlier, the values of physical constants are given in the text only to five significant figures, since this is sufficient for ordinary calculations like the solution of problems and exercises in physical chemistry. Should greater accuracy be required, please refer to Cohen *et al.* (2007).

Exercise 9.1-1

Electrolysis of brine results in the formation of sodium chlorate. Write down the chemical equation that describes this reaction. Hence derive an expression for the mass of ClO_3^- formed.

Answer to Exercise 9.1-1

$$Cl^- + 3H_2O = ClO_3^- + 6H^+ + 6e^-$$

$$\nu_{Cl^-} = -1, \quad \nu_{H_2O} = -3, \quad \nu_{ClO_3^-} = +1, \quad \nu_{H^+} = +6, \quad \nu_{e^-} = +6.$$

Therefore,

$$\Delta m_{ClO_3^-} = M_{ClO_3^-} \left(\frac{Q}{6F} \right) \quad .$$

9.2. Conductivity of Electrolyte Solutions

The conductivity κ of an electrolyte solution of resistivity ρ is

$$\kappa = \frac{1}{\rho} = \frac{l}{RA} = \frac{G}{R} \tag{1}$$

l is the length of the solution between the two electrodes, which are rigid platinum discs of area A, and R is the measured resistance of the solution. $G = l/A$ is the *cell constant*. For any conductivity cell, the geometric factor G can be obtained by measuring the conductivity of certain standard solutions of known conductivity. (For details of conductivity measurements see Robinson and Stokes (1968).) The units of κ are Siemens per metre, $S\,m^{-1}$. κ depends on the amount of the electrolyte (it is often roughly proportional to the concentration c) so that a more useful quantity is the *molar conductivity*

$$\Lambda = \frac{\kappa}{c} \tag{2}$$

Exercise 9.2-1

What are the SI units of Λ?

Consider plots of Λ against $c^{1/2}$, as in Figure 9.1. Here Λ is the molar conductivity of NaCl in water in $S\,cm^2\,mol^{-1}$ at $298.15\,K$ and c is the concentration of the solution in $mol\,dm^{-3}$. For *strong electrolytes* this plot is linear at low concentrations ($0 < c < 0.001\,mol\,dm^{-3}$). Strong electrolytes

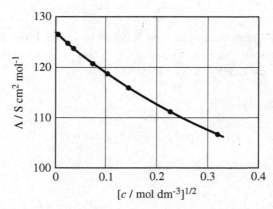

Figure 9.1. Molar conductivity of solutions of NaCl in water at 25°C plotted against $c^{1/2}$, where c is the concentration in $\mathrm{mol\,dm^{-3}}$.

are completely dissociated in solvents of high permittivity (such as water, $\varepsilon_r = 78.42$ at 298.15 K) and the decrease of Λ with increasing concentration is due primarily to ion–ion interactions. Although the conductivity of weak electrolytes is also high at very low concentrations, it falls rapidly with increasing c because weak electrolytes are incompletely dissociated. This makes it hard to determine Λ^0 (or Λ^∞) the value of Λ at $c = 0$ (referred to as infinite dilution) by extrapolation. The dissociation of the strong electrolyte $\mathrm{B}_{\nu_+}\mathrm{C}_{\nu_-}$ is represented by the equation

$$\mathrm{B}_{\nu_+}\mathrm{C}_{\nu_-} = \nu_+\mathrm{B}^{z+} + \nu_-\mathrm{C}^{z-} \tag{3}$$

Charge conservation requires that

$$\nu_+ z_+ + \nu_- z_- = 0 \tag{4}$$

Since strong electrolytes are completely dissociated, at least in dilute solution, the ions migrate independently, so that

$$\Lambda^0 = \nu_+\lambda_+^0 + \nu_-\lambda_-^0 \tag{5}$$

Λ^0 is molar conductivity extrapolated to $c = 0$. Equation (5) is a statement of Kohlrausch's law of the independent migration of ions.

Example 9.2-1

The molar conductivities at infinite dilution Λ^0 of HCl, NaAc (sodium acetate) and NaCl are 425.0, 91.0 and 128.1 $\mathrm{S\,cm^2\,mol^{-1}}$ at 298.15 K. Find

Λ^0 for acetic acid

(5) $\Lambda^0(\text{HAc}) = \Lambda^0(\text{HCl}) + \Lambda^0(\text{NaAc}) - \Lambda^0(\text{NaCl})$

$$= 425.0 + 91.0 - 128.1 = 387.9\,\text{S}\,\text{cm}^2\,\text{mol}^{-1}$$

The weak electrolyte, acetic acid (HAc) is only partially dissociated in aqueous solution

$$\text{HAc} + \text{H}_2\text{O} = \text{H}_3\text{O}^+ + \text{Ac}^- \tag{6}$$

If the extent of dissociation is $\xi < 1$, the equilibrium constant K_c is

(6) $$K_c = \frac{(\xi c)(\xi c)}{(1 - \xi)c} = \frac{\xi^2 c}{(1 - \xi)} \tag{7}$$

Solving (7) for the extent of reaction yields

$$\xi = \frac{K_c}{2c}\left[\left(1 + \frac{4c}{K_c}\right)^{1/2} - 1\right] \tag{8}$$

Only the dissociated ions contribute to the conductivity. At infinite dilution, no further dissociation is possible and so $\Lambda = \Lambda^0$, but at higher concentrations

$$\Lambda = \xi\Lambda^0 \tag{9}$$

(7) $$\frac{1}{\Lambda} = \frac{1}{\Lambda^0} + \frac{c\Lambda}{K_c(\Lambda^0)^2} \tag{10}$$

which is known as the Ostwald dilution law. However, we already have a value for Λ^0 ($387.9\,\text{S}\,\text{cm}^{-2}\,\text{mol}^{-1}$) from Kohlrausch's law. This will not be very accurate because it involved the sum of three different Λ^0's, each determined by an extrapolation procedure, but since ideal behaviour was assumed in (7), it will do for a preliminary estimate of K_c

(7), (9) $$K_c = \frac{\xi^2 c}{(1 - \xi)} = \frac{(\Lambda/\Lambda^0)^2 c}{[1 - (\Lambda/\Lambda^0)]} \tag{11}$$

Exercise 9.2-2

The table below shows the results for the conductivity of dilute solutions of acetic acid, as determined by MacInnes and Shedlovsky. Find the dissociation constant K_c of acetic acid. $c^{\ominus}$ is the standard concentration, $1\,\text{mol}\,\text{dm}^{-3}$. The table shows the results of a calculation using Eq. (11), with the value of Λ^0 from Exercise 9.2-1.

$\Lambda/\mathrm{S\,m^2\,mol^{-1}}$	$c/c^{\diamond}$	Λ/Λ^0	$1-(\Lambda/\Lambda^0)$	$10^5 K_c/\mathrm{mol\,dm^{-3}}$
210.38	0.00002801	0.542356	0.457644	1.800
127.75	0.0001135	0.329337	0.670663	1.836
96.493	0.0002184	0.248757	0.751243	1.799
48.146	0.0010283	0.124120	0.875880	1.809
32.217	0.002414	0.083055	0.916945	1.816
20.962	0.005912	0.054040	0.945960	1.886

The mean value for K_c from this calculation is $1.82 \times 10^{-5}\,\mathrm{mol\,dm^{-3}}$. However, the accuracy of this estimate will be no better than about $\pm 3\%$ because of the neglect of ion interactions in the equilibrium constant (we should be calculating K_a rather than K_c) and the rough value used for Λ^0. An improved calculation will be described in Section 9.7.

Answers to Exercises 9.2

Exercise 9.2-1

The SI units of Λ are

$$\frac{\mathrm{S\,m^{-1}}}{\mathrm{mol\,m^{-3}}} = \mathrm{S\,m^2\,mol^{-1}}$$

Exercise 9.2-2

From the slope and intercept in the plot of data for acetic acid according to Eq. (9.2.10) we find

$$\Lambda^0 = 395.7\,\mathrm{S\,m^2\,mol^{-1}}, \quad K_c = 1.753 \times 10^{-5}$$

9.3. Ionic Mobility and Transport Number

The electrical resistance of a solution is

$$R = \frac{\Delta\phi}{I} \tag{1}$$

where $\Delta\phi$ is the applied potential difference (V) and I the electric current (A)

(1), (9.2.1)
$$\kappa = \frac{I}{A}\frac{l}{\Delta\varphi} = \frac{j}{E} \tag{2}$$

where j is the current density ($\mathrm{A\,m^{-2}}$) and E is the electric field ($\mathrm{V\,m^{-1}}$). The *flux* of species k, that is the number of particles crossing a plane of area $1\,\mathrm{m}^2$ in one second, is

$$J_k = C_k v_k \tag{3}$$

In (3), C_k is the number density (m^{-3}) of particles of species k and v_k is their speed $(\mathrm{m\,s}^{-1})$. The contribution to the current density from species k is therefore

$$j_k = C_k v_k Q_k = c_k v_k z_k \mathrm{F} \tag{4}$$

Exercise 9.3-1

Confirm units in Eq. (4).

The total current density is

(4) $$j = \mathrm{F} \sum_k z_k c_k \tag{5}$$

In solution, c mole of an electrolyte produce $\xi \nu_+ c$ mole of cations and $\xi \nu_- c$ mole of anions. The conductivity κ is therefore

(2), (5) $$\kappa = \frac{j}{E} = \frac{\xi \mathrm{F} c}{E}(\nu_+ z_+ v_+ + \nu_- z_- v_-) \tag{6}$$

Warning! Don't confuse ν_+, ν_-, the stoichiometric numbers of cations and anions (symbol, italic nu) produced by one mole of electrolyte in solution, with v_+, v_- which are the speeds attained by the cations and anions in the electric field of magnitude E. In (6)

(6) $$v_k/|E| = u_k \tag{7}$$

where u_k is the *mobility* of species k. Therefore

(6), (7) $$\Lambda = \frac{\kappa}{c} = \xi \mathrm{F}(\nu_+ |z_+| u_+ + \nu_- |z_-| u_-) \tag{8}$$

For strong electrolytes, $\xi = 1$ but for weak electrolytes, ξ is normally less than one and only reaches one at infinite dilution. In the limit of $c \to 0$, Λ becomes

(8) $$\Lambda^0 = \mathrm{F}(\nu_+ |z_+| u_+^0 + \nu_- |z_-| u_-^0) \tag{9}$$

Because the ions migrate independently, Λ^0 maybe separated into the two terms

(9) $$\lambda_+^0 = F(\nu_+|z_+|u_+^0)$$ (10)

(9) $$\lambda_-^0 = F(\nu_-|z_-|u_-^0)$$ (11)

The *transport number* t_k of species k is defined as the fraction of the total current that is carried by that species. Therefore

(5), (7) $$t_k = \frac{j_k}{j} = \frac{|z_k|c_k u_k}{\sum_k |z_k|c_k u_k}$$ (12)

It follows from the definition of t_k that

$$\sum_k t_k = 1$$ (13)

For a single strong electrolyte

$$c_+ = \nu_+ c, \quad c_- = \nu_- c, \quad \nu_+|z_+| = \nu_-|z_-|$$ (14)

(12), (14) $$t_+ = \frac{|z_+|c\nu_+ u_+}{|z_+|c\nu_+ u_+ + |z_-|c\nu_- u_-} = \frac{u_+}{u_+ + u_-}$$ (15)

(15) $$t_- = \frac{|z_-|c\nu_- u_-}{|z_+|c\nu_+ u_+ + |z_-|c\nu_- u_-} = \frac{u_-}{u_+ + u_-}$$ (16)

(10), (11), (15) $$t_+ = \frac{\lambda_+^0/|z_+|}{(\lambda_+^0/|z_+|) + (\lambda_-^0/|z_-|)} = \frac{\nu_+\lambda_+^0}{\nu_+\lambda_+^0 + \nu_-\lambda_-^0}$$ (17)

(17) $$t_+ = \frac{\nu_+\lambda_+^0}{\Lambda^0}, \quad t_- = \frac{\nu_-\lambda_-^0}{\Lambda^0}$$ (18)

By measuring $t_+(c)$ and $\Lambda(c)$, we can obtain Λ_+^0, Λ_-^0 from (18) and u_+^0, u_-^0 from (15) and (16). Although the measurement of $t_+(c)$ is sufficient, measurement of both $t_+(c)$ and $t_-(c)$ provides a check on experimental accuracy. Some ionic mobilities are given in Table 9.1 and cation transport numbers in Table 9.2. It is important to realize that these data are for the *hydrated* ion in solution. For example, although the crystal radius of Li^+ is the smallest of the alkali metal cations it has the lowest mobility in aqueous solution. This is because of the tightly bound water molecules attached to the Li^+ cation by the strong ion–dipole interaction, which impede the motion of the hydrated Li^+ in the electric field.

Table 9.1. Ionic mobilities in aqueous solution at 298.15 K.

Cation	$u_+/10^{-8}\,\mathrm{m^2\,s^{-1}\,V^{-1}}$	Anion	$u_-/10^{-8}\,\mathrm{m^2\,s^{-1}\,V^{-1}}$
H^+	36.30	OH^-	20.52
K^+	7.62	SO_4^{2-}	8.27
Ba^{2+}	6.59	Cl^-	7.91
Na^+	5.19	NO_3^-	7.40
Li^+	4.01	HCO_3^-	4.61

Table 9.2. Cation transport numbers in aqueous solution at a concentration of $0.01\,\mathrm{mol\,dm^{-3}}$.

Solute	HCl	LiCl	NaCl	KCl	AgNO$_3$	KNO$_3$	$\frac{1}{2}$BaCl$_2$	$\frac{1}{3}$LaCl$_3$
t_+	0.8251	0.3289	0.3918	0.4902	0.4648	0.5084	0.4400	0.4625

9.3.1. *Experimental measurement of transport numbers*

In the *Hittorf* method, one uses an electrolysis cell divided into three compartments. After electrolysis is completed, the contents of each cell are analyzed. At the negative electrode, the reaction is

$$(9.2.3) \qquad B^{z+} + z_+e^- = B \qquad (19)$$

so that $\nu_{B^{z+}} = -1, |\nu_{e^-}| = |z_+|$. As a result of electrolysis, the change in the amount of B^{z+} in the cathode compartment is

$$(9.1.5) \qquad \Delta n_{B^{z+}} = \frac{-1}{|z_+|}\left(\frac{Q}{F}\right) \qquad (20)$$

The amount of B^{z+} gained in the cathode compartment by migration during electrolysis is

$$\Delta n_{B^{z+}} = \frac{t_+}{|z_+|}\left(\frac{Q}{F}\right) \qquad (21)$$

Therefore, the net change in the amount of B^{z+} in the cathode compartment is

$$(20),\ (21) \qquad \Delta n_{B^{z+}} = -\frac{t_-}{|z_+|}\left(\frac{Q}{F}\right) \qquad (22)$$

Consequently, the net change in the amount of electrolyte $B_{\nu_+} C_{\nu_-}$ in the cathode compartment during a Hittorf experiment is

(22) .
$$\Delta n^c = -\frac{t_-}{\nu_+ |z_+|} \left(\frac{Q}{F}\right) \tag{23}$$

Hence, by measuring Q and Δn^c we obtain t_-.

At a reactive anode the reaction is

$$B = B^{z+} + z_+ e^- \tag{24}$$

Therefore the change in the amount of the cation due to *reaction* is

$$\Delta n_{B^{z+}} = \frac{1}{|z_+|} \left(\frac{Q}{F}\right) \tag{25}$$

At the same time

$$\Delta n_{B^{z+}} = \frac{-t_+}{|z_+|} \left(\frac{Q}{F}\right) \tag{26}$$

mole of the cation are lost due to *migration*. Therefore the net change in the anode compartment is

(25), (26)
$$\Delta n^a = \frac{t_-}{\nu_+ |z_+|} \left(\frac{Q}{F}\right) \tag{27}$$

If the anode is an inert electrode, the reaction at the anode is

(9.2.3)
$$A^{z-} = A \left(\text{or} \frac{1}{2} A_2\right) + z_- e^- \tag{28}$$

so that

$$\Delta n_{A^{z-}} = \frac{-1}{|z_-|} \left(\frac{Q}{F}\right) \tag{29}$$

by reaction, and

$$\Delta n_{A^{z-}} = \frac{t_-}{|z_-|} \left(\frac{Q}{F}\right) \tag{30}$$

by migration into the anode compartment. The change in the amount of electrolyte $B_{\nu_+} A_{\nu_-}$ in the anode compartment is, therefore,

(29), (30)
$$\Delta n^a = -\frac{t_+}{\nu_- |z_-|} \left(\frac{Q}{F}\right) \tag{31}$$

In either case, only one compartment need be analyzed, but analysis of both provides a check on the accuracy of the experiment.

In the *moving boundary method* two solutions containing a salt with the same anion, but two different cations, are placed in a vertical conductivity cell of cross-sectional area A so that a horizontal boundary is formed between the two solutions. A constant current is passed through the solution. Because of the applied field, the two cations move upwards. If the lower solution contains $B'_{\nu_+} A_{\nu_-}$, where the mobility of B'_{ν_+} is less than that of B_{ν_+}, i.e.

$$u(B'_{\nu_+}) < u(B_{\nu_+}) \tag{32}$$

the solute in the lower solution will tend to lag behind that in the upper solution as both migrate upwards. But this would make the solution more dilute and therefore of higher resistance and the electric field would increase, thus increasing the velocity of B_{ν_+} (see Eqs. (9.3.1) and (9.3.2)). The net result is that the boundary remains sharp and its progress up the tube can be measured with a suitable telescopic device. Suppose that the boundary moves a distance x as a charge Q is passed through the electrodes. The number of moles of B_{ν_+} transported is

$$n(B_{\nu_+}) = \frac{t_+}{|z_-|}\left(\frac{Q}{F}\right) \tag{33}$$

The volume of solution transported is

$$V = \frac{t_+}{\nu_+ c|z_-|}\left(\frac{Q}{F}\right) = xA \tag{34}$$

$$\text{(34)} \qquad t_+ = \frac{FxA\nu_+ c|z_-|}{Q} \tag{35}$$

A third method of obtaining t_+, from the measurement of the cell potential of a galvanic cell with transport, will be described in Section 9.9.

Answer to Exercise 9.3-1

Units of $C_k v_k Q_k$ are: $\text{m}^{-3}\,\text{m s}^{-1}\,\text{C} = \text{A m}^{-2} = $ units of current density, j_k.
Units of $c_k v_k z_k\,\text{F}$ are: $\text{mol m}^{-3}\,\text{m s}^{-1}\,\text{C mol}^{-1} = \text{A m}^{-2}$.

9.4. Electrochemical Potential

The work done on adding a particle of B with charge $z_B e$, at constant T, p, and the amounts of all other components, to the interior of a phase α in which the electrostatic potential is ϕ^α is $z_B e \phi^\alpha$ and for N_A such particles it is $z_B F \phi^\alpha$. This is therefore the change in the molar Gibbs energy of phase α. The *electrochemical potential* $\tilde{\mu}_B^\alpha$ of B in phase α is defined by

$$\tilde{\mu}_B^\alpha = \left(\frac{\partial G^\alpha}{\partial n_B} \right)_{T,p,n_{C \neq B}} = \mu_B^\alpha + z_B F \phi^\alpha \tag{1}$$

where the Gibbs function $G^\alpha(T, p, n_B, n_C, \ldots, \phi)$ includes the electrical work done as well as the usual chemical effects. The first term on the RS of (1) is the ordinary chemical potential and the second term is the electrochemical contribution arising from the fact that the particles of B are charged. The condition for equilibrium between two phases α and β that both contain the charged species B, C, ... is

$$\tilde{\mu}_B^\alpha = \tilde{\mu}_B^\beta \quad \text{(all B, C \ldots)} \tag{2}$$

In a metal, the electrochemical potential is the electronic energy of the highest occupied level, which is the Fermi level E_F (see Section 10.5). Why have we specified the *interior of the phase* in the first sentence of Section 9.4? A little consideration will convince the reader that when a particle of B is added to a phase α, the particle of B is always added to the interior of α. For if this particle is added to a solid or a liquid and it is placed on top of another particle of B, that particle is now no longer in the surface layer but becomes an interior particle. (It may be that in order to satisfy constraints imposed by crystal structure, the B atom is actually on top of an A atom, but then B can only be added to α by adding a whole unit cell, so that a surface atom of B is certainly covered in this process.) The argument is equally true if the surface consists of several atomic layers. Similarly, it is true for a gas confined in a container of fixed area and volume. The point is not trivial, because the potential χ at the surface of α will be different from that, ϕ, in the interior of α. ϕ is therefore called the *inner potential* and χ is the *surface potential*.

9.5. Electrode Potential

Consider a system S consisting of a piece of metal M (a good electronic conductor) that is attached to a copper wire and then immersed in an electrolyte solution. If an equilibrium of the form

$$0 = \nu_O O + \nu_e e^-(M) + \nu_R R \tag{1}$$

is set up rapidly, then S is called a *reversible reduction-oxidation* (or *redox*) *couple*. In (1), O denotes the oxidized form and R the reduced form of some reactant. *Oxidation* is a chemical process involving the loss of electrons by the substance oxidized. The reverse process is *reduction*. For example, the reduction of iron(III) ions (Fe^{3+}) in an aqueous solution containing a platinum electrode, to iron(II) ions (Fe^{2+}) is described by the equation

$$Fe^{3+}(aq) + e^-(Pt) = Fe^{2+}(aq) \tag{2}$$

Charge compensation in (1) is expressed by

$$\sum_B \nu_B z_B = 0 \tag{3}$$

For example

(3), (1)
$$\nu_O z_O + \nu_e z_e + \nu_R z_R = 0 \tag{4}$$

where z_e is -1.

Exercise 9.5-1

Apply Eq. (3) to the redox couple (2).

The equilibrium condition for (1) is

$$0 = \Delta G$$
$$= \nu_R \tilde{\mu}_R + \nu_O \tilde{\mu}_O + \nu_e \tilde{\mu}_e$$
$$= \nu_R(\mu_R + z_R F \phi^S) + \nu_O(\mu_O + z_O F \phi^S) + \nu_e(\mu_e + z_e F \phi^M)$$
(4) $$= \nu_e z_e F(\phi^M - \phi^S) + (\nu_R \mu_R + \nu_O \mu_O + \nu_e \mu_e) \tag{5}$$

Since $z_e = -1$ and ν_e is a negative number, $\nu_e z_e = |\nu_e|$ which, to save writing, will henceforth be replaced by n, the number of electrons involved in the redox process. For measurement purposes, M is attached to a copper lead. The *electrode potential E* is the potential difference between the Cu lead and the solution

$$E = \phi^{Cu} - \phi^S \tag{6}$$

The condition for equilibrium across the metal junction is

$$\tilde{\mu}_e^{\text{Cu}} = \tilde{\mu}_e^{\text{M}} \tag{7}$$

(7) $$\mu_e^{\text{Cu}} - \mu_e^{\text{M}} = \text{F}(\phi^{\text{Cu}} - \phi^{\text{M}})$$

(6) $$n\text{F}E = n\text{F}(\phi^{\text{Cu}} - \phi^{\text{S}})$$

$$= n\text{F}(\phi^{\text{Cu}} - \phi^{\text{M}} + \phi^{\text{M}} - \phi^{\text{S}})$$

$$= n\text{F}(\phi^{\text{Cu}} - \phi^{\text{M}}) - (\nu_{\text{R}}\mu_{\text{R}} + \nu_{\text{O}}\mu_{\text{O}} + \nu_e\mu_e)$$

$$= n\text{F}(\phi^{\text{Cu}} - \phi^{\text{M}}) - (\nu_{\text{R}}\mu_{\text{R}}^{\ominus} + \nu_{\text{O}}\mu_{\text{O}}^{\ominus} + \nu_e\mu_e)$$

$$+ RT\ln(a_{\text{R}}^{\nu_{\text{R}}} a_{\text{O}}^{\nu_{\text{O}}}) \tag{8}$$

(8) $$E = E^{\ominus} - \frac{RT}{n\text{F}} \ln\left(a_{\text{R}}^{\nu_{\text{R}}} a_{\text{O}}^{\nu_{\text{O}}}\right) \tag{9}$$

(9) $$E = E^{\ominus} - \frac{(\ln 10)RT}{n\text{F}} \log\left(\frac{a_{\text{R}}^{\nu_{\text{R}}}}{a_{\text{O}}^{|\nu_{\text{O}}|}}\right) \tag{10}$$

Equations (9) and (10) are different forms of the *Nernst equation*. Note that the *standard potential* $E^{\ominus}$ includes the contact potential. In (10)

$$\frac{2.3026RT}{n\text{F}} = \frac{2.3026 \times 8.3145 \text{ J K}^{-1}\text{mol}^{-1} \times 298.15 \text{ K}}{n \times 96485\text{C mol}^{-1}} = \frac{0.05916 \text{ V}}{n} \tag{11}$$

Some commonly used electrodes are:

(a) *The hydrogen electrode*

Exercise 9.5-2

When hydrogen gas is bubbled over a Pt electrode immersed in a solution of HCl, the equilibrium

$$\text{H}^+(\text{aq}) + \text{e}^-(\text{Pt}) \rightleftharpoons \frac{1}{2}\text{H}_2(\text{g}) \tag{12}$$

is established. Identify the oxidized and reduced species in reaction (12)

(10), (12) $$E = E^{\ominus} - \frac{RT}{\text{F}} \ln\left(\frac{(\tilde{p}_{\text{H}_2}/\tilde{p}^{\ominus})^{1/2}}{a_{\text{H}^+}}\right) \tag{13}$$

Exercise 9.5-3

Why does a_{e^-} not occur in Eq. (1)?

(b) *Metal/metal ion electrodes*

For example, consider a zinc rod immersed in a solution containing Zn^{2+} ions. The chemical equilibrium is described by

$$Zn^{2+}(aq) + 2e^-(M) \rightleftharpoons Zn(M) \qquad (14)$$

(14), (10)
$$E = E^{\ominus} - \frac{RT}{2F} \ln\left(\frac{1}{a_{Zn^{2+}}}\right) \qquad (15)$$

(c) *Metal/insoluble salt electrodes*

An example of type (c) is a silver wire coated with a layer of AgCl by electrolysis and immersed in a solution containing Cl^- ions

$$AgCl(cr) + e^-(M) \rightleftharpoons Ag(M) + Cl^-(aq) \qquad (16)$$

The AgCl crystal is ionic and made up of Ag^+ and Cl^-. When an Ag^+ ion is reduced to Ag by receiving an electron from the metal, the Ag atom formed is added to the metal and a Cl^- is added to the solution

(10), (16)
$$E = E^{\ominus} - \frac{RT}{F} \ln(a_{Cl^-}) \qquad (17)$$

Answers to Exercises 9.5

Exercise 9.5-1

$$(-1)(3) + (-1)(-1) + (1)(2) = 0$$

Exercise 9.5-2

Molecules of H_2 are adsorbed dissociatively on the Pt and the H atoms lose an electron to the metal to give H^+ ions in solution. Thus, in this redox couple, $H_2(g)$ is oxidized or $H^+(aq)$ is reduced until equilibrium is established.

Exercise 9.5-3

In reactions involving metals, the electron states involved are at, or infinitesimally close to, the Fermi level, and so $a_{e^-} = 1$.

9.6. The Cell Potential of an Electrochemical Cell

The potential difference between the electrode and the electrolyte cannot be measured precisely.[b] We may only measure precisely the potential difference between two phases of the same composition which, in the case of electrochemical cells, are usually copper leads attached to the two electrodes. This potential difference is the *cell potential*. The cell potential was formerly called the electromotive force (emf). However, the cell potential is a potential difference, not a force, and so use of the former name for this quantity is now discouraged by IUPAC.

Since absolute values of the electrode potential E may not be determined precisely, values of E are referred to that of the *standard hydrogen electrode*, that is a hydrogen electrode with the fugacity of the H_2 gas equal to the standard pressure (0.1 MPa) and H^+ at unit activity. This is called the *standard electrode potential $E^\ominus$*. When all substances involved in the reaction in the other half-cell are also at unit activity, the measured cell potential is the *standard cell potential E*. Standard electrode potentials are therefore based on the arbitrarily chosen zero

$$E^\ominus(H^+/H_2) = 0\,\mathrm{V}(\text{all } T) \tag{1}$$

The standard electrode potentials (SEPs) are listed (e.g. Bard *et al.* 1985) as *reduction potentials*, for example

$$Ag^+ + e^- \to Ag, \quad E^\ominus = +0.7991\,\mathrm{V} \tag{2}$$

Example 9.6-1

The following cell was used as an example in Section 1.4

$$Cu(s)|Pt(s)|H_2(g)|HCl(aq)|AgCl(s)|Ag(s)|Cu'(s) \tag{3}$$

The hydrogen gas at constant pressure is bubbling over the platinum electrode, maintaining a saturated layer of adsorbed hydrogen.

According to the 1953 *Stockholm convention* (Cohen *et al.* 2007, p. 73) in a cell diagram like (3) the *positive* electrode (that is, the one which must be connected to the positive terminal of a potentiometer if a balance is to

---

[b]This statement is true because we require an accuracy of 1 mV or better. The potential difference between an electrode and a solution can be measured *approximately* with an accuracy of about ± 50 mV using a method devised by Gomer and Tryson (1977).

be achieved) is on the *right*. The *cell potential* E is then given, in sign and magnitude, by

$$E = \phi^R - \phi^L = \phi^{Cu'} - \phi^{Cu} \tag{4}$$

The *cell reaction* is

RS $$AgCl(s) + e^-(Ag) \rightleftharpoons Ag(Ag) + Cl^-(aq) \tag{5}$$

LS $$\frac{1}{2}H_2(g) \rightleftharpoons H^+(aq) + e^-(Ag) \tag{6}$$

Since the electrode on the RS is positive, electrons are being consumed there (reduction). The LS electrode is negative, so that electrons are being produced there (oxidation). The balanced cell reaction is (5) + (6)

$$AgCl(cr) + \frac{1}{2}H_2(g, \tilde{p} = p^\circ) \rightleftharpoons Ag(s) + H^+(aq) + Cl^-(aq) \tag{7}$$

for which

$$\Delta G = \mu_{Ag} + \mu_{H^+} + \mu_{Cl^-} - \frac{1}{2}\mu_{H_2} - \mu_{AgCl} \tag{8}$$

$$= -FE = w' \tag{9}$$

the electrical work done on the cell, reversibly at constant T and p. Notice that for this cell (see (5) and (6)) $n = 1$ and that in general (9) is to be replaced by

$$\Delta G = -nFE \tag{10}$$

where n is the magnitude of the stoichiometric number of electrons appearing in each half-reaction like (5) and (6).

Substituting for the chemical potentials in (8) gives

(9) $$-FE = -FE^\circ + RT\ln(a_{H^+}a_{Cl^-}) \tag{11}$$

since the activity of pure solid phases is unity and the hydrogen gas is also at unit activity. The *mean ionic activity* $a_\pm$ for HCl is given by

$$(a_\pm)^2 = a_+ a_- \tag{12}$$

(11), (12) $$E = E^\circ - \frac{2RT}{F}\ln a_\pm \tag{13}$$

The above cell is an example of a cell *without transport* since both electrodes are in contact with the same solution. Equation (13) could have been derived by applying the Nernst equation (9.2.18) to each electrode. The cell potential is

$$E = E^{\mathrm{R}} - E^{\mathrm{L}} \tag{14}$$

$$E = E^{\ominus}(\mathrm{AgCl|Ag, Cl^-}) - \frac{RT}{F} \ln(a_{\mathrm{Cl-}})$$

$$- E^{\ominus}(\mathrm{H^+|H_2}) + \frac{RT}{F} \ln\left(\frac{1}{a_{\mathrm{H^+}}}\right) \tag{15}$$

$$(15) \qquad E = E^{\ominus} - \frac{2RT}{F} \ln a_{\pm} \tag{16}$$

which is the same equation as (13). $a_{\pm}$ is the mean ionic activity for a 1:1 electrolyte, in this case HCl. (For the general case, see (9.7.4).) In order to evaluate the standard cell potential from (16) we need information about activity coefficients in electrolyte solutions. Since strong electrolytes are completely dissociated into ions, the interactions that cause deviations from ideality are assumed to be primarily electrostatic in nature.

9.7. Activity Coefficients in Electrolyte Solutions

The strong electrolyte $\mathrm{B}_{\nu_+}\mathrm{C}_{\nu_-}$ is dissociated into B^{z+} and C^{z-} ions according to the equation

$$\mathrm{B}_{\nu_+}\mathrm{C}_{\nu_-} = \nu_+\mathrm{B}^{z+} + \nu_-\mathrm{C}^{z-} \tag{1}$$

The chemical potential of $\mathrm{B}_{\nu_+} + \mathrm{C}_{\nu_-}$ is therefore

$$\begin{aligned}
\mu(\mathrm{B}_{\nu_+}\mathrm{C}_{\nu_-}) &= \nu_+\mu_+ + \nu_-\mu_- \\
&= \nu_+\mu_+^{\ominus} + \nu_+RT\ln a_+ + \nu_-\mu_-^{\ominus} + \nu_-RT\ln a_- \\
&= \nu_+\mu_+^{\ominus} + \nu_+RT\ln(\gamma_+ m_+) + \nu_-\mu_-^{\ominus} + \nu_-RT\ln(\gamma_- m_-) \quad (2) \\
&= \nu_+\mu_+^{\ominus} + \nu_-\mu_-^{\ominus} + \nu RT\ln m_{\pm} + \nu RT\ln\gamma_{\pm} \tag{3}
\end{aligned}$$

where

$$\nu = \nu_+ + \nu_-, \quad m_{\pm}^{\nu} = m_+^{\nu_+}m_-^{\nu_-}, \quad \gamma_{\pm}^{\nu} = \gamma_+^{\nu_+}\gamma_-^{\nu_-} \tag{4}$$

$m_{\pm}$ is the *mean ionic molality* and $\gamma_{\pm}$ is the *mean ionic activity coefficient*. An activity coefficient describes the behaviour of a real system compared to some ideal reference state. The above activity coefficients are referenced to Henry's law, molality basis.

The theory of activity coefficients of the ions in an electrolyte solution was given by Debye and Hückel (1923). The explanation of this theory is somewhat lengthy, so those readers who prefer to accept the final result without worrying about the detailed arguments may skip to Eq. (20). But first a few words about the solvent water. The HOH molecule is non-linear with a bond angle of 105°, OH bond lengths of 0.97 Å and an H−H distance of 1.54 Å. The melting and boiling temperatures and the entropy of vaporization of water are all comparatively high (compared to other covalent hydrides), indicating strong interactions between water molecules, as is to be expected from their polar nature. Even the purest water has a residual conductivity, indicating small concentrations of H^+ and OH^- ions, due to the ionization of water.

Exercise 9.7-1

Write down the chemical equation for the ionization of water and the expression for the ionization constant of water, K_w. Given that the numerical value of K_w is 1.008×10^{-14} at 25°C, calculate the hydrogen ion activity in pure water at this temperature.

A strong electrolyte $B_{\nu_+} C_{\nu_-}$ is considered to be completely dissociated into ions. Only interionic interactions are considered. It is surmised that a particular ion B^{z^+}, which for brevity we call ion j and which is used as the origin of the coordinate system, will be surrounded by an "ion atmosphere" of + and − ions, but with a net charge of $-z_+$ in order to balance the charge on the central ion, j. Each of the ions in the ion atmosphere is itself surrounded by its own atmosphere. The ions are in a constant state of motion, so properties like charge density and electric potential at a point (that is, within a volume element at that point) mean their time-averaged values.

If the electric potential at a distance r from our reference ion j is $\phi_j(r)$, then the electric potential energy of an ion i at this distance r from j is $z_i e \phi_j(r)$. The potential $\phi_j(r)$ is related to the charge density ρ by Poisson's equation (A7.3.5)

$$\nabla^2 \phi_j(r) = -\frac{\rho}{\varepsilon} \tag{5}$$

In the spherical symmetry prevailing

(5)
$$\nabla^2 \phi_j(r) = \frac{1}{r^2} \frac{d}{dr}\left(r^2 \frac{d\phi_j(r)}{dr}\right) = -\frac{\rho}{\varepsilon} \tag{6}$$

The condition that the solution be electrically neutral is

$$\sum_{i=1}^{s} C_i z_i = 0 \tag{7}$$

where C_i is the number density of ions of species i. Since there is no net charge on the system, the total charge in the ion atmosphere must balance that on the central ion

$$\int_a^{\infty} 4\pi \sum_{i=1}^{s} \rho_i dr = -z_j e \tag{8}$$

a is the distance of closest approach to ion j and is assumed to be independent of i.

Assuming a Boltzmann distribution, the charge density of ions at any point (that is, within a volume element at that point) is

$$\rho = \sum_{i=1}^{s} C_i z_i e \exp\left(-\frac{z_i e \phi_j(r)}{kT}\right) \tag{9}$$

Expand the exponential and retain only the first three terms. The first term is zero (why?) and the third term also vanishes for 1:1 electrolytes (show this). The sole remaining term, which should hold best for an extremely dilute solution, is

$$\rho = -\sum_{i=1}^{s} C_i z_i e \left(\frac{z_i e \phi_j}{kT}\right) \tag{10}$$

This formula is usually called the *linearized Boltzmann distribution for the charge density.*

Exercise 9.7-2

Show that for a symmetrical (1:1, or 2:2 etc.) electrolyte the linear distribution (10) yields $\rho = -2 C_i z_i e \left(z_i e \phi_j / kT\right)$

(6), (9) $$\frac{1}{r^2} \frac{d}{dr}\left(r^2 \frac{d\phi_j}{dr}\right) = \frac{e^2}{\varepsilon kT} \sum_{j=1}^{s} C_i z_i^2 \phi_j = \kappa^2 \phi_j \tag{11}$$

which defines κ.

Exercise 9.7-3

Check units in Eq. (11). What are the units of κ?

The substitution

$$u = \phi_j r \tag{12}$$

reduces (11) to

$$\frac{d^2 u}{d r^2} = \kappa^2 u \tag{13}$$

which has the general solution

$$u = A \exp(-\kappa r) + B \exp(\kappa r) \tag{14}$$

The potential $\phi_j = u/r$ remains finite at large r, so that $B = 0$. With this boundary condition satisfied, the electroneutrality condition gives

(14), (8) $$A \kappa^2 \varepsilon \int_a^\infty r \exp(-\kappa r) dr = -z_j e \tag{15}$$

Integration by parts yields

$$A = \frac{z_j e \exp(\kappa a)}{\varepsilon (1 + \kappa a)} \tag{16}$$

so that the potential is

(16), (14) $$\phi_j = \frac{z_j e \exp(\kappa a) \exp(-\kappa r)}{\varepsilon (1 + \kappa a) r} \tag{17}$$

Equation (17) is the expression derived by Debye and Hückel for the potential due to an ion j at a distance r. The approximations made are (i) the use of a single distance of closest approach a, even though a solution may contain cations and anions with quite different radii, and (ii) use of the linear distribution (9). Alternative distribution functions are discussed by Robinson and Stokes (1968).

The potential due to the ion atmosphere at any point at a distance r from j is obtained from (17) by subtracting the contribution from ion j, giving

(17) $$\varphi_{j,atm}(r) = \frac{z_j e \exp(\kappa a) \exp(-\kappa r)}{\varepsilon (1 + \kappa a) r} - \frac{z_j e}{\varepsilon r} \tag{18}$$

This expression holds down to $r = a$, when any other ion would be in contact with ion j. At this radius

(18) $$\varphi_{j,atm}(a) = -\frac{z_j e \kappa}{\varepsilon (1 + \kappa a)} \tag{19}$$

In fact, since the potential due to the ion atmosphere cannot change within ion j, this expression (19) holds for $0 < r \leq a$.

Starting from an initially ideal state as far as ion interactions are concerned, the work done in establishing the ion atmosphere w' is the product of the charge on the ion and the potential due to the ion atmosphere. Therefore, the presence of an ion atmosphere around each ion in a solution containing one mole of ions of type j changes the molar Gibbs energy by

$$\Delta G^{\mathrm{el}} = w' = -\frac{z_{\mathrm{j}}^2 \mathrm{e}^2 \kappa N_{\mathrm{A}}}{2\varepsilon(1 + \kappa a)} = RT\ln\gamma_j \qquad (20)$$

Exercise 9.7-4

Explain the origin of the factor $\frac{1}{2}$ in Eq. (20).

Single-ion activity coefficients may not be determined experimentally, so formula (20) must be expressed in terms of the mean ionic activity coefficient. For a single electrolyte

$$(20) \qquad\qquad \ln\gamma_{\pm} = -\frac{|z_+ z_-|\mathrm{e}^2}{2\varepsilon kT}\frac{\kappa}{1 + \kappa a} \qquad\qquad (21)$$

$$(11) \qquad\qquad \kappa = \left(\frac{8\pi N_{\mathrm{A}}\mathrm{e}^2}{\varepsilon kT}\right)^{\frac{1}{2}} I_{\mathrm{c}}^{1/2} \qquad\qquad (22)$$

where I_c is the *ionic strength*, defined by

$$I_{\mathrm{c}} = \frac{\sum_{\mathrm{B}} N_{\mathrm{B}} z_{\mathrm{B}}^2}{2N_{\mathrm{A}}V} = \sum_{\mathrm{B}}\frac{1}{2}c_{\mathrm{B}}z_{\mathrm{B}}^2 \quad \text{(concentration basis)} \qquad (23a)$$

N_A is the Avogadro constant, N_B, $N_C \ldots$ the amounts of the ionic species B, C $\ldots$ and c_B is the concentration of B in mol dm^{-3}

$$(21),\ (22) \qquad\qquad \log\gamma_{\pm} = -\frac{\alpha|z_+ z_-|I_{\mathrm{c}}^{1/2}}{1 + \beta a I_{\mathrm{c}}^{1/2}} \qquad\qquad (24)$$

where

$$\alpha = \left(\frac{2\pi N_{\mathrm{A}}}{10^3}\right)^{1/2}\frac{\mathrm{e}^2}{2.3026(\varepsilon kT)^{3/2}} = \frac{1.8246 \times 10^6}{(\varepsilon_{\mathrm{r}}\,T/\mathrm{K})^{3/2}}\mathrm{mol}^{-1/2}\mathrm{m}^{3/2} \qquad (25)$$

$$\beta = \left(\frac{8\pi N_{\mathrm{A}}\mathrm{e}^2}{10^3\varepsilon\,kT}\right)^{1/2} = \frac{50.29 \times 10^{10}}{(\varepsilon_{\mathrm{r}}T/\mathrm{K})^{1/2}}\mathrm{mol}^{-1/2}\mathrm{m}^{1/2} \qquad (26)$$

Exercise 9.7-5

Check the units of κ, α and β.

On the molality basis, the ionic strength is

$$I_m = \sum_B \frac{1}{2} m_B z_B^2 \tag{23b}$$

Exercise 9.7-6

What is the relation between I_c and I_m? Evaluate α and β for water at 298.15 K when the ionic strength I_m is $1 \, \text{mol} \, \text{kg}^{-1}$.

At low values of the ionic strength ($I_m < 0.01 \, \text{mol} \, \text{kg}^{-1}$) the term containing β may be neglected, giving

$$(24) \qquad\qquad \log \gamma_\pm = -\alpha |z_+ z_-| I_m^{1/2} \tag{27}$$

which is called the *Debye–Hückel limiting law*. It predicts that $\log \gamma_\pm$ should be a linear function of the square root of the ionic strength, with α given by (25). There is abundant experimental evidence that this is so (see, for example, Figure 9.2). At higher concentrations, plots of $\log \gamma_\pm$ against $I_m^{1/2}$ become concave upwards due, in part, to the finite size of the ions, which is neglected in (27). Attempts to account for this by (24), called the extended Debye–Hückel law, have not always been successful, due to the approximations inherent in the Debye–Hückel model, notably the assumption of a single value for a (the collision diameter of anions and cations), the neglect of ion–solvent interaction, the formation of ion-pairs and covalent complexes and the linearization of the Boltzmann equation. However, the Debye–Hückel model remains a valuable aid in interpreting the properties of ionic solutions, particularly in very dilute solutions and even in dilute solutions ($0.01 < I_m/\text{mole} \, \text{kg}^{-1} < 0.1$) provided a is used as an adjustable parameter.

The presence of the ion atmosphere affects transport properties in ionic solutions. Because a small majority of ions in the ion atmosphere are moving in the opposite direction to the central ion, there is a net flow of solvent in the opposite direction to the central ion, tending to retard the motion of the central ion. This is called the *electrophoretic effect*. When an electric field is applied, the ions do not move towards the appropriate electrode at a constant velocity. Instead, their forward progress is erratic, due to continuous collisions with solvent molecules and other ions. If the net distance moved towards the electrode in a time interval t is d then d/t is the

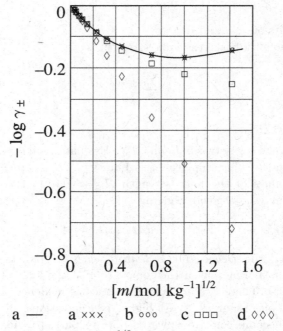

a —   a ×××   b ∘∘∘   c ▢▢▢   d ◇◇◇

Figure 9.2.   Plots of log $\gamma_{\pm}$ against $I_m^{1/2}$ illustrating the validity of the Debye–Hückel limiting law in dilute solution. (a) Experimental data for NaCl, below $0.1\,m$ from Shedlovsky (1950), at and above $0.1\,m$, Robinson (1945); (b) Eq. (27); (c) Eq. (24); (d) an empirical equation due to Guggenheim (1935) in which βa is set equal to 1 and a term bI_m added to the RS of (24), giving

$$\log \gamma_{\pm} = -\frac{\alpha|z_+ z_-|(\rho I_m)^{1/2}}{1 + (\rho I_m)^{1/2}} + bI_m$$

b is an adjustable parameter, set here, to -0.055.

drift velocity in the electric field. Before the electric field is applied, the ion distribution about the central ion is spherically symmetric, but when the central ion moves, the equilibrium ion distribution becomes distorted and it takes a finite time τ (called the *relaxation time*) for the ion atmosphere to recover its spherical symmetry. The distorted ion atmosphere exerts an electrostatic drag on the ions motion. The combined effects were analyzed by Onsager, who derived the equation

$$\Lambda_i = \Lambda_i^0 - (\sigma + \theta\Lambda_i^0)\sqrt{c} \qquad (28)$$

which is known as the Debye–Hückel–Onsager (DHO) equation (Fuoss and Onsager 1957). c is the concentration of ionized solute in mol dm^{-3} and Λ_i is the contribution to the molar conductivity from ionic species i. For a weak electrolyte, replace c by ξc and Λ by Λ/ξ, where ξ is the extent of ionization. For a 1:1 electrolyte, (28) becomes

$$\Lambda = \Lambda^0 - (2\sigma + \theta\Lambda^0)\sqrt{c} \qquad (29)$$

Equation (29) gives a good representation (to within 0.04%) of the conductivity of 1:1 electrolytes below 10^{-2} mol dm^{-3} and the limiting slope of $\Lambda(\sqrt{c})$ is predicted extremely well for a number of electrolytes, solvents and temperatures.

For weak electrolytes

(29) $$\Lambda/\xi = \Lambda^0 - (2\sigma + \theta\Lambda^0)\sqrt{\xi c} \qquad (30)$$

Apart from predicting the behaviour of $\log \gamma_\pm$ and of Λ in very dilute solutions, the Debye–Hückel theory predicts successfully the consequences of ionic interactions on equilibrium constants. For example, the dissociation of acetic acid may be represented by the equation

$$H_2O + HAc = H_3O^+ + Ac^- \qquad (9.2.6)$$

The equilibrium constant for this reaction is

$$K_a = \frac{\gamma_\pm^2 \xi^2 c^2}{(1-\xi)c} \qquad (31)$$

In dilute solution the activity coefficient of undissociated HAc can be replaced by one, and the mean ion activity coefficient of the ionic species can be calculated from the Debye–Hückel theory.

Consider the equilibrium of a sparingly soluble salt in contact with excess solvent

$$B_{\nu_+} C_{\nu_-}(s) = \nu_+ B^{z+} + \nu_- C^{z-} \qquad (32)$$

At equilibrium

$$\mu^s = \nu_+\mu_+ + \nu_-\mu_- \qquad (33)$$

(32) $$K_{sp} = \exp\left[\frac{-(\mu^{\ominus} - \mu^s)}{RT}\right] = \frac{(\gamma_+ m_+)^{\nu_+}(\gamma_- m_-)^{\nu_-}}{(m^{\ominus})^{\nu}}$$

$$= \gamma_\pm^{\nu}\,(\nu_+)^{\nu_+}(\nu_-)^{\nu_-}(m/m^{\ominus})^{\nu} \qquad (34)$$

K_{sp} is called the *solubility product*

$$(34) \qquad \log(m/m^\ominus) = \nu^{-1} \log K_{sp} - \nu^{-1}(\nu_+ \log \nu_+ + \nu_- \log \nu_-)$$

$$+ \frac{\alpha|z_+ z_-|I_c^{1/2}}{1 + \beta a I_c^{1/2}} \tag{35}$$

Equation (34) shows that the solubility m is increased by adding an indifferent electrolyte which increases the ionic strength. For a 1:1 electrolyte

$$(35) \qquad \log(m/m^\ominus) = \frac{1}{2} \log K_{sp} + \frac{\alpha I_c^{1/2}}{1 + \beta a I_c^{1/2}} \tag{36}$$

Answers to Exercises 9.7

Exercise 9.7-1

$$H_2O + H_2O = H_3O^+ + OH^-$$

$$K_w = a_{H^+} a_{OH^-} = 1.008 \times 10^{-14}$$

$$a_{H^+} = 1.004 \times 10^{-7}$$

We write H^+ in the above equations for brevity, despite the fact that in aqueous solution the proton is attached to a water molecule.

Exercise 9.7-2

In Eq. (11)

Units of LS $= V\,m^{-2}$; units of RS $= \dfrac{C^2V}{C}\dfrac{m}{C}\dfrac{m^{-3}V}{V} = V\,m^{-2} =$ units of LS

Therefore the units of κ are m^{-1}. κ^{-1} is the *Debye length*.

Exercise 9.7-3

$$\rho = \sum_{i=1}^{s} C_i z_i e - \sum_{i=1}^{s} C_i z_i e \left(\frac{z_i e \phi_j}{kT}\right) + \sum_{i=1}^{s} \frac{C_i z_i e}{2} \left(\frac{z_i e \phi_j}{kT}\right)^2 - \cdots \tag{37}$$

The first term on the RS of (2) is zero because of the electroneutrality condition. The third term is quadratic in ϕ_j in general but for the special case of a single electrolyte of symmetrical valence (BC, B_2C_2, etc.), when $C_B = C_C$, $z_C = -z_B$ and the coefficient of ϕ_j^2 is consequently zero.

Exercise 9.7-4

The expression (20), without the factor $N_A/2$ correctly (within the approximations made in its derivation), gives the electrical work done in creating the ion atmosphere around a particular ion, referred to a state in which the ions are distributed uniformly. If we were to multiply by N_A to obtain the molar Gibbs energy, this result would be too large by a factor of two since every ion would have been counted twice, once as a central j ion and once as a member i of the ion atmosphere around another ion. Therefore the correct multiplying factor is $N_A/2$.

Exercise 9.7-5

Units of $\kappa = \left(8\pi N_A e^2/10^3 \varepsilon\, kT\right)^{1/2} I_c^{1/2}$ are: $\left(\mathrm{mol}^{-1}\mathrm{C}^2/\mathrm{CV}^{-1}\mathrm{m}^{-1}\mathrm{VC}\right)^{1/2}$
$(\mathrm{mol\,m}^{-3})^{1/2} = \mathrm{m}^{-1}$

Units of $\alpha = \left(2\pi N_A/10^3\right)^{1/2} e^3/2.3026(\varepsilon\, kT)^{3/2}$ are: $\mathrm{mol}^{-1/2}\mathrm{C}^3/$
$(\mathrm{CV}^{-1}\mathrm{m}^{-1}\mathrm{VC})^{3/2}(\mathrm{mol\,m}^{-3})^{1/2} = 1$

Or, units of $\alpha = (1.8246 \times 10^6 \times I_c^{1/2})/(\varepsilon_r\, T/\mathrm{K})^{3/2} \times \mathrm{mol}^{-1/2}\,\mathrm{m}^{3/2}$ are: $\mathrm{mol}^{-1/2}\mathrm{m}^{3/2}\mathrm{mol}^{1/2}\mathrm{m}^{-3/2} = 1$

Units of $\beta = \left(8\pi N_A e^2/10^3 \varepsilon\, kT\right)^{1/2} = \left(\mathrm{mol}^{-1}\mathrm{C}^2/\mathrm{CV}^{-1}\mathrm{m}^{-1}\mathrm{VC}\right)^{1/2} = \mathrm{mol}^{-1/2}\,\mathrm{m}^{1/2}$

and of $\beta a I_c^{1/2}$ are: $\mathrm{mol}^{-1/2}\,\mathrm{m}^{1/2}\,\mathrm{m\,mol}^{1/2}\,\mathrm{m}^{-3/2} = 1$.

Exercise 9.7-6

The SI units of I_m are $\mathrm{mol\,kg}^{-1}$ and of I_c are $10^3\,\mathrm{mol\,m}^{-3}$. Therefore

$$\rho I_m = I_c \tag{38}$$

where ρ is the density of solvent in $\mathrm{mol\,dm}^{-3}$. Checking units:

SI units of LS are: $\mathrm{kg\,dm}^{-3}\,\mathrm{mol\,kg}^{-1} = \mathrm{mol\,dm}^{-3}$ or $10^3\,\mathrm{mol\,m}^{-3} =$ units of RS.

Interpolation of the data in Fernandez *et al.* (1997) yields a value of 78.42 ± 0.04 for the permittivity of water at $298.15\,\mathrm{K}$. The large temperature coefficient of the permittivity of water (about $-0.4\,\mathrm{K}^{-1}$ near $298.15\,\mathrm{K}$) may be the reason for the differences in the value of ε_r found by various workers

$$\alpha = \frac{1.8246 \times 10^6 \times (\rho I_m/\text{mol m}^{-3})^{1/2}}{(78.42 \times 298.16)^{3/2}} = 0.5104 \tag{39}$$

$$(26) \quad \beta/\text{mol}^{-1/2}\,\text{m}^{1/2} = \frac{50.29 \times 10^{10}}{(\varepsilon_r T/\text{K})^{1/2}} = \frac{50.29 \times 10^{10}}{(78.42 \times 298.16)^{1/2}}$$

$$= 0.3277 \times 10^{10} \tag{40}$$

9.8. The Standard Cell Potential

$$(9.6.16) \qquad\qquad E = E^{\ominus} - \frac{2RT}{F}\ln(\gamma_{\pm}m/m^{\ominus}) \tag{1}$$

$$(1) \qquad E + \left(\frac{2RT}{F}\ln(m/m^{\ominus})\right) = E^{\ominus} - \frac{2RT}{F}\ln(\gamma_{\pm}) \tag{2}$$

According to the theory of Debye and Hückel, in a dilute solution of a single 1:1 electrolyte, $\gamma_{\pm}$ is proportional to the square root of m. Thus $E^{\ominus}$ may be determined by plotting the LS of (2) against $m^{1/2}$ and extrapolating to $m = 0$.

Two further examples of cells without transport follow. The first has two gas electrodes in which the gas is at different pressures at the anode and cathode. An example is

$$\text{Pt(s)}|\text{H}_2(\tilde{p}_1)|\text{HCl}(m)|\text{H}_2(\tilde{p}_2)|\text{Pt(s)} \tag{3}$$

$$(9.6.13) \qquad E = E^{\ominus} - \frac{RT}{F}\ln\left\{\left(\frac{\tilde{p}_2}{p^{\ominus}}\right)^{1/2}\frac{1}{a_{\text{H}^+}}\right\}$$

$$- \left[E^{\ominus} - \frac{RT}{F}\ln\left\{\left(\frac{\tilde{p}_1}{p^{\ominus}}\right)^{1/2}\frac{1}{a_{\text{H}^+}}\right\}\right] \tag{4}$$

$$(4) \qquad\qquad E = -\frac{RT}{2F}\ln\left(\frac{\tilde{p}_2}{\tilde{p}_1}\right) \tag{5}$$

The second example consists of two cells like (9.6.3) connected in series, but with the cell on the right reversed.

$$\text{Pt(s)}|\text{H}_2(p)|\text{HCl}(m_1)|\text{AgCl(s)}|\text{Ag(s)}|$$
$$- |\text{Ag(s)}|\text{AgCl(s)}|\text{HCl}(m_2)|\text{H}_2(p)|\text{Pt(s)} \tag{6}$$

where $m_2 > m_1$. The two cells are mirror images except for the different molalities of HCl. The cell potential is

$$(9.6.13) \qquad E = -\left[E^{\ominus} - \frac{2RT}{F}\ln(a_\pm)_2\right] + \left[E^{\ominus} - \frac{2RT}{F}\ln(a_\pm)_1\right]$$

$$= \frac{2RT}{F}\ln\left\{\frac{(a_\pm)_2}{(a_\pm)_1}\right\} \tag{7}$$

which is greater than 0, since $m_2 > m_1$.

Example 9.8-1

Calculate the cell potential of the cell

$$Zn|ZnSO_4(m_1)||CuSO_4(m_2)|Cu \tag{8}$$

at 298.15 K. From tables of activity coefficients

	$m/\text{mol kg}^{-1}$	$\gamma_\pm$
$CuSO_4$	0.100	0.16
$ZnSO_4$	1.00	0.045

The standard potentials for the cell half-reactions are:

$$\text{RS} \qquad Cu^{2+} + 2e^- = Cu, \quad E^{\ominus} = 0.337\,\text{V} \tag{9}$$

$$\text{LS} \qquad Zn^{2+} + 2e^- = Zn, \quad E^{\ominus} = -0.763\,\text{V} \tag{10}$$

The balanced cell reaction[c] (RS–LS) is

$$(9),\ (10) \qquad Cu^{2+} + SO_4^{2-} + Zn = Zn^{2+} + SO_4^{2-} + Cu \tag{11}$$

The standard cell potential is

$$E^{\ominus} = E^{\ominus}(\text{RS}) - E^{\ominus}(\text{LS}) = 0.337 - (-0.763)\,\text{V} = 1.100\,\text{V} \tag{12}$$

The Nernst equation for the cell potential is

$$E = E^{\ominus} - \frac{RT}{2F}\ln\left\{\frac{a_{Zn^{2+}}a_{SO_4^{2-}}}{a_{Cu^{2+}}a_{SO_4^{2-}}}\right\} \tag{13}$$

$$= E^{\ominus} - \frac{RT}{2F}\ln\left\{\frac{a_\pm^2(ZnSO_4, m_1)}{a_\pm^2(CuSO_4, m_2)}\right\} \tag{14}$$

---

[c]According to the cell convention, *reduction* occurs at the *right-hand* electrode.

where

$$a_\pm = \gamma_\pm (\nu_+^{\nu_+} \nu_-^{\nu_-})^{1/\nu} m = \gamma_\pm m \tag{15}$$

Using the data supplied for $\gamma_\pm$ and m

(14), (15) $\qquad E = 1.100 - 0.05916 \log \left(\dfrac{0.045}{0.016} \right) \text{V} = 1.073 \,\text{V}$

In an electrolytic cell *without transport*, either the two electrodes are immersed in the same solution, as in the above example, or they are immersed in separate solutions which are connected by a *salt bridge*. A salt bridge consists of a tube, often in the shape of an inverted u, containing a strong solution of an electrolyte (such as KCl) which has the property that t_+ is as nearly equal as possible to t_-. This tube is terminated by two sintered glass discs that minimize the flow of electrolyte into the two compartments that contain the electrodes of the electrolytic cell, yet provide an electrical connection between the two compartments that is of low resistivity. The presence of a salt bridge is indicated in a cell diagram by a double vertical line.

9.9. pH

The definition of pH (Buck *et al.* 2002) is

$$\text{pH} = -\log a_{\text{H}+} = -\log(m_{\text{H}+} \gamma_{m,\text{H}+}/m^{\ominus}) \tag{1}$$

$a_{\text{H}+}$ is the H^+ activity in aqueous solution, H^+ (aq) and $\gamma_{m,\text{H}+}$ is the activity coefficient of H^+ (aq) on the molality basis at molality $m_{\text{H}+}$. Thus the symbol pH in (1) is to be interpreted as the operation $-\log a_{\text{H}+}$.

Exercise 9.9-1

What is the pH of pure water at 25°C and at 37°C, at which temperature $K_{\text{w}} = 2.1 \times 10^{-14}$?

Answer to Exercise 9.9-1

From Exercise 9.7-1, $a_{\text{H}+} = 1.004 \times 10^{-7}$ and therefore the pH of pure water at 25°C is 7.00. At 37°C the pH of pure water is 6.84.

The primary method for the measurement of pH involves the measurement of the cell potential of the *Harned cell*

$$\text{Pt(s)}|\text{H}_2\text{(g)}|\text{Buffer S, Cl}^-\text{(aq)}|\text{AgCl(s)}|\text{Ag(s)} \tag{2}$$

A buffer is a mixture of a weak acid and a salt of that weak acid (for example phthallic acid and potassium phthalate). These mixtures provide solutions of stable and reproducible pH and so make suitable standards. The pH of S is approximately

$$\text{pH} = pK_c + \log(c_{\text{salt}}/c_{\text{acid}}) \tag{3}$$

pK_c is $-\log K_c$, where K_c is the equilibrium constant for the dissociation of the acid. Thus solutions S that approximate any desired pH can be obtained. However, the pH calculated from (3) will not be exact, because of lack of knowledge of precise values for the activity coefficients. Instead, the pH of S is calculated from the cell potential of (2), which is

$$E = E^{\ominus} + (RT \ln 10/\text{F}) \log[(m_{\text{H}^+}\gamma_{\text{H}^+}/m^{\ominus})(m_{\text{Cl}^-}\gamma_{\text{Cl}^-}/m^{\ominus})] \tag{4}$$

where $E^{\ominus}$ is the standard potential of the AgCl|Ag electrode

$$(4) \qquad -\log((a_{\text{H}^+}\gamma_{\text{Cl}^-}) = \frac{E - E^{\ominus}}{RT \ln 10/\text{F}} + \log(m_{\text{Cl}^-}/m^{\ominus}) \tag{5}$$

Measurements of E are made as a function of the molality of Cl$^-$ and the LS of (5) obtained by extrapolation to $m_{\text{Cl}^-} = 0$. γ_{Cl^-} is obtained from the Debye–Hückel theory (Bates and Guggenheim 1960). This procedure yields the pH of standard solutions, which may then be used to calibrate a glass electrode. For further details on the measurement of pH, consult Cohen *et al.* (2007).

9.10. Thermodynamic Functions of Ions in Aqueous Solution

Tabulated data for $\Delta G_{\text{f}}^{\ominus}$, $\Delta H_{\text{f}}^{\ominus}$, $S^{\ominus}$ and $C_p^{\ominus}$ are based on the *convention* that, in systems of uniform temperature, $\mu^{\ominus}$, $\Delta H_{\text{f}}^{\ominus}$, $S^{\ominus}$ and $C_p^{\ominus}$ for H$^+$(aq, $\gamma = 1$, $m = m^{\ominus}$) are *zero at any temperature*.

Example 9.10-1

Consider the cell

$$Cd|Cd^{2+}(a=1)||H^+(a=1)|H_2(\tilde{p}=p^{\ominus})|Pt \qquad (1)$$

for which $E^{\ominus} = 0.403\,\text{V}$. The cell half-reactions are

RS $\qquad\qquad\qquad H^+ + e^- = \frac{1}{2}H_2(g) \qquad\qquad\qquad (2)$

LS $\qquad\qquad\qquad Cd = Cd^{2+} + 2e^- \qquad\qquad\qquad (3)$

The balanced cell reaction is

(1), (3) $\quad 2H^+ + Cd + 2e^- = H_2(g) + Cd^{2+} + 2e^- \qquad (4)$

(4) $\qquad -nFE^{\ominus} = \Delta G^{\ominus} = \mu^{\ominus}_{Cd^{2+}} + \mu^{\ominus}_{H_2} - \mu^{\ominus}_{Cd} - 2\mu^{\ominus}_{H^+}$

$$= \mu^{\ominus}_{Cd^{2+}}, \quad \text{by conventions} \qquad (5)$$

(5) $\qquad \mu^{\ominus}_{Cd^{2+}} = -nFE^{\ominus} = -2 \times 96485\,\text{C mol}^{-1} \times 0.403\,\text{V}$

$$\qquad\qquad\qquad\qquad\qquad\qquad\qquad\qquad (6)$$

$$= -77.74\,\text{kJ mol}^{-1}$$

Since $-nFE$ is ΔG for the cell reaction

$$nF\left(\frac{\partial E}{\partial T}\right)_p = -\left(\frac{\partial \Delta G}{\partial T}\right)_p = \Delta S \qquad (7)$$

$$\Delta H = \Delta G + T\Delta S$$

(7) $\qquad\qquad\qquad = -nFE + nFT\left(\frac{\partial E}{\partial T}\right)_p \qquad (8)$

Example 9.10-2

The standard cell potential $E^{\ominus}$ of the cell

$$Pt|H_2(\tilde{p}^{\ominus})|H_2SO_4(aq)|Hg_2SO_4(c)|Hg(liq) \qquad (9)$$

is $0.61515\,\text{V}$ at $298.16\,\text{K}$ and its temperature coefficient at $298.15\,\text{K}$ is $-8.23 \times 10^{-4}\,\text{V K}^{-1}$. What is the balanced cell reaction for this cell? Calculate the standard Gibbs function change, standard entropy change and standard enthalpy change for this reaction.

RS $\qquad$ $Hg^{2+} + 2e^- = Hg$ $\hfill$ (10)

LS $\qquad$ $H_2(g) = 2H(ad) = 2H^+ + 2e^-$ $\hfill$ (11)

(10), (11) $\qquad$ $\Delta G^{\ominus} = -nFE^{\ominus} = -2 \times 96\,485 \times 0.61515$

$$= -118\,705 \text{ J mol}^{-1}$$

$$\Delta S^{\ominus} = nF \left(\frac{\partial E}{\partial T} \right)_p = -2 \times 96\,485 \times 8.23 \times 10^{-4}$$

$$= -158.8 \text{ J K}^{-1} \text{mol}^{-1}$$

$$\Delta H^{\ominus} = \Delta G^{\ominus} + T\Delta S^{\ominus} = -118\,705 - 47\,348 \text{ J mol}^{-1}$$

$$= -163\,050 \text{ J mol}^{-1}$$

9.10.1. *Standard states in biochemistry*

Most reactions in biological systems occur at a pH near 7. It therefore makes little sense to continue using a standard state of pH = 1 when the system of interest is a biological one. So in biochemistry, the standard state for H^+ is pH 7, that is

$$- \log a_{H^+} = 7 \hfill (12)$$

Using this standard state

$$\mu^{\text{std}} = \mu^{\ominus} - 2.303 \times 7 \times RT \log a_{H^+} = \mu^{\ominus} - 16.1\,RT \hfill (13)$$

At the time of writing, there was no IUPAC-recommended symbol for this standard state, which is therefore denoted by μ^{std} in Eq. (13).

9.11. Electrochemical Cells with Transport

In the electrochemical cells considered so far, in no case has there been contact between two solutions of the same solute at different concentrations, $c_2 > c_1$. A system containing such a junction is not at equilibrium, and in these cells, transport of the solute from the solution of higher concentration to that of lower concentration occurs, setting up a concentration gradient. The transport of matter because of a concentration gradient is called *diffusion*. Diffusion is an example of a process during which the system is not at equilibrium and entropy is being created (recall Chapter 2)

$$dS = d_e S + d_i S = \frac{dq}{T} + d_i S, \quad d_i S > 0 \tag{1}$$

Classical thermodynamics based on the four laws (zero to three) deals only with systems in equilibrium, so that processes are restricted to those called reversible or quasi-static processes in which the system is always arbitrarily close to equilibrium and $d_i S = 0$. An introduction to the thermodynamics of non-equilibrium processes will be given in Chapter 12. However, in order to complete our discussion of electrochemical cells, we shall in this section derive a formula for the cell potential by assuming that the irreversible process of diffusion is superimposed upon the reversible transfer of charge through the liquid junction in which the concentration is varying *continuously*.

Consider the cell

$$Ag|AgCl(s)|HCl(m_1)|bridge|HCl(m_2)|AgCl(s)|Ag \tag{2}$$

in which the two solutions of HCl are of different concentrations ($m_2 > m_1$). These two solutions are connected by a bridge solution, in which the composition of Cl^- varies *continuously* from m_1 (on the left) to m_2 (on the right). There is thus no discontinuity in concentration. If the bridge were absent, contact between the two solutions of different concentration would give rise to the diffusion of solute from that of higher concentration to the one of lower concentration, an irreversible process. Let us suppose that the mobility of the cation exceeds that of the anion. Then the faster cations tend to leave behind the slower moving anions, thus setting an electrical double layer. The potential difference across this double layer is the *liquid junction potential*. This potential does not continue to increase without limit because the separation of the ions in the electrical double layer gives rise to an electric field which slows down the faster moving cations, thus giving rise to a *steady state*. If the bridge solution is formed by the interdiffusion of two solutions of HCl of molalities m_1 and m_2 the condition of continuity of concentration will be met. Consider two planes in the bridge solution that are normal to flow of solute from left to right. Since the concentration of solute differs only infinitesimally between the two planes, the passage of an infinitesimal electric current will be reversible and the methods of equilibrium thermodynamics are applicable. Of course, there is an irreversible diffusion process going on simultaneously, but this can be regarded as superimposed upon, and independent of the reversible electrical process.

Suppose a charge of one Faraday is passed through the cell from left to right. At each plane in the bridge solution the change in G is

$$(t_+/z_+)\mathrm{d}\mu_+ + (t_-/-z_-)(-\mathrm{d}\mu_-) \tag{3}$$

Therefore, the change in G occurring because of the passage of one Faraday through the cell from one electrode, through the bridge solution to the other electrode, is

$$\int_{m_1}^{m_2} [(t_+/z_+)\,\mathrm{d}\mu_+ + (t_-/z_-)\mathrm{d}\mu_-] \tag{4}$$

Associated with this process, the chemical reactions

$$\mathrm{Ag(s)} + \mathrm{Cl}^-(\mathrm{aq}, m_1) = \mathrm{AgCl(s)} \tag{5}$$

$$\mathrm{AgCl(s)} = \mathrm{Ag(s)} + \mathrm{Cl}^-(\mathrm{aq}, m_2) \tag{6}$$

occur at the electrodes. The net cell reaction $(5) + (6)$ is

$$(1/z_-)\mathrm{Cl}^-(\mathrm{aq}, m_1) = (1/z_-)\mathrm{Cl}^-(\mathrm{aq}, m_2) \tag{7}$$

which is associated with an increase in G of

$$(1/z_-)[\mu(\mathrm{Cl}^-, \mathrm{aq}, m_2) - \mu(\mathrm{Cl}^-, \mathrm{aq}, m_1)] \tag{8}$$

Since $\Delta G = -nEF$, the cell potential is given by

(4), (8) $\qquad -EF = (1/z_-)[\mu(\mathrm{Cl}^-, \mathrm{aq}, m_2) - \mu(\mathrm{Cl}^-, \mathrm{aq}, m_1)]$

$$+ \int_{m_1}^{m_2} [(t_+/z_+)\mathrm{d}\mu_+ + (t_-/z_-)\mathrm{d}\mu_-] \tag{9}$$

In this cell (2), $z_+ = 1$ and $z_- = -1$, but the charge numbers of the ions have been retained for the sake of generality.

The liquid junction potential in this steady state is

(9) $\qquad \phi_j = -\dfrac{RT}{F} \displaystyle\int_{m_1}^{m_2} \left(\dfrac{t_+}{z_+}\mathrm{d}\ln a_+ + \dfrac{t_-}{z_-}\mathrm{d}\ln a_- \right) \tag{10}$

If we neglect the small variation of the transport numbers with concentration, the RS of Eq. (10) may be integrated to yield

$$\phi_j = -\frac{RT}{F}\left(\frac{t_+}{z_+}\ln\frac{(a_+)_2}{(a_+)_1} + \frac{t_-}{z_-}\ln\frac{(a_-)_2}{(a_-)_1}\right)$$ (11)

The cell potential of (2) is then, on setting $t_- = 1 - t_+$ and $z_+ = -z_- = 1$

$$E = -\frac{RT}{F}\left[\left(\ln\frac{(a_-)_2}{(a_-)_1}\right) + \left(t_+\ln\frac{(a_+)_2}{(a_+)_1} - (1 - t_+)\ln\frac{(a_-)_2}{(a_-)_1}\right)\right]$$

$$= -\frac{RT}{F}t_+\ln\left\{\frac{(a_+)_2(a_-)_2}{(a_+)_1(a_-)_1}\right\}$$

$$= -\frac{2RT}{F}t_+\ln\left\{\frac{(a_\pm)_2}{(a_\pm)_1}\right\}$$ (12)

By measuring both E and the cell potential of (9.8.6), which is

$$E = \frac{2RT}{F}\ln\left\{\frac{(a_\pm)_2}{(a_\pm)_1}\right\}$$ (9.8.7)

t_+ can be calculated. For the cell potential of the most general type of cell with transport see Taylor (1927) and Guggenheim (1930).

9.12. Some Applications

9.12.1. *Batteries*

The reader will have appreciated by now that there are two types of electrochemical cells. In a *galvanic cell*, chemical reactions are used to produce an electrical potential. In an *electrolytic* cell, an electric current is used to produce chemical reactions. A rechargeable battery is an electrochemical cell that is capable of performing both these functions with high efficiency. The most familiar example is the lead acid accumulator, consisting of Pb plates (the electrodes) immersed in a strong aqueous solution of H_2SO_4. When discharged by using the battery as a source of electrical current, it is a galvanic cell in which the following reactions are occurring:

At the positive pole

$$PbO_2(s) + 4H^+(aq) + SO_4^{2-}(aq) + 2e^-(Pb) \rightarrow 2H_2O(l) + PbSO_4(s)$$ (1)

and at the negative pole

$$Pb(s) + SO_4^{2-}(aq) \rightarrow PbSO_4(s) + 2e^-(Pb) \tag{2}$$

During recharge (by connecting the battery to a source of electric current) the direction of these reactions is reversed. The standard cell potential is

$$E_{cell} = E_1 - E_2 = 1.685\,V - (-0.356)\,V = 2.04\,V \tag{3}$$

Lithium ion batteries are of particular interest for device technology because of their high cell potential (3.7 V) and relatively low mass. The cell reactions when charging are:

positive pole $\qquad C + nLi^+ + ne^- \rightarrow CLi_n$ $\qquad\qquad\qquad$ (4)

negative pole $\qquad LiCoO_2(s) \rightarrow Li_{1-n}CoO_2(s) + nLi^+ + ne^-$ $\qquad$ (5)

so that the net reaction is

$$LiCoO_2(s) + C(gr) = Li_{1-n}CoO_2 + CLi_n, \quad E^{\ominus} \approx -3.7\,V \tag{6}$$

Thus the charged battery contains non-stoichiometric lithium cobalt oxide and graphite intercalated with Li. During discharge, the reactions (4)–(6) are reversed.

9.12.2. *Fuel cells*

In a fuel cell a fuel such as hydrogen (or methanol) is oxidized to water (or water + CO_2) by oxygen and the chemical energy liberated directly as electricity. Because of their high energy density, fuel cells are likely to have a major impact on device technology and as power sources for transportation. In this rapidly advancing field it is impossible to give an up-to-date account in a comprehensive text book on thermodynamics intended for undergraduates. (For many details on fuel cell technology see Vielstich *et al.* 2003.) My brief account will therefore be limited to one example, the *proton exchange membrane fuel cell.* It depends on the ability of a thin membrane of a perfluorosulfonic acid polymer to act as a proton conductor. The membrane contains channels lined with ionizable $-SO_3H$ groups, along which the hydrogen ions migrate. The electrodes have channels that facilitate the gases reaching the catalyst-covered polymer. The surfaces of the membrane are coated with a Pt or Pt alloy catalyst to facilitate the half-cell chemical reactions, which are:

positive pole $\quad\frac{1}{2}O_2(g) + 2H^+(polymer) + 2e^-(Pt') \to H_2O(liq)$ $\quad$ (1)

negative pole $\quad\quad\quad H_2(g) \to 2H^+(polymer) + 2e^-(Pt)$ $\quad\quad$ (2)

net reaction $\quad\quad\quad\frac{1}{2}O_2(g) + H_2(g) \to H_2O(liq)$ $\quad\quad$ (3)

Problems 9

9.1 Calculate the mean ionic molality $(m_\pm)$, the mean ion activity $(a_\pm)$, the activity (a) of the electrolyte and the ionic strength (I_m) for the following solutions:

Solute	$m/\text{mol kg}^{-1}$	$\gamma_\pm$
$AgNO_3$	0.100	0.72
$CuSO_4$	0.050	0.21
H_2SO_4	0.010	0.544
$In_2(SO_4)_3$	0.100	0.035

9.2 A conductivity cell has plane parallel electrodes of area $3.142\times10^{-4}\,\text{m}^2$ and the distance between the electrodes is $8.908 \times 10^{-2}\,\text{m}$. The cell contains a 1:1 electrolyte BC of concentration $0.1000\,\text{mol dm}^{-3}$. The transport number of the B^+ ion is 0.4898. When a potential difference of $6.943\times10^{-2}\,\text{V}$ is applied between the electrodes, the current flowing through the solution is $2.850 \times 10^{-4}\,\text{A}$. Find:

  (i) the electric field between the electrodes;
  (ii) the current density;
  (iii) the electrical conductivity of the solution of BC;
  (iv) the molar conductivity of the solution;
  (v) the flux of B^+ ions in mol m^{-2} s^{-1};
  (vi) the speed of the B^+ ions;
  (vii) the mobility of the B^+ ions.

9.3 The transport number of Na^+ in aqueous solutions of NaCl of concentration $20.0\,\text{mol m}^{-3}$ was measured by Longsworth (1932) by the moving boundary method. x denotes the position of the boundary. The cross-sectional area of the tube $A = 0.115\,\text{cm}^2$ and the current $I = 1.6001\,\text{mA}$.

Find t_+ from Longsworth's data:

x/cm	0	1.00	6.00	10.00
t/s	0	344	2070	3453

9.4 The solubility m of $AgIO_3$ in solutions of KNO_3 of molality m_1 is given in the following table:

$10^3 m_1$/mol kg^{-1}	0	1.301	3.252	6.053	14.10
$10^4 m$/mol kg^{-1}	1.761	1.813	1.863	1.908	1.991

Plot $m/m^\ominus$ against $I_m^{1/2}$ and hence determine K_{sp}, the solubility product of $AgIO_3$. Calculate $\gamma_\pm$ for $AgIO_3$ at each of the above concentrations and compare your results with the predictions of the Debye–Hückel limiting law.

9.5 The cell potential of

$$Pt|H_2(\tilde{p}^\ominus)|HCl(m^\ominus)|AgCl(cr)|Ag$$

is 0.2333 V at 298.15 K. Find $\gamma_\pm$ for the HCl solution.

9.6 Deduce the balanced cell reaction for the electrochemical cell

$$Ag|AgI(cr)|I^-(a=1)|Ag^+(a=1)|Ag$$

Find the cell potential and the solubility product of AgI at 298.15 K. Necessary standard electrode potentials are $E^\ominus_{Ag^+|Ag} 0.7991$ V and $E^\ominus_{AgI|Ag} = -0.1518$ V.

9.7 The potential of cell (a) with transport

(a) $Pt|H_2(\tilde{p}^\ominus)|HCl$ (aq, $a_\pm = 0.009048$)|HCl (aq, $a_\pm = 0.01751$)| $H_2(\tilde{p}^\ominus)|Pt$ is $E_a = 0.02802$ V, and that of the cell

(b) $Pt|H_2(\tilde{p}^\ominus)|HCl$ (aq, $a_\pm = 0.009048$)||HCl(aq, $a_\pm = 0.01751$)| $H_2(\tilde{p}^\ominus)|Pt$ is $E_b = 0.01696$ V. Calculate the liquid junction potential in cell (a) and the transport number of H$^+$. Comment on the value of t_+.

9.8 The molar conductivity Λ of nitric acid in methanol at 25°C as a function of concentration c is given in the following table:

$10^4 c$/mol dm^{-3}	0.966	1.814	3/075	4.730	9.163
Λ/S cm^2 mol^{-1}	178.7	165.2	151.0	137.6	116.3

Neglecting interionic effects find the limiting molar conductivity and the dissociation constant of nitric acid in methanol.

9.9 Start from

$$\Lambda = \xi\Lambda^0 - (A + B\Lambda^0)(\Lambda/\Lambda^0)\sqrt{\xi c} \tag{2}$$

Let $\xi = \Lambda S/\Lambda^0$, so that (2) becomes

$$\Lambda = \Lambda S - (A + B\Lambda^0)(\Lambda/[\Lambda^0]^{3/2})\sqrt{\Lambda S c} \tag{3}$$

Divide through by Λ and rearrange, giving

$$S - ZS^{1/2} - 1 = 0 \tag{4}$$

where

$$Z = (A + B\Lambda^0)(\Lambda c)^{1/2}/(\Lambda^0)^{3/2} \tag{5}$$

Solve (4) to obtain S and hence ξ as a power series in Z. (**Hint**: (4) is a quadratic equation in $S^{1/2}$.) Write down the expression for K_a that includes interionic effects through the factor S and rearrange this equation in a form similar to the Ostwald dilution law.

9.10 The resistance of a conductivity cell (cell constant $0.7096\,\text{cm}^{-1}$) when filled with $0.01\,\text{M}$ acetic acid is $4372\,\text{ohm}$ at $25°\text{C}$. Given that the molar conductivities at infinite dilution and $25°\text{C}$ of the hydrogen and acetate ions are 349.8 and $40.9\,\text{S}\,\text{cm}^2\,\text{mol}^{-1}$, calculate the molar conductivity of the acetic acid solution and the dissociation constant of acetic acid.

9.11 Calculate the potential energy of interaction between a positive ion and a negative ion, each of charge equal in magnitude to the elementary charge $1.6022 \times 10^{-19}\,\text{C}$ at a distance apart of $3.00 \times 10^{10}\,\text{m}$, (a) in vacuum and (b) in a medium of relative permittivity 78.42. Comment on the significance of your results.

10

Fermi–Dirac and Bose–Einstein Statistics

Note to the reader: The subject matter in Section 10.6 and Section 10.8 inevitably requires that these two sections contain large numbers of equations. Their detailed study could well be postponed for a second reading or until a graduate course.

10.1. Ensembles

The aim of statistical mechanics is the calculation of the properties of macroscopic systems from the properties of the individual molecules that comprise the system. When the focus of interest is the calculation of the thermodynamic properties of a system, we usually use the term "statistical thermodynamics" (Fowler and Guggenheim 1939). In Chapter 3, we presented a brief introduction to statistical thermodynamics. In this, thermodynamic properties were identified with the most probable distribution over quantum states. The first thing to be settled is what kind of averaging procedure is to be used. Systems of this size contain an exceedingly large number of quantum states and it is therefore difficult to imagine an average over time being accomplished in the comparatively short time (in the macroscopic sense) available to complete a physical measurement (such as that of the wavelength of a spectral line, for example). This conceptual difficulty may be removed by averaging over a very large number N_e of replicas of the original thermodynamic system, called an *ensemble*. Each of these replicas in the ensemble duplicates the thermodynamic system and its surroundings. For example, if a thermodynamic system is described by the variables (T, V, N) then an ensemble of these systems is a very large number of identical (in the macroscopic sense) replicas of this system all in thermal contact with one another and initially in contact with a large heat bath (thermostat) at temperature T. The N_e systems are then disconnected from the heat bath and adiabatically isolated, and so maintain the average temperature T. Focus attention on one particular ensemble member. Then the remaining $N_e - 1$ members function as a heat reservoir for the selected system. (It is necessary for the ensemble as a whole to be isolated in order that postulate II below shall be applicable.) We use N as an abbreviation for the whole set $\{N_i\}$ so that a state of the system is described by stating its temperature, volume and composition. An ensemble constructed from such a system is called a *canonical ensemble* (*canonical* = according to a rule, standard, fundamental, as in the "equations of motion in canonical form" = Hamilton's equations of motion).

10.1.1. *Postulates of statistical thermodynamics*

I The macroscopic time average of a thermodynamic variable M (such as $U, V, p, \ldots$) is equal to the ensemble average of M in the limit $N_e \to \infty$. (This is called the ergodic hypothesis.)

II Each of the large number of quantum states that correspond to *the same state* of an isolated, macroscopic thermodynamic system is represented in the ensemble and they occur with equal probability. This postulate of the equal *a priori* probability of the quantum states of a thermodynamic state of a macroscopic system was used in Section 3.5 in the derivation of the Boltzmann distribution.

A canonical ensemble consists of replicas of a closed thermodynamic system of specified temperature, volume and composition; a *microcanonical ensemble* consists of replicas of an isolated thermodynamic system of specified energy, volume and composition; and a *grand canonical ensemble* consists of replicas of an open thermodynamic system of specified temperature, volume and chemical potential (or absolute activity) of each component.

10.2. The Canonical Ensemble

We must now look more closely at how a canonical ensemble would be constructed. In order to calculate an ensemble average of some particular property, this property must have a definite value in each quantum state that corresponds to the macroscopic thermodynamic state specified. Properties that satisfy this condition, like energy, volume and pressure, are called *mechanical properties.* But there are other thermodynamic properties, like temperature and entropy, called *thermal properties*, which do not have a definite value in each quantum state. The collection of replicas of a particular thermodynamic system that comprise a canonical ensemble consists of systems of composition N and volume V, all in thermal contact with one another. The whole ensemble is insulated from its environment in order that postulate II will be applicable to the whole ensemble. Within the canonical ensemble, when we focus attention on one particular system, the remaining $N_e - 1$ systems function as a heat reservoir for our system. The same is true for each system in the ensemble so that thermal equilibrium is established throughout the ensemble, which therefore has a uniform temperature T. The derivation of the Boltzmann distribution now proceeds as in Chapter 3, the "system" being the whole ensemble of replicas at temperature T, volume $N_e V$ and composition $N_e N$.

All the N_e systems in the ensemble have the same set of energy states (because they all have the same N and V). Let n_i be the number of systems in the ensemble that have an energy E_i. The probability that a particular system chosen at random from the canonical ensemble will have energy E_i is

$$P_i = \frac{\langle n_i \rangle}{N_e} \tag{1}$$

where $\langle n_i \rangle$ is the mean value of n_i, which we identify with n_i^* the value of n_i in the most probable distribution of the $\{n_i\}$ over the systems in the ensemble. The justification for this step is the extreme sharpness of the distribution. The number of states of the ensemble for a given distribution is

$$\Omega = \frac{N_e!}{\prod_i n_i!} \tag{2}$$

As before, we identify the most probable distribution with the maximum value of Ω. On using Stirling's approximation

$$(2) \qquad \ln \Omega = N_e \ln N_e - \sum_i n_i \ln n_i \tag{3}$$

Setting $c = 0$ in (3.10.18), that is, assuming the absence of metastable phases and ignoring any entropy contributions from isotopic mixing and from the degeneracy of the nuclear and electronic ground states by our choice of entropy zero

$$(3.9.18) \qquad S = k \ln Q + kT \frac{\partial \ln Q}{\partial T} \tag{4}$$

where the partition function Q is

$$Q = \sum_i \exp[-E_i(V,N)/kT] \tag{5}$$

$$(4) \qquad S = k \ln Q + kT \frac{1}{Q} \left(\frac{\partial Q}{\partial T} \right)_{V,N}$$

$$(5) \qquad = k \ln Q - \frac{1}{Q} \sum_i [E_i(V,N)/T] \exp[-E_i(V,N)/kT]$$

$$= -k \sum_i P_i \ln P_i \tag{6}$$

where

$$P_i(T,V,N) = \frac{\exp[-E_i(V,N)/kT]}{Q} = \frac{\exp[-E_i(V,N)/kT]}{\sum_i \exp[-E_i(V,N)/kT]} \qquad (7)$$

is the probability that a system is in any particular state E_i. Thus the molecular basis of the thermodynamic properties of systems described by the thermodynamic potential $F(T,V,N)$ is established.

10.3. The Microcanonical Ensemble

A microcanonical ensemble may be derived from a canonical ensemble by selecting from it only those systems which have (very nearly) the same energy and then isolating each system. The ensemble is then a collection of isolated thermodynamic systems described by the thermodynamic variables (U,T,V). This is probably an opportune moment to remind the reader that we are using E for the energy of the quantum state of a macroscopic system (and in particular E_i for the ith quantum state) and U for the thermodynamic internal energy of a system. We identify U with $\langle E \rangle$, the ensemble average of E, which is calculated by equating it to the most probable value of E. This we are entitled to do because of the sharpness of the energy distribution.

The microcanonical ensemble is a degenerate canonical ensemble in which all members of the ensemble, $\Omega(E,V,N)$ in number, have the same energy (within the very fine limits set by the natural fluctuations in E). The probability P_i of a state having the energy E is therefore the same for all members of the ensemble and equal to $1/\Omega$. U is therefore equal to E and

$$(10.2.3) \qquad S(U,V,N) = -k\sum_i P_i \ln P_i = -k\Omega[(1/\Omega)\ln(1/\Omega)]$$

$$= k\ln\Omega(E,V,N) \qquad (1)$$

which is the Boltzmann equation for the entropy and the route to the other thermodynamic potentials. From the fundamental equation

$$dS = \frac{1}{T}dU + \frac{p}{T}dV - \sum_B \frac{\mu_B}{T}dN_B \qquad (2)$$

$$(1),\,(2) \qquad \left(\frac{\partial \ln\Omega}{\partial E}\right)_{V,N} = \frac{1}{kT} \qquad (3)$$

(1), (2)
$$\left(\frac{\partial \ln \Omega}{\partial V}\right)_{E,N} = \frac{p}{kT} \tag{4}$$

. (1), (2)
$$\left(\frac{\partial \ln \Omega}{\partial N_{\mathrm{B}}}\right)_{E,V,N_{\mathrm{C} \neq \mathrm{B}}} = -\frac{\mu_{\mathrm{B}}}{kT} \tag{5}$$

10.4. The Grand Canonical Ensemble

The grand canonical ensemble resembles the canonical ensemble except that the walls of the replica systems are permeable to the molecular species labelled B, C, ... Therefore, the independent variables are T, V, μ, where μ stands for the set of chemical potentials $\{\mu_{\mathrm{B}}, \mu_{\mathrm{C}} \ldots\}$. The ensemble is initially in a heat reservoir which is also a source of all the chemical species B, C, ... and remains there until thermal and chemical equilibrium are established. It is then adiabatically isolated and removed from the heat reservoir. The whole ensemble can now be regarded as an isolated giant system to which postulate II may be applied.

Let $n_i(N)$ denote the number of systems that contain N molecules and have energy $E_i(V, N)$ where, as usual in this chapter, N denotes the whole set $\{N_{\mathrm{B}}, N_{\mathrm{C}}, \ldots\}$. The total number of quantum states in the super system is

$$\Omega_{\mathrm{t}}(n) = \frac{\left[\sum_{i,N} N_i(N)\right]!}{\prod_{i,N} n_i(N)!} \tag{1}$$

The $n_i(N)$ distributions satisfy the conditions

$$\sum_{i,N} n_i(N) = N_{\mathrm{e}} \tag{2}$$

$$\sum_{i,N} n_i(N) E_i(V, N) = E_{\mathrm{tot}} \tag{3}$$

$$\sum_{i,N} n_i(N) N = N_{\mathrm{tot}} \tag{4}$$

Multiply (2), (3) and (4) by the undetermined multipliers α, β and γ; then Lagrange's method (see Section 3.5) yields for the most probable distribution

$$n_{\mathrm{i}}^*(N) = \exp(-\alpha)\exp[-\beta E_i(V, N)]\exp(-\gamma N) = \langle n_i(N)\rangle \tag{5}$$

Eliminating α by summing over i and using the fact that the total number of distributions is N_e

(5) $$\langle n_i(N)\rangle = \frac{N_e \exp[-\beta E_i(V, N)]\exp(-\gamma N)}{\sum_{i,N}\exp[-\beta E_i(V, N)]\exp(-\gamma N)}$$ (6)

Again, because of the sharpness of the distribution, we have in (5) identified the mean value of the distribution with its most probable value. The probability that a system in the ensemble will contain N molecules and have energy $E_i(V, N)$ is

(6) $$P_i(N) = \frac{\langle n_i(N)\rangle}{N_e} = \frac{\exp(-\beta E_i)\exp(-\gamma N)}{\sum_{i,N}\exp(-\beta E_i)\exp(-\gamma N)}$$ (7)

The thermodynamic energy U is equal to the ensemble average $\langle E\rangle$ where

$$\langle E\rangle = \sum_{i,N} P_i(N)E_i(V, N)$$ (8)

The number of particles in the thermodynamic system is the ensemble average

$$N = \langle N\rangle = \sum_{i,N} P_i(N)N$$ (9)

and the pressure in the thermodynamic system is

$$p = \langle p\rangle = -\sum_{i,N} P_i(N)\left(\frac{\partial E}{\partial V}\right)_N$$ (10)

(8) $$d\langle E\rangle = -\sum_{i,N} P_i(N)dE_i + E_i dP_i$$ (11)

$$= \sum_{i,N} P_i(N)\left(\frac{\partial E}{\partial V}\right)_N dV + E_i dP_i$$ (12)

Exercise 10.4-1

Why is E in (12) apparently a function of V only?

Define the *grand partition function* Ξ by

$$\Xi = \sum_{i,N} \exp(-\beta E_i)\exp(-\gamma N)$$ (13)

$$= \sum_N Q(T, V, N)\exp(-\gamma N)$$ (14)

(7), (12), (13)
$$-d \sum_{i,N} P_i(N) \ln P_i(N) = \beta d\langle E \rangle$$

$$+ \beta \langle p \rangle dV + \gamma d\langle N \rangle \qquad (15)$$

On comparing (15) with the thermodynamic equation

$$T dS = dU + p dV - \mu dN \qquad (16)$$

we conclude that

$$-k \sum_{i,N} P_i(N) \ln P_i(N) = S, \quad \beta = 1/kT, \quad \gamma/\beta = -\mu \qquad (17)$$

(15)
$$d(pV) = S dT + p dV + N d\mu \qquad (18)$$

which is the fundamental equation for the thermodynamic potential $pV(T, V, \mu)$

(17), (13)
$$S = kT \left(\frac{\partial \ln \Xi}{\partial T} \right)_{V,\mu} + k \ln \Xi \qquad (19)$$

(13)
$$N = kT \left(\frac{\partial \ln \Xi}{\partial \mu} \right)_{V,T} \qquad (20)$$

(13)
$$p = kT \left(\frac{\partial \ln \Xi}{\partial V} \right)_{T,\mu} = kT \frac{\ln \Xi}{V} \qquad (21)$$

Exercise 10.4-2

Express the equation of state for a real gas in terms of the grand partition function.

Exercise 10.4-3

Express the average energy as a function of the chemical potential and the grand partition function.

Answers to Exercises 10.4

Exercise 10.4-1

Because the sum over N in (12) removes the expected dependence on N.

Exercise 10.4-2

(21) $$pV = kT \ln \Xi \tag{22}$$

Exercise 10.4-3

$$\langle E \rangle = U = G + TS - pV$$

(19), (22) $$= \mu N + kT^2 \frac{\partial \ln \Xi}{\partial T} \tag{23}$$

10.5. Fermi–Dirac and Bose–Einstein Statistics

In Chapter 3 and Section 10.1, the only quantum-mechanical property made use of was the existence of eigenstates with state functions ψ_i and energy ε_i. Since the number of quantum states available was much larger than the number of particles to be accommodated, any restrictions on the number of particles allowed per quantum state were unimportant. However, in some systems, notably the free electrons in a metal, the *Pauli exclusion principle* (PEP) restricts the number of electrons permitted to occupy a quantum state to either zero or one. The same would be true for any other particle of spin $1/2$. The PEP is that the complete state function for a system of particles of spin $1/2$ must be antisymmetric with respect to the interchange of any two particles. It is a statement of how Nature behaves and may not be derived from other fundamental principles. Particles of spin $1/2$ (and generally any odd power of $1/2$) are therefore called *fermions* and obey Fermi–Dirac statistics. In contrast, particles or quasi-particles of spin $0, 1, 2, \ldots$ are called *bosons* and obey Bose–Einstein statistics, with the result that the number of particles (or quasi-particles) that can be accommodated in one quantum state is unlimited. Examples are *photons*, *phonons* and the isotope He^4.

The antisymmetric requirement for an n-electron system may be met by using for an n-electron state function, the determinant

$$\Psi_{\mathrm{FD}} = C \begin{vmatrix} \psi_1(1) & \psi_1(2) & \psi_1(3) & \cdots & \psi_1(n) \\ \psi_2(1) & \psi_2(2) & \psi_2(3) & \cdots & \psi_2(n) \\ \psi_3(1) & \psi_3(2) & \psi_3(3) & \cdots & \psi_3(n) \\ \cdots & \cdots & \cdots & \cdots & \cdots \\ \psi_n(1) & \psi_n(2) & \psi_n(3) & \cdots & \psi_n(n) \end{vmatrix} \tag{1}$$

where C is a normalization constant. The same list of products but with all signs positive (indicated by the prime on the determinant) is a Bose–Einstein state function for a $n-$electron system

$$\Psi_{\mathbf{BE}} = C' \begin{vmatrix} \psi_1(1) & \psi_1(2) & \psi_1(3) & \cdots & \psi_1(n) \\ \psi_2(1) & \psi_2(2) & \psi_2(3) & \cdots & \psi_2(n) \\ \psi_3(1) & \psi_3(2) & \psi_3(3) & \cdots & \psi_3(n) \\ \cdots & \cdots & \cdots & \cdots & \cdots \\ \psi_n(1) & \psi_n(2) & \psi_n(3) & \cdots & \psi_n(n) \end{vmatrix}' \tag{2}$$

Denote the grand partition function for a single particle (sub-system) by ξ. The particles are indistinguishable but the quantum states are distinguishable. In FD statistics, there are only the two possibilities, $n_i = 0$, when the state i is empty, and $n_i = 1$ when it is occupied. In FD statistics therefore

$$\Xi(T, V, \mu) = \prod_i \xi_i(T, V), \quad \xi_i = 1 + \exp[-E_i(V)/kT]\lambda,$$

$$\lambda = \exp(\mu/kT) \tag{3}$$

The first term in ξ_i corresponds to $n_i = 0$ and the second one to $n_i = 1$. $\exp[-E_i(V)/kT]$ is the factor q in the partition function of one particle in the energy state E_i. The average number of particles is

$$(10.4.20) \qquad \langle N \rangle = \lambda \left(\frac{\partial \ln \Xi}{\partial \lambda} \right)_{T,V} = \lambda \sum_i \left(\frac{\partial \ln \xi_i}{\partial \lambda} \right)_{T,V} = \sum_i n_i$$

$$= \sum_i \frac{\exp[-E_i/kT]\lambda}{1 + \exp[-E_i/kT]\lambda}$$

$$= \sum_i \frac{1}{\exp[(E_i - \mu)/kT] + 1} \tag{4}$$

Equation (4) is the FD distribution law. At $T = 0\,\mathrm{K}$, for constant μ and finite E_i, all the energy states are occupied below $E_i = \mu$ and empty above $E_i = \mu$, so that the FD distribution function is a step function. The sharp corners of this step function become rounded off as T increases and electrons are excited thermally from filled states to empty states. The highest occupied level at $0\,\mathrm{K}$ is called the *Fermi level*. The energy required to take an electron from the Fermi level to a point well outside a metal is the *work function*, Φ. So far in this section we have ignored the possibility that the electric potential ϕ of the phase in question might be non-zero. If $\phi \neq 0$, then the chemical potential μ_e must be replaced by the electrochemical potential $\tilde{\mu}_e$.

Exercise 10.5-1

Let $f(E)$ be the probability that the state E is occupied. Sketch the form of the function $df(E)/dE$ at room temperature.

Exercise 10.5-2

Two different metals are brought into electrical contact. What happens?

In *Bose–Einstein statistics*

$$\Xi(T, V, \mu) = \prod_i \xi_i(T, V) \tag{5}$$

$$\xi_i = 1 + \exp[-E_i(V)/kT]\lambda + \exp[-2E_i(V)/kT]\lambda^2 + \cdots$$

$$= \frac{1}{1 - \exp[-E_i(V)/kT]} \tag{6}$$

The series in (6) is a geometric progression which is convergent if the common ratio

$$\exp[-E_i/kT]\lambda < 1 \tag{7}$$

The condition (7) holds for all values of E_i and in particular for the ground state, the energy of which is usually taken as zero, in which case

$$(7) \qquad\qquad\qquad\qquad \lambda < 1 \tag{8}$$

$$(10.4.20) \qquad \langle N \rangle = \lambda \left(\frac{\partial \ln \Xi}{\partial \lambda} \right)_{T,V} = \sum_i \frac{1}{\exp[(E_i - \mu)/kT] - 1} \tag{9}$$

Equation (9) is the BE (Bose–Einstein) distribution law. Notice that it differs from the FD distribution law only in the sign before unity in the denominator

$$(3) \qquad\qquad \Xi = \prod_i \frac{1}{1 - \exp[(\mu - E_i)/kT]} \tag{10}$$

$$(10) \qquad\qquad \ln \Xi = -\sum_{i=0}^{\infty} \ln[1 - \exp\{(\mu - E_i)/kT\}] \tag{11}$$

on differentiating (3.4.2) with respect to E, the density of states is

$$N_E \equiv \frac{dN(\leq E)}{dE} = (2s+1)2\pi(2m/h^2)^{3/2}\,VE^{1/2}dE \qquad (12)$$

(s is the spin quantum number). Replacing the sum in (11) by an integral gives

(11), (12) $\quad \ln\Xi = -2\pi(2s+1)(2m/h^2)^{3/2}V$

$$\times \int_0^\infty E^{1/2}\ln[\{1 - \exp\{(\mu - E)/kT\}]dE \qquad (13)$$

which can only be evaluated in the limiting cases[a] of *weak degeneracy*, $\exp(\mu/kT) \ll 1$, or *strong degeneracy*, $\exp(\mu/kT) \cong 1$ (provided the energy zero has been chosen to coincide with the lowest energy level).

The energy and pressure are given by

(Ex. 10.4-3) $\quad E = \mu N + kT^2\dfrac{\partial \ln \Xi}{\partial T} = kT^2\dfrac{\partial}{\partial T}\left[-\dfrac{\mu N}{kT} + \ln \Xi\right] \qquad (14)$

(10.4.21) $\quad p = kT\left(\dfrac{\partial \ln \Xi}{\partial V}\right)_{T,\mu} = kT\dfrac{\partial}{\partial V}\left[-\dfrac{\mu N}{kT} + \ln \Xi\right] \qquad (15)$

On substituting

$$x^2 = E/kT \qquad (16)$$

(13) $\quad \ln \Xi = -4\pi(2s+1)(2mkT/h^2)^{3/2}V$

$$\times \int_0^\infty x^2 \ln\{1 - \exp\{(\mu/kT) - x^2\}dx \qquad (17)$$

(9) $\quad N = -\displaystyle\sum_{i=0}^\infty \dfrac{1}{\exp[(E_i - \mu)/kT]}$

$$= 4\pi(2s+1)(2mkT/h^2)^{3/2}V$$

$$\times \int_0^\infty \dfrac{x^2 dx}{\exp[(-\mu/kT) + x^2] - 1} \qquad (18)$$

---

[a]It should be made clear that the words degenerate and degeneracy have two distinct meanings. Firstly, when two different states have the same energy, they are said to be degenerate. This is the common meaning in quantum mechanics. For example the three p-states in the hydrogen atom have the same energy and are thus degenerate. But a degenerate Fermi gas means one for which the Fermi energy is large in comparison with kT.

Equations (17) and (18) show that $\ln \Xi$ and μ/kT depend on $VT^{3/2}$ and therefore that the Helmholtz function F depends on $VT^{5/2}$.

Exercise 10.5-3

How can this be when T appears in the integral, both directly and as x?

$$(10.4.22) \qquad\qquad -\mu N + kT\ln \Xi = -G + pV = -F \qquad\qquad (19)$$

$$\frac{\partial F}{\partial T} = \frac{5}{2}VT^{3/2}\frac{\partial F}{\partial(VT^{5/2})} \qquad\qquad (20)$$

$$\frac{\partial F}{\partial V} = T^{5/2}\frac{\partial F}{\partial(VT^{5/2})} \qquad\qquad (21)$$

Therefore

$$E = kT^2\frac{\partial F}{\partial T} = \frac{3}{2}VT^{5/2}\frac{\partial F}{\partial(VT^{3/2})} \qquad\qquad (22)$$

$$p = kT\frac{\partial F}{\partial V} = kT^{5/2}\frac{\partial F}{\partial(VT^{3/2})} \qquad\qquad (23)$$

$$(22),\ (23) \qquad\qquad pV = \frac{2}{3}E \qquad\qquad (24)$$

as for an ideal gas.

Answers to Exercises 10.5

Exercise 10.5-1

At room temperature, some electrons have been excited from occupied states into empty states just above the Fermi level, thus rounding off the step function (Figure 10.2). The derivative of this function is a narrow Gaussian-like function centered on $E = \mu$.

Exercise 10.5-2

The condition for equilibrium is $\tilde{\mu}_e^\alpha = \tilde{\mu}_e^\beta$. Suppose that $\Phi^\beta > \Phi^\alpha$. Then when contact is made, $\tilde{\mu}_e^\alpha > \tilde{\mu}_e^\beta$ and electrons will flow from phase α to phase β until equilibrium is established.

Exercise 10.5-3

A definite integral is a number, not a function. Therefore the dependence on T and V comes from the terms outside the integral.

10.6. The Ideal Bose–Einstein Gas

Since we wish to study quantum effects in a gas composed of bosons we choose helium, because these effects should be observable most clearly in a gas composed of light molecules with weak intermolecular forces.

Exercise 10.6-1

What kind of statistics should be used for D, H, H_2, D_2, He^4?

Exercise 10.6-2

For a monatomic ideal gas, $pV = NkT$ and $E = \frac{3}{2}NkT$ so that $pV = \frac{2}{3}E$. Would you expect the same result to hold for an ideal gas of bosons?

We consider a gas of indistinguishable, independent particles which have *symmetric* state functions, spin s and a non-zero rest mass. Let n_i be the number of particles with energy E_i so that

$$\sum_i n_i = N, \quad \sum_i n_i E_i = E \tag{1}$$

Unlike the FD case, there are no restrictions on the values of the n_i which may have any value that satisfies (1). The BE distribution law is

$$(10.5.9) \qquad n_i = \frac{1}{\exp[\{E_i - \mu\}/kT] - 1} \tag{2}$$

The grand partition function is

$$\Xi(T, V, \mu) = \prod_i \xi_i(T, V) \tag{3}$$

where

$$\xi_i = 1 + \lambda \exp[-E_i(V)/kT] + \lambda^2 \exp[-2E_i(V)/kT] + \cdots$$
$$= \frac{1}{1 - \lambda \exp[-E_i/kT]} \tag{4}$$

The series is a geometric progression and therefore convergent only if

$$\lambda \exp[-E_i(V)/kT] < 1 \tag{5}$$

for all values of i, including the ground state E_0 which is usually set equal to zero

$$(3),\ (4) \qquad \ln \Xi = \sum_{i=0}^{\infty} \ln[1 - \exp(\mu - E_i)/kT] \tag{6}$$

Exercise 10.6-3

How many states are there with energy between E and $E + \mathrm{d}E$?

Replacing the sum in (6) by an integral, we have, for Bose–Einstein statistics

$$(6) \qquad pV = kT\ln\Xi = -2\pi kT(2s+1)(2m/h^2)^{3/2}V$$

$$\times \int_0^\infty E^{1/2}\ln[1 - \exp(\mu - E)/kT]\mathrm{d}E \qquad (7)$$

Equation (7) can be evaluated provided the energy zero has been taken as that of the lowest quantum state. We are concerned, therefore, with the range $0 \leq \lambda \leq 1$.

Exercise 10.6-4

Prove that λ can never be greater than unity.

We consider a gas of non-interacting boson particles, which is an approximate model for He^4 in which the intermolecular forces are relatively weak. The phase diagram of liquid He^4 shows that there are two liquid forms, He^I and He^{II}. The viscosity of He^I is not unusual but He^{II} shows a remarkable fluidity and is referred to as a superfluid. Helium is a liquid at temperatures very close to 0 K under its own vapour pressure and indeed under pressures as high as $2.5\,\mathrm{MN\,m^{-2}}$. Heat capacity measurements reveal a λ-type transition at $2.17\,\mathrm{K}$. This is the He^I–He^{II} transition, since the viscosity of the liquid below the transition is extremely small.

We are at first concerned only with λ close to one (the *strong degeneracy* case). This means that the Fermi energy E_F is much greater than kT. Consequently, states with energy lower than E_F will be almost entirely occupied while those with energy greater than E_F will be almost entirely vacant. The BE distribution function is

$$(2) \qquad N = \sum_{i=0}^\infty \frac{1}{\exp[(E_i - \mu)/kT] - 1} \qquad (8)$$

The number of particles in all the excited states is

$$N_{\mathrm{ex}} = \frac{V}{4\pi}\left(\frac{2m}{\hbar^2}\right)^{3/2}\int_0^\infty \frac{E^{1/2}}{\lambda^{-1}\exp(E/kT)}\mathrm{d}E \qquad (9)$$

Exercise 10.6-5

Why does the integral in (5) not include the particles in the ground state?

Setting $E_0 = 0$, the number of particles in the ground state at low temperatures is

$$(10.5.9) \qquad n(0,T) = \frac{1}{\exp[(-\mu)/kT] - 1} \qquad (10)$$

At $T = 0\,\mathrm{K}$ there would be no particles in excited states. Thus in a system of Bose–Einstein particles, all these particles accumulate in the ground state as T approaches $0\,\mathrm{K}$, a phenomenon known as *Einstein condensation*. (Note that Eqs. (11) and (12) occur in the solution to Problem (10.5).)

The number of particles in excited states is

$$N_{ex}(T) = \frac{V}{4\pi^2} \left(\frac{2m}{\hbar^2}\right)^{3/2} \int_0^\infty \frac{E^{1/2}}{\lambda^{-1}\exp(E/kT) - 1}\,dE \qquad (13)$$

or, on setting $x = E/kT$

$$N_{ex}(T) = \frac{V}{4\pi^2} \left(\frac{2m}{\hbar^2}\right)^{3/2} (kT)^{3/2} \int_0^\infty \frac{x^{1/2}}{\lambda^{-1}\exp(x) - 1}\,dx \qquad (14)$$

Following Hill (1962) expand $[(1 - \lambda\exp(-x)]^{-1}$ and integrate term by term giving

$$N_{ex}(T) = \left(\frac{2\pi mkT}{h^2}\right)^{3/2} V F_{3/2}(\alpha) \qquad (15)$$

where

$$\alpha = -\mu/kT, \quad \exp(-\alpha) = \lambda \qquad (16)$$

and the function

$$F(\alpha) = \sum_{j\geq1} j^{-\sigma}\exp(-j\alpha), \quad F_\sigma(0) = \sum_{j\geq1} j^{-\sigma} = \zeta(\sigma) \qquad (17)$$

$\zeta(\sigma)$ is the Riemann zeta function, with $\sigma = 3/2$, so that $\zeta(1.5) = 2.612$ (Jahnke and Emde 1945).

Weak degeneracy

With

$$y^2 = E/kT \qquad (18)$$

the total number of particles is

(2), (18) $$N = 4\pi(2s+1)\left(\frac{2mkT}{h^2}\right)^{3/2} V \int_0^\infty \frac{y^2 dy}{\exp[-(\mu/kT)+y^2]-1}$$

(19)

As a first approximation, take $\exp(-\mu/kT)$ to be independent of y. Then

(19) $$N = 4\pi(2s+1)\left(\frac{2mkT}{h^2}\right)^{3/2} V \exp(\mu/kT)$$

$$\times \int_0^\infty y^2 \exp(-y^2) dy$$

$$= \pi(2s+1)\left(\frac{2mkT}{h^2}\right)^{3/2} V \exp(\mu/kT)\pi^{1/2} \qquad (20)$$

(20) $$\exp(\mu/kT) = \frac{1}{2s+1}\left(\frac{N}{V}\right)\left(\frac{h^2}{2\pi mkT}\right)^{3/2} \qquad (21)$$

which is a first approximation for μ. To obtain a second approximation, use

(19) $$N = 4\pi(2s+1)\left(\frac{2mkT}{h^2}\right)^{3/2} V \int_0^\infty \frac{y^2 \exp[(\mu/kT)-y^2]dy}{1-\exp[(\mu/kT)-y^2]} \qquad (22)$$

and expand the denominator, giving

$$N = -4\pi(2s+1)\left(\frac{2mkT}{h^2}\right)^{3/2} V \int_0^\infty y^2 e^{\{(\mu/kT)-y^2\}}$$

$$\times [1+e^{\{(\mu/kT)-y^2\}}+e^{2\{(\mu/kT)-y^2\}}+\cdots]dy \qquad (23)$$

(23) $$N = (2s+1)\left(\frac{2\pi mkT}{h^2}\right)^{3/2} V \left[e^{\mu/kT}+\frac{e^{2\mu/kT}}{2^{3/2}}+\frac{e^{3\mu/kT}}{3^{3/2}}+\cdots\right]$$

(24)

Define

$$z = \frac{1}{2s+1}\frac{N}{V}\left(\frac{h^2}{2\pi mkT}\right)^{3/2} \qquad (25)$$

(24), (25) $$z = e^{\mu/kT}+\frac{z^2}{2^{3/2}}+\cdots \qquad (26)$$

(26) $$e^{\mu/kT} = z - \frac{z^2}{2^{3/2}}+\cdots \qquad (27)$$

We may now obtain the thermodynamic functions in this approximation. Since

$$-\ln(1-t) = \sum_{n=1}^{\infty} \frac{t^n}{n} \tag{28}$$

$$\ln \Xi = 4\pi(2s+1)\left(\frac{2mkT}{h^2}\right)^{3/2} V$$

$$\times \int_0^{\infty} y^2 \ln[1 - e^{(\mu/kT)-y^2}]dy \tag{29}$$

$$= \frac{4N}{z\pi^{1/2}} \sum_{n=1}^{\infty} e^{\mu n/kT} \int_0^{\infty} y^2 e^{-ny^2} dy$$

$$= \frac{N}{z} \sum_{n=1}^{\infty} \frac{e^{\mu n/kT}}{n^{5/2}} \tag{30}$$

(27)

$$\ln \Xi = \frac{N}{z}\left[e^{\mu/kT} + \frac{e^{2\mu/kT}}{2^{5/2}} + \cdots\right] \tag{31}$$

(31)

$$U = \frac{3}{2}NkT\left[1 - \frac{z}{2^{5/2}} + \cdots\right] \tag{32}$$

$$S = \frac{U-F}{T} = \frac{U-\mu N}{T} + k\ln \Xi \tag{33}$$

$$= Nk\left[\frac{5}{2} - \ln z - \frac{z}{2^{7/2}} + \cdots\right] \tag{34}$$

Answers to Exercises 10.6

Exercise 10.6-1

Electrons, protons and neutrons all have spin $1/2$ and therefore D and He^3, which have an odd number of these particles, have state functions which are antisymmetric, while H, H_2, D_2, He^4, which have an even number of these fundamental particles, have symmetric state functions.

Exercise 10.6-2

Since BE statistics become MB statistics when $(E_i - \mu)/kT$ is sufficiently large, yes one would expect the same ideal gas relationship to hold, as indeed turns out to be the case.

Exercise 10.6-3

The number of states with energy between E and $E + dE$ is

(10.6.2) $$(2s + 1)2\pi(2m/h^2)^{3/2}\,VE^{1/2}dE$$

Exercise 10.6-4

The number of molecules in the ground state $(E_1 = 0)$ is $\lambda/(1 - \lambda)$ and hence $\lambda \leq 1$, which means that $\mu < 0$.

Exercise 10.6-5

The integrand in (5) is the density of states and because of the factor $E^{1/2}$, this is zero when $E = 0$.

10.7. The Ideal Fermi–Dirac Gas

In an ideal gas, the number of translational states with energy $\leq E$ is

(3.4.2) $$N(\leq E) = \frac{4\pi}{3}\left(\frac{2mE}{h^2}\right)^{\frac{3}{2}}V \tag{1}$$

Therefore the density of states is

(1) $$N_E dE = \frac{dN(\leq E)}{dE}dE = 2\pi\left(\frac{2m}{h^2}\right)^{\frac{3}{2}}VE^{1/2}dE \tag{2}$$

For the electrons in a metal, there is a spin-degeneracy of two because each may have a spin of $\pm^1/_2$ in the usual units of $h/2\pi$. Therefore the density of states in (2) must be multiplied by 2 to give

$$N_E dE = 4\pi\left(\frac{2m}{h^2}\right)^{\frac{3}{2}}VE^{1/2}dE \tag{3}$$

The energy zero chosen is that of an electron at rest in the gas phase well outside the metal. Consequently, inside the metal free electrons from the valence shell of each metal atom are subjected to a constant (negative) potential of the order of several eV due to the attraction of the positive metal cores. (In the free-electron model this potential is taken as constant as a first approximation. In reality, the core potential is periodic, but this makes the situation much more difficult, so we content ourselves here with the "free-electron" approximation which is adequate for our purpose.) It is convenient to shift the zero on the potential scale so that $\phi = 0$ inside the metal. Then

$$\langle N \rangle = 4\pi \left(\frac{2m}{h^2}\right)^{\frac{3}{2}} V \int_0^\infty \frac{E^{1/2}}{\exp[(E-\mu)/kT]+1} dE \qquad (4)$$

For $T = 0\,\mathrm{K}$, $\langle N \rangle = 1$ for $E < \mu$ and $\langle N \rangle = 0$ for $E > \mu$, and so

$$(4) \qquad \langle N \rangle = 4\pi \left(\frac{2m}{h^2}\right)^{\frac{3}{2}} V \int_0^\mu E^{1/2} dE = \frac{8\pi}{3}\left(\frac{2m\mu}{h^2}\right)^{\frac{3}{2}} V \qquad (5)$$

For given electron density $\langle N \rangle/V$, all states are filled up to the Fermi energy

$$(5) \qquad \mu = \frac{h^2}{8m}\left(\frac{3\langle N \rangle}{\pi V}\right)^{\frac{2}{3}} \qquad (6)$$

Exercise 10.7-1

Explain why the electronic contribution to the heat capacity of a metal is much less than that expected for a classical free-electron gas.

The (kinetic) energy at $T = 0\,\mathrm{K}$ is

$$\langle E \rangle = -\sum_{i=0}^{\infty} n_i E_i = 4\pi \left(\frac{2m}{h^2}\right)^{\frac{3}{2}} V \int_0^\mu E^{3/2} dE = \frac{3N\mu}{5} \qquad (7)$$

Notice that the ground state energy was taken as zero in this calculation.

Exercise 10.7-2

Explain why the zero-point energy is so high in this system.

Answers to Exercises 10.7

Exercise 10.7-1

The valence electrons in a metal are free in the sense that each electron is no longer associated with one particular atom but is free to roam throughout the space occupied by the metal atoms, thus accounting for the high electronic conductivity of metals. However, only the electrons near the Fermi level can be excited thermally into vacant excited states. The large majority of the electrons may not be excited thermally because there are no empty states available and the PEP prevents each energy state from being occupied by more than two electrons with opposed spins.

Exercise 10.7-2

Because of the Pauli exclusion principle, μ is of the order of a few electron volts and hence such systems have a high zero-point energy.

10.8. Temperature Dependence of Thermodynamic Functions for a FD gas

For the ideal FD gas

(10.6.6)
$$\ln \Xi = \sum_{i=0}^{\infty} \ln[1 + \exp(\mu - E_i)/kT] \qquad (1)$$

(Note the sign change between BE and FD statistics.) As for the ideal BE gas, μ/kT and $\ln \Xi$ depend on V and T through the factor $VT^{3/2}$ so that again

$$pV = \frac{2}{3}E \qquad (2)$$

As for the BE case, $\ln \Xi$ is most easily evaluated for weak and strong degeneracy.

Again, as for the BE gas

(10.6.27)
$$e^{\mu/kT} = z + \frac{z^2}{2^{3/2}} + \cdots \qquad (3)$$

where

(10.6.25)
$$z = \frac{1}{2s+1} \frac{N}{V} \left(\frac{h^2}{2\pi mkT} \right)^{3/2} \qquad (4)$$

Therefore

(10.6.31)
$$pV = kT\ln \Xi = NkT \left[1 + \frac{z}{2^{5/2}} + \cdots \right] \qquad (5)$$

(10.6.33)
$$U = \frac{3}{2}NkT \left[1 + \frac{z}{2^{5/2}} + \cdots \right] \qquad (6)$$

(10.6.34)
$$S = Nk \left[\frac{5}{2} - \ln z + \frac{z}{2^{7/2}} + \cdots \right] \qquad (7)$$

Other thermodynamic functions may be found from (5), (6) and (7).

Strong degeneracy

The calculation of the temperature dependence of the FD thermodynamic functions under conditions of strong degeneracy is rather lengthy.

In FD statistics, the probability of a state having an energy between E and $E + \mathrm{d}E$ being occupied is

(10.5.4)
$$f(E)\mathrm{d}E = \frac{\mathrm{d}E}{\exp\{(E - \mu)/kT\} + 1} \tag{8}$$

Consider the integral

$$I = \int_0^\infty f(E)\frac{\mathrm{d}F}{\mathrm{d}E}\mathrm{d}E \tag{9}$$

where $F(E)$ is a function with the property $F(0) = 0$.

Because of the properties of $f(E)$ (see Section 10.4), $\mathrm{d}f/\mathrm{d}E$ has an appreciable value only near $E = \mu$. Expand $F(E)$ in a Taylor series about μ, giving

$$F(E) = \sum_{n=0}^\infty \frac{(E - \mu)^n}{n!}\left(\frac{\mathrm{d}^n F}{\mathrm{d}E^n}\right)_{E=\mu} \tag{10}$$

Since

$$\frac{\mathrm{d}f}{\mathrm{d}E} = -\frac{1}{kT}\frac{\exp\{(E - \mu)/kT\}}{[\exp\{(E - \mu)/kT\} + 1]^2} \tag{11}$$

(3), (4), (5) $\quad I = \dfrac{1}{kT}\displaystyle\int_0^\infty \sum_{n=0}^\infty \frac{(E - \mu)^n}{n!}\frac{\exp\{(E - \mu)/kT\}}{[\exp\{(E - \mu)/kT\} + 1]^2}\left(\frac{\mathrm{d}^n F}{\mathrm{d}E^n}\right)_\mu$

$$= \sum_{n=0}^\infty \frac{(kT)^n}{n!}\left(\frac{\mathrm{d}^n F}{\mathrm{d}E^n}\right)_\mu \int_{-\infty}^\infty z^n e^{-z}(1 + e^{-z})^{-2}\mathrm{d}z \tag{12}$$

Exercise 10.8-1

Show that $e^z(1 + e^z)^{-2} = e^{-z}(1 + e^{-z})^{-2}$, as used in the above derivation.

Expanding $(1 + e^{-z})^{-2}$ gives

(12)
$$I = \sum_{n=0}^\infty \frac{(kT)^n}{n!}\left(\frac{\mathrm{d}^n F}{\mathrm{d}E^n}\right)_\mu \int_{-\infty}^\infty z^n e^{-z}\sum_{k=0}^\infty k(-1)^{k-1}e^{-kz}\,\mathrm{d}z$$

$$= F(\mu) + 2\sum_{n=1}^\infty (kT)^{2n}\left(\frac{\mathrm{d}^{2n}F}{\mathrm{d}E^{2n}}\right)_\mu\left(\sum_{k=1}^\infty \frac{(-1)^{k-1}}{k^{2n}}\right) \tag{13}$$

Generally, only the first term is used, giving

$$\int_0^\infty f(E)\frac{\mathrm{d}F}{\mathrm{d}E}\mathrm{d}E = F(\mu) + \frac{\pi^2}{6}(kT)^2\left(\frac{\mathrm{d}^2 F}{\mathrm{d}E^2}\right)_\mu \tag{14}$$

which is Sommerfeld's integral (9). The thermodynamic functions may now be evaluated

$$N = \int_0^\infty N_E(E)f(E)dE = 2\pi(2s+1)\left(\frac{2m}{h^2}\right)^{3/2}V\int_0^\infty E^{1/2}f(E)dE$$

$$= \frac{4\pi}{3}(2s+1)\left(\frac{2m}{h^2}\right)^{3/2}V\mu^{3/2}\left[1+\frac{\pi^2}{8}\left(\frac{kT}{\mu}\right)^2\right] \tag{15}$$

At $T = 0\,\mathrm{K}$,

$$N = \frac{4\pi}{3}(2s+1)\left(\frac{2m}{h^2}\right)^{3/2}V\mu(0)^{3/2} \tag{16}$$

(Note that Eqs. (17), (20) and (21) occur in the solutions to Problems 10.)

$$E = \int_0^\infty EN_E(E)f(E)dE = 2\pi(2s+1)\left(\frac{2m}{h^2}\right)^{3/2}V\int_0^\infty E^{3/2}f(E)dE$$

$$= (2s+1)\frac{4\pi}{5}\left(\frac{2m}{h^2}\right)^{3/2}V\mu^{5/2}\left[1+\frac{5\pi^2}{8}\left(\frac{kT}{\mu(0)}\right)^2+\cdots\right] \tag{18}$$

(16), (17), (18)

$$E = \frac{3N}{5}\mu(0)\left[1-\frac{\pi^2}{12}\left(\frac{kT}{\mu(0)}\right)^2+\cdots\right]\left[1+\frac{5\pi^2}{8}\left(\frac{kT}{\mu(0)}\right)^2+\cdots\right]$$

$$= \frac{3N}{5}\mu(0)\left[1+\frac{5\pi^2}{12}\left(\frac{kT}{\mu(0)}\right)^2+\cdots\right] \tag{19}$$

Answer to Exercise 10.8-1

Multiply numerator and denominator of the RS by e^{z^2}, giving

$$\frac{e^{z^2}}{e^{z^2}}\times\frac{e^{-z}}{(1+e^{-z})^2}=\frac{e^z}{(e^z+1)^2}=\mathrm{LS}$$

Problems 10

10.1 Use Lagrange's method of undetermined multipliers to derive Eq. (10.4.5).

10.2 Calculate the Fermi energy of (a) Ag and (b) Cu at 298 K given that the electron concentration at this temperature in Ag is $5.85\times10^{22}\,\mathrm{cm}^{-3}$ and that in Cu is $8.45\times10^{22}\,\mathrm{cm}^{-3}$.

10.3 Show that the He^3 atom is a fermion. At temperatures approaching 0 K, the density of liquid He^3 is $0.81\,\mathrm{g\,cm^{-3}}$. Calculate the Fermi energy of He^3.

10.4 Find the molar heat capacity at constant volume and the molar Helmholtz energy for an ideal Bose–Einstein gas.

10.5 Find the chemical potential for an ideal Bose–Einstein gas as T approaches 0 K.

10.6 Show that the Fermi energy μ decreases slightly with increasing T.

10.7 Find expressions for the heat capacity C_V and the entropy S of an ideal FD gas in the strong degeneracy limit.

11

Thermodynamics of Solids

11.1. Symmetry and the Physical Properties of Crystals

That symmetry can be useful in a discussion of the physical properties of crystals is shown by Neumann's principle that the symmetry of any physical property of a crystal must include at least that of the point group of the crystal. We begin with a brief description of tensors. It will be more convenient in this section to replace the notation x, y, z by x_1, x_2, x_3. Symmetry operations transform points in space so that under a proper or improper rotation A, $P(x_1, x_2, x_3)$ is transformed into $P'(x'_1, x'_2, x'_3)$. The result of applying A may be represented by

$$|x'_1 \ x'_2 \ x'_3 >= A|x_1 \ x_2 \ x_3 > \tag{1}$$

where the ket brackets denote a column matrix. (If necessary, please see Appendix A1 of Jacobs (2005) for a description of matrices and matrix notation, sufficient for the present purpose.) Equation (1) may be re-written more concisely as

$$x'_i = a_{ij} x_j \tag{2}$$

in which the Einstein summation convention implies a sum over repeated indices. Since the a_{ij} are real, Eq. (2) describes an orthogonal transformation (Jacobs 2005, pp. 60–61) with

$$AA^T = E, \quad \det A = \pm 1 \tag{3}$$

Here, $+1$ applies to a proper rotation and -1 to an improper rotation (that is a proper rotation followed by inversion). When two symmetry operators B and A are applied successively, so that $P \rightarrow P' \rightarrow P''$

$$x''_i = a_{ij} x'_j = a_{ij} b_{jk} x_k = c_{ik} x_k \tag{4}$$

The symmetry operator $C = AB$ is represented by the matrix $C = [c_{ik}] = AB$.

319

If T is invariant under all proper and improper rotations then T is a scalar, or tensor of rank zero, written $T(0)$. If T is invariant under proper rotations but changes sign on inversion, then it is a pseudoscalar with the property $IT(0) = -T(0)$. Pseudoscalars are also called axial tensors of rank 0 and denoted by the symbol $T(0)^{\text{ax}}$. If T has three components $\{T_1 \; T_2 \; T_3\}$ that transform like the coordinates of a point P, that is, like the components of the position vector $\boldsymbol{r}$

$$T_i' = a_{ij}T_j, \quad i,j = 1,2,3 \tag{5}$$

then T is a polar vector or tensor of rank 1, $T(1)$. If

$$T_i' = \pm a_{ij}T_j, \quad i,j = 1,2,3 \tag{6}$$

(where the $+$ sign applies to proper rotations $\{R\}$ and the $-$ sign to improper rotations $\{IR\}$), then T is a pseudovector or axial vector or axial (or pseudo-) tensor of rank 1, $T(1)^{\text{ax}}$. The products $u_i v_j$ of the components of two vectors $\boldsymbol{u}$, $\boldsymbol{v}$ transform as

$$u_i' v_j' = a_{ik}a_{jl}u_k v_l \tag{7}$$

The $\{a_{ik}a_{jl}\}$ are the set of components of a tensor of rank 2. These definitions can be extended to tensors of rank n. If a $T(2)$ has

$$T_{ij} = T_{ji}, \quad \forall \, i,j \tag{8}$$

then it is symmetric, with only six independent components. But if

$$T_{ij} = -T_{ji}, \quad \forall \, i,j \tag{9}$$

then it is antisymmetric, with the diagonal elements T_{ii} zero and therefore only three independent components. The symmetry of a tensor is a feature of the physical property it represents and consequently symmetry operations may impose additional relations between tensor components.

Exercise 11.1-1

Write down the defining equations, analogous to (5), for a tensor of rank 4 and a tensor of rank n.

Physical properties of crystals are often represented by symmetric tensors of rank 2, in which case a single index notation introduced by Voigt is useful. (The second row below contains the contracted form.)

$$T_{ijkl} \quad ij \text{ or } kl \quad 11 \quad 22 \quad 33 \quad 23 \text{ or } 32 \quad 31 \text{ or } 13 \quad 12 \text{ or } 21$$

$$T_{pq} \quad p \text{ or } q \quad 1 \quad 2 \quad 3 \quad 4 \quad 5 \quad 6 \qquad (10)$$

Note, however, (i) that the $\{T_{pq}\}$ do not form a second rank tensor, so that symmetry transformations must be carried out using the four-index notation T_{ijkl}; (ii) the contraction of T_{ijkl} to T_{pq} may be accompanied by the introduction of numerical factors, as happens, for example, when $T(4)$ is the elastic stiffness (Nye 1957).

Answer to Exercise 11.1-1

A $T(4)$ has $3^4 = 81$ components like T_{pqrs}, which transform like

$$T'_{ijkl} = a_{ip} a_{jq} a_{kr} a_{ls} T_{pqrs}$$

where i, j, k, l, p, q, r, $s = 1, 2,$ or 3.
A $T(n)$ has 3^n components that transform like

$$T'_{ijk\ldots} = a_{ip} a_{jq} a_{kr} \ldots T_{pqr\ldots}$$

where $i, j, k, \ldots p, q, r_{\ldots} = 1, 2, 3$ and there are n subscripted indices on T and T'.

11.2. Stress and Strain

Thermodynamics, as developed in Chapters 1–10 of this book, is applicable to gases, liquids and solids that are not subjected to any stress other than a uniform pressure. We now consider the changes that are necessary when a solid is subjected to anisotropic stresses and, as a consequence, develops strains.

A solid body is strained when the relative distance and orientation of two points is changed. Consider two points in the solid Q and R (see Figure 11.1) such that the vector distance OQ is r and that OR is $r + dr$. Now suppose that because of an applied stress the points Q and R are displaced to Q' and R', such that the distance Q'R' is dr'. Let QQ' be denoted by $s(r)$. Then RR' $= s(r + dr)$. We limit considerations to *small deformations* so that the strain $s(r)$ is a continuous function of r and the change in the displacement $dr' - dr$ depends only on the first derivatives

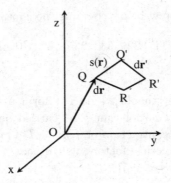

Figure 11.1. A strained crystal.

of the components of $s(r)$ with respect to position. Then

$$
dr' = dr + dr \cdot
\begin{bmatrix}
\dfrac{\partial s_x(r)}{\partial x} & \dfrac{\partial s_y(r)}{\partial x} & \dfrac{\partial s_z(r)}{\partial x} \\[2ex]
\dfrac{\partial s_x(r)}{\partial y} & \dfrac{\partial s_y(r)}{\partial y} & \dfrac{\partial s_z(r)}{\partial y} \\[2ex]
\dfrac{\partial s_x(r)}{\partial z} & \dfrac{\partial s_y(r)}{\partial z} & \dfrac{\partial s_z(r)}{\partial z}
\end{bmatrix}
\tag{1}
$$

$$
= dr + dr \cdot S(r)
\tag{2}
$$

$s(r)$ is the strain and it is a vector function of position with components $s(x)$, $s(y)$, $s(z)$. Its derivative with respect to position $S(r)$ has nine components and is a second rank tensor. The multiplication in Eq. (2) is row by column and yields the change in the small distance dr that results from straining the crystal.

A translation of the crystal does not change its thermodynamic state. In this case, $dr' = dr$ and

$$
S(r) = 0 \qquad \text{(translation)}
\tag{3}
$$

A rotation of the crystal also leaves its thermodynamic state unchanged. In this case $S(r)$ is not zero though it must be antisymmetric. Any second order tensor may be rewritten as the sum of symmetric and antisymmetric parts. The antisymmetric part represents a pure rotation and will be discarded. The symmetric part is the strain tensor Σ.

Exercise 11.2-1

Show that $S(r)$ may be written as the sum of symmetric and antisymmetric parts. (**Hint:** Recall that $S(r)$ is represented by a 3×3 matrix.)

Exercise 11.2-2

Write out the matrix Σ showing the nine components of the strain tensor explicitly.

So if the coordinate axes are embedded in the system and therefore rotated with the system, the antisymmetric part of S becomes zero and

$$dr' = dr + dr \cdot \Sigma \tag{6}$$

To ensure this simplification, locally-rotated coordinates will always be used. The diagonal element Σ_{xx} is the fractional change in length of an element of the solid which is initially parallel to the x-axis and is therefore called the *linear dilation* in the x direction (and similarly for the y and z directions).

To understand the physical meaning of the off-diagonal elements, consider the angle between two elements $(dl)\mathbf{e}_1$ and $(dl)\mathbf{e}_2$ that are initially along the x and y axes. In the strained state $(dl)\mathbf{e}_1$ becomes

$$(dl)\mathbf{e}_1 + (dl)\mathbf{e}_1 \cdot \Sigma = (dl)(1 + \Sigma_{xx})\mathbf{e}_1 + (dl)\Sigma_{xy}\mathbf{e}_2 + (dl)\Sigma_{xz}\mathbf{e}_3 \tag{7}$$

Similarly, $(dl)\mathbf{e}_2$ becomes

$$(dl)\mathbf{e}_2 + (dl)\mathbf{e}_2 \cdot \Sigma = (dl)\Sigma_{xy}\mathbf{e}_1 + (dl)(1 + \Sigma_{yy})\mathbf{e}_2 + (dl)\Sigma_{yz}\mathbf{e}_3 \tag{8}$$

Let ψ_{xy} be the angle between these two elements in the strained state. Then

$$(7), (8) \quad \cos\psi_{xy} = \Sigma_{xy}(1 + \Sigma_{xx}) + \Sigma_{xy}(1 + \Sigma_{yy}) + \Sigma_{xz}\Sigma_{yx} = 2\Sigma_{xy} \tag{9}$$

to first order in the strain components. Since ψ_{xy} differs only slightly from $\pi/2$ we define the small angle φ_{xy} by

$$\varphi_{xy} = (\pi/2) - \psi_{xy} \tag{10}$$

and find

$$(10) \qquad \cos\psi_{xy} = \sin\varphi_{xy} = \varphi_{xy} \quad \text{(to first order in } \varphi_{xy}) \tag{11}$$

The off-diagonal element Σ_{xy} in the strain tensor is thus equal to *half* the decrease in the angle between two line elements that are initially along the

x and y axes in the unstressed state. The other off-diagonal elements Σ_{xz} and Σ_{yz} have similar meanings.

Because of the symmetry of the strain tensor there are only six independent components and it is customary to adopt the notation (*cf.* (1.1.10))

$$\Sigma_1 = \Sigma_{xx}; \quad \Sigma_2 = \Sigma_{yy}; \quad \Sigma_3 = \Sigma_{zz}$$

$$\Sigma_4 = 2\Sigma_{yz}; \quad \Sigma_5 = 2\Sigma_{zx}; \quad \Sigma_6 = 2\Sigma_{xy} \tag{12}$$

Σ_1 through Σ_6 are referred to as the *strain components*, though they are not the components of the strain tensor because of the factors of two introduced in their definition.

Answers to Exercises 11.2

Exercise 11.2-1

Let $S^{\mathbf{T}}$ denote the transpose of S. Then

$$S = \frac{1}{2}[S + S^{\mathbf{T}}] + \frac{1}{2}[S - S^{\mathbf{T}}] = \Sigma + \frac{1}{2}[S - S^{\mathbf{T}}] \tag{4}$$

where Σ is the symmetric part of S. Σ is therefore the *strain tensor*.

Exercise 11.2-2

$$\Sigma = \begin{bmatrix} \Sigma_{xx} & \Sigma_{xy} & \Sigma_{xz} \\ \Sigma_{xy} & \Sigma_{yy} & \Sigma_{yz} \\ \Sigma_{xz} & \Sigma_{yz} & \Sigma_{zz} \end{bmatrix}$$

$$= \begin{bmatrix} \dfrac{\partial s_x}{\partial x} & \dfrac{1}{2}\left(\dfrac{\partial s_x}{\partial y} + \dfrac{\partial s_y}{\partial x}\right) & \dfrac{1}{2}\left(\dfrac{\partial s_x}{\partial z} + \dfrac{\partial s_z}{\partial x}\right) \\ \dfrac{1}{2}\left(\dfrac{\partial s_x}{\partial y} + \dfrac{\partial s_y}{\partial x}\right) & \dfrac{\partial s_y}{\partial y} & \dfrac{1}{2}\left(\dfrac{\partial s_y}{\partial z} + \dfrac{\partial s_z}{\partial y}\right) \\ \dfrac{1}{2}\left(\dfrac{\partial s_x}{\partial z} + \dfrac{\partial s_z}{\partial x}\right) & \dfrac{1}{2}\left(\dfrac{\partial s_y}{\partial z} + \dfrac{\partial s_z}{\partial y}\right) & \dfrac{\partial s_z}{\partial z} \end{bmatrix} \tag{5}$$

11.3. Thermodynamics of Stress and Strain

At ordinary pressures, the $p\,dV$ term is negligibly small for solids and it is omitted from the fundamental equation for the energy, which therefore becomes

$$dU = TdS + \sum_{i=1}^{6} T_i d(V_0\Sigma_i) + \sum_B \mu_B dN_B \qquad (1)$$

The strain components Σ_i are intensive properties but become extensive properties on multiplication by V_0, the volume of the unstrained crystal. The T_i are the *stress components* and are defined by

$$(1) \qquad\qquad T_i = \frac{1}{V_0}\left(\frac{\partial U}{\partial \Sigma_i}\right)_{S,\Sigma_{j\neq i}} \qquad (2)$$

Since the strain is represented by a symmetric second-order tensor, Σ, so is the stress.

Exercise 11.3-1

Clarify the physical meaning of the stress components T_{xx} and T_{xy}.

As U is a homogeneous function of degree 1, Euler's theorem (Appendix A1) yields

$$(1) \qquad\qquad U = TS + \sum_{i=1}^{6} T_i(V_0\Sigma_i) + \sum_B \mu_B N_B \qquad (3)$$

The Gibbs–Duhem equation for a strained system is

$$(1), (3) \qquad SdT + \sum_{i=1}^{6}(V_0\Sigma_i)dT_i + \sum_B N_B\, d\mu_B = 0 \qquad (4)$$

Since (1) is a complete differential, various Maxwell relations follow, for example

$$V_0\left(\frac{\partial \Sigma_i}{\partial T}\right)_{T_i} = \left(\frac{\partial S}{\partial T_i}\right)_{T,T_{j\neq i}} \qquad (5)$$

$$\frac{1}{V_0}\left(\frac{\partial \mu_B}{\partial \Sigma_i}\right)_{T,\mu_{C\neq B},\Sigma_{j\neq i}} = \left(\frac{\partial T_i}{\partial N_B}\right)_{T,\Sigma_i,\mu_{C\neq B}} \qquad (6)$$

$$\left(\frac{\partial \Sigma_i}{\partial T_j}\right)_{T,T_{i\neq j}} = \left(\frac{\partial \Sigma_j}{\partial T_i}\right)_{T,T_{j\neq i}} \qquad (7)$$

11.3.1. *Elastic coefficients*

The *isothermal elastic stiffness coefficients* are defined by

$$c_{ij} = \left(\frac{\partial T_i}{\partial \Sigma_j} \right)_{T, \Sigma_i \neq j} \quad , \quad i, j = 1, \ldots 6 \tag{8}$$

Take constant composition as understood in this section. The Maxwell relation (7) tells us that the array of c_{ij} is symmetric, $c_{ij} = c_{ji}$. This reduces the number of independent coefficients from 36 to 21. The *isothermal elastic compliance coefficients* are defined by

$$\kappa_{ij} = \left(\frac{\partial \Sigma_i}{\partial T_j} \right)_{T, T_i \neq j} \quad , \quad i, j = 1, \ldots 6 \tag{8}$$

Again, the matrix $\kappa = [\kappa_{ij}]$ is symmetric. The isothermal compliance coefficients are analogous to the isothermal compressibility κ of a fluid, and the isothermal elastic stiffness coefficients are analogous to the isothermal bulk modulus of a fluid.

The 21 elastic stiffness coefficients are further reduced in number by crystal symmetry. For example, consider cubic symmetry. The 48 symmetry operators which leave a cube (or octahedron) in an indistinguishable configuration are R and IR, where $R \in \{E, 8C_3, 6C_4, 3C_2, 6C_2'\}$. E is the identity, I the inversion operator, C_n a rotation about an n-fold axis of symmetry, and C_2' a rotation about a two-fold axis that is normal to a principle axis. For further details on group theory see Jacobs (2005) or Cotton (1963). Indistiguishability with respect to the cyclic change $x \rightarrow y \rightarrow z \rightarrow x$ (which results from the presence of a three-fold axis along the [111] direction) requires that $c_{11} = c_{22} = c_{33}$.

It is physically obvious that if an object (crystal) has cubic symmetry, there cannot be any difference between the x, y and z directions, which are chosen arbitrarily. For the same reason, there cannot be any difference between c_{12}, c_{13}, and c_{23}. Now consider a reflection in the mirror plane zx which transforms y into $\bar{y}$.

$$c_{41} = \frac{\partial T_{xy}}{\partial \Sigma_{xx}} = \frac{\partial T_{x\bar{y}}}{\partial \Sigma_{xx}} = -\frac{\partial T_{xy}}{\partial \Sigma_{xx}} = -c_{41}$$

Therefore

$$c_{41} = c_{51} = c_{61} = 0$$

Again, since the x, y, and z axes are indistinguishable, the c matrix is reduced to

$$
\begin{bmatrix}
c_{11} & c_{12} & c_{12} & 0 & 0 & 0 \\
c_{12} & c_{11} & c_{12} & 0 & 0 & 0 \\
c_{12} & c_{12} & c_{11} & 0 & 0 & 0 \\
0 & 0 & 0 & c_{44} & 0 & 0 \\
0 & 0 & 0 & 0 & c_{44} & 0 \\
0 & 0 & 0 & 0 & 0 & c_{44}
\end{bmatrix}
\tag{9}
$$

in which there are only three independent elastic stiffness coefficients c_{11}, c_{12} and c_{44}. Although cubic symmetry is the highest symmetry that a crystalline substance may possess, amorphous materials may be isotropic and therefore invariant under any rotation. In this case the stiffness matrix is

$$
\begin{bmatrix}
2\mu_L + \lambda_L & \lambda_L & \lambda_L & 0 & 0 & 0 \\
\lambda_L & 2\mu_L + \lambda_L & \lambda_L & 0 & 0 & 0 \\
\lambda_L & \lambda_L & 2\mu_L + \lambda_L & 0 & 0 & 0 \\
0 & 0 & 0 & \mu_L & 0 & 0 \\
0 & 0 & 0 & 0 & \mu_L & 0 \\
0 & 0 & 0 & 0 & 0 & \mu_L
\end{bmatrix}
\tag{10}
$$

The two independent constants μ_L and λ_L are the Lamé constants of the material.

If all the operators in a symmetry group can be generated from a sub-set of the group elements, the elements comprising this sub-set are called the group generators. Therefore we need only examine the restrictions on the group generators due to symmetry.

Example 11.3-1

Determine the non-zero elements of the elasticity tensor for a crystal of D_4 symmetry.

The group D_4 consists of the operators $\{E, \ 2C_4, \ C_2, \ 2C_2', \ 2C_2''\}$. The generalized form of Hooke's law ("*ut tensio sic vis*") is

$$
\sigma_{ij} = c_{ijkl}\, \varepsilon_{kl}
\tag{11}
$$

Table 11.1.   Table (a) shows the transforms of x_1, x_2, x_3 under the operators C_{4z} and C_{2x}. Tables (b) and (c) give, in suffix notation, the transforms c'_{pq} of c_{pq} under the operators C_{4z} and C_{2x}. Table (d) gives the upper half of the symmetric matrix $[c_{pq}]$.

(a)	x'_1	x'_2	x'_3
C_{4z}	x_2	$-x_1$	x_3
C_{2x}	x_1	$-x_2$	$-x_3$

(b) C_{4z}	p, q	ij	$i'j'$	p', q'						
	1	11	22	2	22	12	23	25	−24	−26
	2	22	11	1		11	13	15	−14	−16
	3	33	33	3			33	35	−34	−36
	4	23	13	5				55	−45	−56
	5	13	−23	−4					−44	−46
	6	12	−12	−6						66

(c) C_{2x}										
	1	11	11	1	11	12	13	14	−15	−16
	2	22	22	2		22	23	24	−25	−26
	3	33	33	3			33	34	−35	−36
	4	23	23	4				44	−45	−46
	5	13	−13	−5					55	56
	6	12	−12	−6						66

(d) $[c_{pq}]$					
11	12	13	0	0	0
	11	13	0	0	0
		33	0	0	0
			44	0	0
				44	0
					66

where both the stress σ and the strain ε are symmetric T_2's. It will be sufficient to study the transformation of x_1, x_2, x_3 under the operators C_{4z} and C_{2x}. The results x'_1, x'_2, x'_3 are shown in Table 11.1 (a) and the remainder of the working is shown in parts (b), (c) and (d) of Table 11.1. A similar procedure may be followed for other point groups and tensors representing other physical properties.

The matrix c_{pq} has to satisfy the requirements of symmetry, established in Table 11.1, parts (b) and (c). For example, $c_{11} = c_{22}$. Since c_{36} transforms into $-c_{36}$ it must be zero. Also $c_{16} = -c_{26} = -c_{16} = 0$. In this way we construct the matrix c_{pq} shown in (d).

Answer to Exercise 11.3-1

T_{xx} is the force per unit area acting in the x direction on a plane normal to the x-axis. It is therefore called a *normal stress*. T_{xy} is the force per unit area acting in the y direction on a plane normal to the x-axis. It is therefore a *shear stress*. It should now be clear that T_{yx} must be equal to T_{xy}, to avoid rotations.

11.4. Point Defects in Ionic Crystals

Ionic crystals like NaCl or MgO have high cohesive energies, as shown by their high fusion temperatures. The ions are located at specific lattice sites determined by the crystal structure of the material, for example the face-centered cubic structure of NaCl. If the structure of a real crystal was perfect, with every lattice site correctly occupied, then we would predict that ion transport in the crystal would be extremely slow, requiring an activation energy approaching that of the cohesive energy of the crystal. Yet it has been known since the time of Pierre and Marie Curie that ionic crystals do conduct an electric current. Investigation of the ionic conductivity of crystals, for example NaCl (Hooton and Jacobs 1990), KCl (Acuña and Jacobs 1980) and RbCl (Jacobs and Vernon 1997) show that plots of $\log \kappa T$ against $10^3 K/T$ consist of two approximately linear sections. The traditional interpretation is that the ion current is carried by defects in the crystal lattice. At high temperatures vacancies are formed in the crystal lattice in thermodynamic equilibrium. These are *intrinsic* defects, called Schottky defects (Figure 11.2). No matter how carefully purified, KCl crystals, for example, contain impurities and every divalent cation impurity like Ca^{2+} must be accompanied by a vacant site on the cation sub-lattice, in order to preserve electroneutrality. It is these naturally occurring cation vacancies that are responsible for the *extrinsic* ionic conductivity of crystals at low temperatures. This explanation of the conductivity curve is easily verified by intentional doping of the crystals of a uni-univalent crystal with divalent cation impurity. In some other types of crystals, Frenkel defects (Figure 11.2) are favoured over Schottky defects, for example (with the interstitial ion in parentheses) the silver halides (Ag^+) (e.g. Corish and Jacobs 1972; Jacobs, Corish and Catlow 1980), lithium oxide (Li^+) and barium fluoride (F^-).

Figure 11.3(a) shows the dependence of the ionic conductivity κ of KCl on temperature, plotted as $\log \kappa T$ against T^{-1}. While the two segments of the conductivity curve predicted by the simple Schottky defect model are clearly present, it is clear on large-scale plots ($1\,m \times 1\,m$, 100–200 points) that in KCl both segments are curved and not linear. (For similar investigations of the ionic conductivity of some other ionic crystals see, for NaCl, Hooton and Jacobs 1990; for RbCl, Jacobs and Vernon 1995;

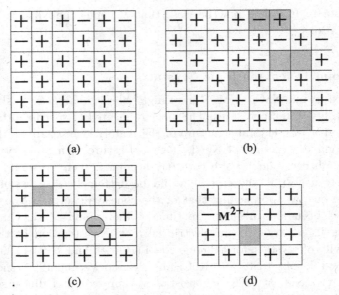

Figure 11.2.   Point defects in crystal structures (schematic). (a) Perfect uni-univalent crystal structure; (b) a Schottky defect; (c) a Frenkel defect; (d) divalent cation impurity and charge-compensating cation vacancy.

for AgCl, Corish and Jacobs 1972; and for BaF_2, Jacobs and Ong 1980.) The conductivity curve (for alkali halides) can be analyzed in terms of a defect model that allows for contributions from both anion and cation vacancies and anion interstitials and includes defect interactions calculated from the Debye–Hückel–Onsager theory. (Cation interstitials were also included in some analyses but the residual sum of squares was no better than if cation interstitials were omitted.) This model has a number of unknown parameters, which are determined by a combination of theoretical calculation and non-linear least squares fitting, e.g. BaF_2 (Jacobs and Ong 1962). Examples of parameter values are given in Table 11.1. These show that although the formation energy of Schottky defects in alkali halides is less than that of interstitials, the Arrhenius energy for the motion of interstitials is not so much greater than that for vacancy motion that their contribution to the conductivity can be neglected.

While Schottky defects are clearly in the majority in KCl and RbCl, it was found necessary to allow also for Frenkel defects on the anion sublattice in order to obtain a consistent set of parameters that fitted the conductivity data of pure KCl, $KCl:Sr^{2+}$ and $KCl:SO_4^{2-}$ (and likewise pure

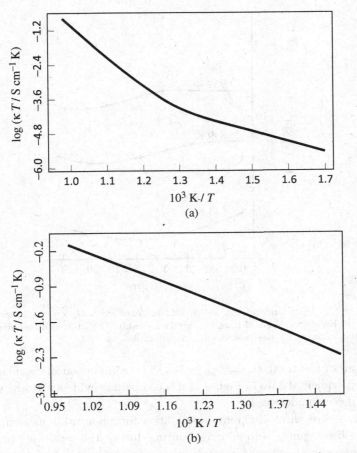

Figure 11.3. Dependence on temperature of the ionic conductivity κ of crystals (a) of pure KCl and (b) of KCl containing 100 ppm of $SrCl_2$ (after Acuña and Jacobs 1980). Please see the original paper for plots showing the experimental points, which are too numerous to be shown here.

RbCl, RbCl:Sr^{2+} and RbCl:S^{2-}). Perhaps the most striking evidence for the presence of anion interstitials in KCl lies in Fuller's (1966) measurements of the diffusion coefficient D_a of Cl^- in KCl:Sr^{2+}. As shown in Figure 11.4, at constant temperature D_a at first decreases with added Sr^{2+}, since this increases the concentration of cation vacancies and therefore decreases the concentration of anion vacancies (because of the Schottky equilibrium). However, at higher concentrations of Sr^{2+} the decreasing concentration of anion vacancies increases the concentration of anion interstitials and

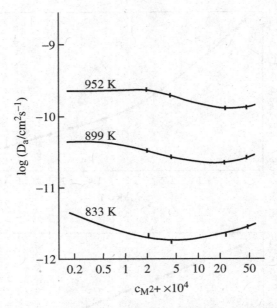

Figure 11.4.   Dependence of the anion diffusion coefficient D_a on the concentration of Sr^{2+} in $KCl:Sr^{2+}$. Vertical bars, experimental results (Fuller 1966); continuous line, calculated, allowing for the existence of interstitial anions.

therefore of the total $D_a = D_{av} + D_{ai}$. The comparison of calculated D_a with experimental data in Figure 11.4 is compelling evidence for the validity of the defect model used.

There are three different possibilities for interstitial motion, which are: a *direct* jump into a neighbouring interstitial position, or either collinear or non-collinear *interstitialcy* motion, in which an interstitial replaces a lattice ion on the same sub-lattice, knocking that lattice ion into another interstitial site. Since the calculated activation energies of the three interstitial jumps are too similar for their individual contributions to be assessed, only the collinear interstitialcy mechanism was assumed in the data analysis.

11.5. Thermodynamics of Ionic Crystals with Point Defects

The idea that point defects, even vacant sites, could be treated as chemical species in thermodynamic equilibrium is due to C. Wagner and W. Schottky. Suppose that a uni-univalent ionic crystal like KCl is in thermal equilibrium at a high temperature. Then it inevitably contains Schottky defects, cation vacancies (cv) and anion vacancies (av) which obey the quasi-chemical equilibrium

$$0 = cv + av \tag{1}$$

(1) $$K_S = \exp(-g_S/kT) = a_{cv}a_{av} = \gamma_{cv}\gamma_{av}c_{cv}c_{av} \tag{2}$$

Equation (1) describes the formation of a Schottky defect pair in a pure uni-univalent crystal. The standard state is pure, perfect crystal. K_S is the equilibrium constant for the quasi-chemical process (1), g_S is the Gibbs energy change accompanying the formation of the Schottky defect, $a_r = \gamma_r c_r$ is the activity, γ_r the activity coefficient and c_r the concentration (site-fraction) of defect species r. A possible mechanism is the formation of a cation vacancy–anion vacancy pair at the surface. It is supposed that a pair of surface ions leave their normal lattice sites and add on to a surface step (Figure 11.2b). The vacancies then diffuse into the crystal. A certain fraction of the vacancy pairs (vp) will survive, the remainder dissociating into single, isolated vacancies (Figure 11.2b) according to the equilibrium

$$vp = cv + av \tag{3}$$

(3) $$K_p = \exp(-g_p/kT) = \frac{a_{cv}a_{av}}{a_p} = \frac{\gamma_{cv}\gamma_{av}c_{cv}c_{av}}{\gamma_p c_p} \tag{4}$$

It may seem strange at first that a vacancy, which is the absence of an ion from a particular crystal site, can have a chemical potential, but remember that our energy zero was the perfect crystal (see Eq. (1)) and therefore each vacancy has energy, its energy of formation. A vacancy will also perturb the vibrational frequencies of nearby ions and thus has a vibrational entropy.

The equilibria (1) and (3) involve only the pure crystal. Every uni-univalent crystal inevitably contains some divalent cation impurity, e.g. Ca^{2+} in NaCl. To ensure electroneutrality, each M^{2+} in solution must be accompanied by a cation vacancy, either separated from or associated with an M^{2+} (Figure 11.2d). The number of free cv due to the presence of divalent cation impurity is determined by the equilibrium

$$M^{2+} + cv = M^{2+}cv \tag{5}$$

The notation used in (5) implies that the cv on the LS of (5) is free and independently mobile whereas that on the RS is bound into a complex of lower mobility. If the degree of association of cation vacancies is p_c

(5) $$K_{ac1} = \frac{p_c}{(1 - p_c)c_{cv}\gamma_M\gamma_{cv}} = 12\exp(-g_{ac1}/kT) \tag{6}$$

Exercise 11.5-1

γ_r for the complex has been set equal to one. Justify this step.

The factor 12 arises from configurational degeneracy and contributes to the association entropy. The rest of the association entropy is vibrational and arises because the vibrational frequencies of nearby ions are perturbed differently by the presence of the cation vacancy, the M^{2+} ion and the complex. Since the calculated energy of formation of nn complexes and of nnn complexes is similar, there is no way of telling which (if either) predominates. Most likely both are present and quoted association parameters refer to a kind of average.

K_{ac1} is the equilibrium constant for the association reaction (5), subscript 1 is used to indicate that a nearest-neighbour (nn) complex has been assumed (Figure 11.2d). Similarly, in a uni-univalent crystal doped with a divalent anion impurity A^{2-} (such as SO_4^{2-} or S^{2-}) an anion vacancy av will be attracted to a nearest-neighbour anion site to form an associated complex (e.g. Acuña and Jacobs 1980).

$$A^{2-} + av = A^{2-}av \tag{7}$$

(4) $$K_{aa1} = \frac{p_a}{(1 - p_a)c_{av}\gamma_A\gamma_{av}} = 12\exp(-g_{aa1}/kT) \tag{8}$$

The subscripts aa1 on the equilibrium constant K identify the equilibrium under discussion as (8) with the vacancy in the nearest neighbour position, though as in cation vacancy complexes we cannot be certain about the precise location of the vacancy.

From accurate measurements of conductivity and/or diffusion, the various defect parameters can be determined by a combination of theoretical calculation and non-linear least squares fitting (e.g. Acuña and Jacobs 1980). For the theoretical calculation of defect enthalpies and entropies, reviews are available (e.g. Harding 1990, Jacobs 1992, Leslie 1985).

Thermodynamics of a crystal with defects

In general, $G(T, p)$ where $p(T, V)$. Therefore

$$\left(\frac{\partial G}{\partial T}\right)_V = \left(\frac{\partial G}{\partial T}\right)_p + \left(\frac{\partial G}{\partial p}\right)_T\left(\frac{\partial p}{\partial T}\right)_V \tag{9}$$

For a change ΔT in the temperature of the crystal

(9)
$$(\Delta G)_V \approx (\Delta G)_p + (\partial G/\partial p)_T (\Delta p)_V \tag{10}$$

which is accurate to first order in $(\Delta p)_V$. The second-order correction is negligible. Since $G = F + pV$,

$$(\Delta G)_V = (\Delta F)_V + V(\Delta p)_V \tag{11}$$

and therefore, neglecting the small terms in $(\Delta p)_V$

(11)
$$(\Delta G)_p \cong (\Delta F)_V \tag{12}$$

The calculations are actually done at constant lattice parameter, so that the energy calculated is ΔU^a, but this is readily converted to the corresponding energy at constant volume by means of the equation

$$\Delta U^V = \Delta U^a + pV^a \tag{13}$$

The correction pV^a is actually negligible at all except geological pressures. We have then, from calculations, the energy and entropy *changes* on forming a defect at constant volume, and therefore the Helmholz energy change

$$(\Delta F)^V = (\Delta U)^V - T(\Delta S)^V \tag{14}$$

Most experiments, however, are carried out at constant pressure, so that we need ΔG^p, ΔH^p, and ΔS^p, the Gibbs function change, enthalpy change and entropy change for the same process, at constant pressure. The only practical way of calculating the temperature dependence of U and S is to use the *quasiharmonic approximation* which attributes their temperature dependence entirely to the thermal expansion of the crystal

$$U(V,T) = U(V(T)) \tag{15}$$

$$S(V,T) = S(V(T)) \tag{16}$$

The energy and entropy can therefore be calculated at a series of temperatures using experimental values of the lattice constant. If we plot $U^V(v)$, where v is the unit cell volume, against v, it is found that U^v decreases smoothly and often linearly with v, so that

$$U^v(v) = U^v(v_0) + (\partial u/\partial v)_T (v - v_0) \tag{17}$$

where $(\partial u/\partial v)_T$ is negative. v_0 is the (extrapolated) volume of the unit cell at $T = 0\,\mathrm{K}$. The thermal expansivity is independent of T for the simple materials examined in detail so far (Figure 11.5).

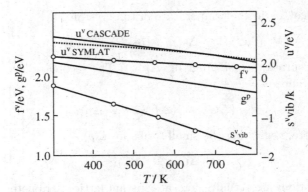

Figure 11.5. Energy of formation of a Schottky defect in NaCl calculated using both of the two main theoretical methods, the crystallite method (CASCADE) and the method that uses periodic boundary conditions (SYMLAT, the dotted line). The agreement between results from these two different methods is reassuring. The temperature dependence was obtained using the quasiharmonic approximation. Also shown is a comparison of the calculated values of f^v, the Helmholtz energy of formation of a Schottky defect in NaCl, with g^p, the experimental Gibbs energy of formation of a Schottky defect obtained from ionic conductivity measurements.

(9)
$$U^v(v) = U^v(v_0) + \alpha T v (\partial U/\partial v)_T \tag{18}$$

where α is the expansivity. Also

$$H^p(T) = U^V(T) - \alpha T V (\partial F^V/\partial V)_T \tag{19}$$

so that

$$H^p(T) = U^v(T = 0\,\mathrm{K}) + \alpha T^2 V \left(\frac{\partial S^V}{\partial V}\right)_T \tag{20}$$

The second term on the RS of (12) is relatively small and generally less than 5% of the first term. This is the reason why comparisons of calculated values of $U^V(T = 0\,\mathrm{K})$ with experimental values of $H^p(T)$ are nevertheless meaningful, despite the temperature difference. Figure 11.5 shows the results of calculations of the energy of formation of a Schottky defect in NaCl using both of the two main theoretical methods, the crystallite method (CASCADE) and the method that uses periodic boundary conditions (SYMLAT). The agreement between results from these two quite different codes for two methods that use different approximations is reassuring.

CASCADE uses the crystallite method (originated by Mott and Littleton) in which the simulated crystal is divided into two regions.

In Region I, which contains the defect and about 100–200 other ions, the ionic interactions are calculated exactly, using interionic potentials that have been tested by calculating properties of the perfect crystal, including the phonon dispersion (e.g. Jacobs and Vernon 1998). Region II, which is the rest of the crystal, is treated as a dielectic continuum. The interaction between Regions I and II is allowed for by calculating the interactions between the ions of Region I and those in the inner part of Region II. The finite size of Region I can be allowed for by plotting the calculated defect energy against the reciprocal of n, the number of ions in Region I and extrapolating to $n^{-1} = 0$. In SYMLAT, the defect and a number of surrounding ions are regarded as a giant unit cell which, reproduced by translational symmetry, generates the whole crystal. Periodic boundary conditions are used, as in lattice dynamics (Born and Huang 1954). Again, the finite size of the unit cell can be allowed for by repeating the calculation with unit cells of various sizes and using an extrapolation technique.[a]

In a similar manner, it is possible to calculate the (vibrational) entropy of various defects and defect processes. (See Jacobs (1992) for a review of entropy calculations.) The references in Table 11.2 are to the experimental work. These papers include references to defect calculations for these materials.

Answer to Exercise 11.5-1

It has been assumed that the primary cause of non-ideality in these very dilute solid solutions is electrostatic in origin. Therefore, the Debye-Hückel approximation is used for charged species, and activity coefficients of neutral species have been neglected.

11.6. Semiconductors

The solids we have been talking about in the last section were crystals with wide band gaps. When NaCl is formed from Na and Cl_2 the 3s electron from a Na atom is donated to a dissociated Cl atom to complete the 3p shell, resulting in two closed shell Na^+ and Cl^- ions and the stable ionic compound NaCl. The empty 3s states from the Na form a band of empty

---

[a]The computer codes CASCADE and SYMLAT were written by Dr M. Leslie, then at the SERC Laboratory, Daresbury, UK. I am much indebted to Dr Leslie for advice on the use of these codes.

Table 11.2. Calculated values of defect energies at $T = 0\,\mathrm{K}$ and experimental defect enthalpies from ionic conductivity measurements at elevated temperatures (see Eq. (12)). The defect energies and enthalpies are given in eV ($1\,\mathrm{eV} = 1.6019 \times 10^{-19}\,\mathrm{J} = 16.019\,\mathrm{aJ}$).

Formation	NaCl[a]		KCl[b]		RbCl[c]		AgCl[d]	
	Calc.	Expt.	Calc.	Expt.	Calc.	Expt.	Calc.	Expt.
Schottky defect	2.32	2.41	2.49	2.50	2.52	2.52		
Cation Frenkel defect	3.50		3.61		3.46		1.36	1.47
Anion Frenkel defect	4.33		3.71		3.48			
Vacancy pair	1.61		1.57		1.58			
migration								
Cation vacancy	0.66	0.65	0.66	0.67	0.66	0.66		
Anion vacancy	0.71	0.77	0.66	0.85	0.67	0.72		
Arrhenius energies								
Cation vacancy	1.82	1.86	1.91	1.92	1.92	1.92		
Anion vacancy	1.87	1.98	1.87	1.92	1.93	1.98		
Cation interstitial	2.28		2.37		1.94			
Anion interstitial	2.57		2.44		1.93			
Cation jump into pair	2.32	2.35	2.32	2.65	2.38	2.39		
Anion jump into pair	2.29	2.37	2.33	2.39	2.38	2.39		
association								
(nn complex)								
$Sr^{2+}cv$	−0.64	−0.64	−0.62	−0.65	−0.60	−0.59		
$Ca^{2+}cv$	−0.59	−0.61	−0.59	−0.59	−0.63			
$S^{2-}av$	−0.63	−0.75			−0.65	−0.87		

[a] Hooton and Jacobs (1990).
[b] Acuña and Jacobs (1980).
[c] Jacobs and Vernon (1998).
[d] Corish and Jacobs (1972), Catlow, Corish and Jacobs (1979).

electronic states called the *conduction band.* This is separated by an energy gap several electron volts in width from another band of electronic states which is formed from filled 3p states of the Cl atoms, and which is therefore called the *full band.* The upper band is normally empty because of the large band gap in NaCl.

The Group IV elements germanium and silicon are covalent compounds and in the solid state the electrons from the valence shell form a filled band of electronic states. These materials are therefore insulators at low temperatures when pure. However, the band gap to the next empty band is sufficiently small (Table 11.3) for electrons to be excited thermally from the full band into the conduction band. Such materials are called *semiconductors* because they have an electronic conductivity less than that of a metallic conductor, but higher than that of an insulator. Each electron

Table 11.3. Band gap energies E_f in eV at 300 K (unless otherwise stated) for a variety of crystalline semiconductors.

Material	E_f/eV	Material	E_f/eV
Si	1.11	SiC(hex)	3.0 (0 K)
Ge	0.744	PbS	0.286
InSb	0.27	ZnO	3.436
InP	1.27	ZnS	3.6
GaP	2.25	Cu_2O	2.172 (0 K)
GaAs	1.43	TiO_2	3.03 (0 K)

excited from the valence band into the conduction band leaves behind a vacant site that had been originally occupied by an electron. This vacant site is mobile since it can be filled by an electron from a neighbouring site. In an electric field the random motion of these electrons becomes directed and consequently the unoccupied electron site moves towards the negative electrode and contributes to the total current. It is therefore helpful to think of this mobile electron vacancy as a positively charged "hole" in the valence band. This positively charged pseudo-particle (which in reality is a vacant electronic state in the otherwise filled valence band) is called a *positive hole* and represented by the letter p. An electron promoted to the conduction band is denoted by n. The promotion of an electron to the CB (conduction band) can be represented by the quasichemical equation

$$0 = n + p \tag{1}$$

The phenomenon just described is that of intrinsic semiconductivity, because it is a property of the pure material. However, if a small amount of a Group V element (say Bi) were to be added to pure Ge then each Bi atom would contribute an extra electron. These electrons can only be accommodated in the conduction band. The procedure of adding an impurity is called *doping* and the result is *extrinsic semiconductivity*.

Since doping with an electron donor like Bi results in excess electrons in the conduction band, this is referred to as n-type behaviour. Similarly, doping a Group IV semiconductor like Si with a Group III element like In results in a deficiency of electrons on the full band, that is the formation of positive holes. This is referred to as p-type doping. These various possibilities are shown in the energy level diagram in Figure 11.6.

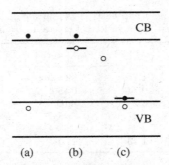

Figure 11.6.   Energy levels in semiconductors. In addition to the formation of holes in the valence band and mobile electrons in the conduction band, there exists the possibility of the elevation of electrons from donor levels close to the conduction band and the formation of mobile holes in the valence band by the promotion of electrons from the full (valence) band into acceptor levels located just above the full band.

Thermodynamics of semiconductors

In intrinsic semiconductors electrons are promoted thermally from the full band to the conduction band

$$(1) \qquad c_n c_p = (np/N^2) = K = A \exp(-E_f/kT) \qquad (2)$$

c_n, c_p are the concentrations of electrons and holes, N is the total number of molecules in the crystal and E_f the energy required to promote an electron from the full band to the conduction band. In the exponent of the equilibrium constant we expect to see a Gibbs energy, but the electron and hole have only translational energy, so that there is no configurational or vibrational entropy involved in Eq. (2). Similarly there is no volume change accompanying (2) and so the exponent is just the energy of formation of a mobile (separated) electron hole pair divided by kT. Both electron and hole are mobile and contribute to the (electronic) conductivity, so that we expect a conductivity of

$$\kappa = \kappa_0 \exp(-E_f/kT) \qquad (3)$$

where κ_0 will depend both on the number of charge carriers and their mobilities. This exponential dependence of conductivity on T^{-1} had been observed at high temperature but at low temperatures the increase of κ with T is less rapid than expected from (3). The magnitude of κ also varies greatly from sample to sample. This behaviour suggests strongly the involvement of impurities acting as donors or acceptors and this suggestion is easily verified by doping with suitable donors or acceptors.

Problems 11

11.1 Show that when a crystal is strained, the diagonal elements of the strain tensor Σ are the fractional changes in length along the x, y, and z axes.

11.2 Consider the distortion of a cube of volume V_0 of sides $(dl)\mathbf{e}_1$, $(dl)\mathbf{e}_2$, $(dl)\mathbf{e}_3$. In the stressed crystal the cube is distorted into a parallelopiped. Calculate the volume of the parallelopiped and the volume dilation, that is the fractional increase in the volume of the cube during stress.

11.3 Prove that in cubic symmetry $c_{11} = c_{22} = c_{33}$.

11.4 Obtain the number of independent elastic compliance coefficients for a substance which has the point group T_d.

11.5 In AgCl and AgBr cation Frenkel defects predominate, whereas in KCl Schottky defects predominate. There is evidence in KCl that anion Frenkel defects are formed in sufficient numbers to affect the anion diffusion coefficient. But no evidence was found in NaCl for cation Frenkel defects. Rationalize these facts in terms of the electronic structure of cations and anions in these crystals.

11.6 The conductivity of KCl crystals is increased by adding small amounts (of the order of 100 ppm) $CaCl_2$ to the KCl before melting and crystal growth. Would you expect the density of these $KCl:Ca^{2+}$ crystals to be greater or less than that of a crystal of pure KCl?

12

Thermodynamics of Non-equilibrium Systems

12.1. Reprise

The First Law of Thermodynamics describes the conservation of energy

$$dU = dq + dw \tag{1}$$

where dU is the infinitesimally small change in the internal energy of a closed system which occurs in time t during an infinitesimal thermodynamic process, dq is the heat absorbed in this process during t and

$$dw = \sum_{k=1}^{N_X} \boldsymbol{X}_k^0 \cdot d\boldsymbol{x}_k \tag{2}$$

is the work done on the system during t. $\boldsymbol{X}_k^0$ denotes a generalized force acting on the system and $\boldsymbol{x}_k$ is the conjugate displacement (that is the range over which the force $\boldsymbol{X}_k^0$ acts). Examples of various kinds of thermodynamic forces were given in Section 1.5. Let $\boldsymbol{X}_k$ denote the corresponding force exerted by the system on its surroundings. Then if

$$X_k = X_k^0, \quad \forall k \tag{3}$$

the system is in mechanical equilibrium, but if

$$X_k \neq X_k^0, \quad \text{at least one } k \tag{4}$$

then a thermodynamic process will occur in which the system changes spontaneously until (3) is satisfied (provided there are no constraints that prevent this occuring). Such processes are termed natural processes. But if

$$X_k = X_k^0 \pm dX_k, \quad \forall k \tag{5}$$

where $dX_{\mathbf{k}}$ is $\ll X_k$, then the thermodynamic process is described as reversible (since it may be reversed by a slight alteration in the conditions that maintain the sign of dX_k). But if (4) is true for at least one k,

343

then the system is not in equilibrium and the processes that occur are non-equilibrium processes.

For example, consider a system comprising a single gas that is confined in a vertical cylinder by a piston and some device (a collection of weights will do) for altering the pressure p^0 exerted on the gas. When the pressure of the gas p is equal to the external pressure p^0, the system is in equilibrium, but if p differs from p^0, the gas expands or contracts until equilibrium is established. The work done on the gas is $-p^0 \, dV$, which becomes $-pdV$ if the process is reversible.

The First Law of Thermodynamics tells us how to calculate the changes in the energy of a system when it is changed either by the absorption of heat or by the performance of work, but it tells us nothing about the direction of thermodynamic change. For this we need the Second Law: there exists an extensive function of state called the entropy[a] S; in a natural process the net entropy change in a system and its surroundings

$$\mathrm{d}S_{\mathrm{net}} = \mathrm{d}S + \mathrm{d}S_0 \geq 0 \tag{6}$$

In (6), the inequality applies to natural, irreversible, non-equilibrium processes and the equality to reversible processes.

The entropy change in the system comprises two terms

$$\mathrm{d}S = \mathrm{d}_{\mathrm{i}}S + \mathrm{d}_{\mathrm{e}}S \tag{7}$$

where $\mathrm{d}_{\mathrm{i}}S$ is due to irreversible processes going on within the system and $\mathrm{d}_{\mathrm{e}}S$ to the absorption of heat $\mathrm{d}q$ from, and to the exchange of matter with, its surroundings. Examples of adiabatic natural processes were given in Chapter 1. Provided a system is in thermal equilibrium with its surroundings $(T = T_0)$ the isothermal absorption of heat is reversible and

$$\mathrm{d}_{\mathrm{e}}S = \frac{\mathrm{d}q}{T} \quad \text{(reversible process, closed system)} \tag{8}$$

From the First and Second Laws we thus obtain the fundamental equation for a closed phase (FECP)

$$\mathrm{d}U = T\mathrm{d}S - pdV \tag{9}$$

Exercise 12.1-1

What restrictions apply to Eq. (4)?

---

[a]From the Greek $\varepsilon \nu \tau \rho \omega \pi \eta$ = evolution.

If there are several generalized forces $\mathbf{X}_k$, N_X in number, operating on the system then the FECP becomes

$$dU = TdS + \sum_{k=1}^{N_X} \mathbf{X}_k \cdot d\mathbf{x}_k \tag{10}$$

Equation (10) is the FECP for the thermodynamic potential U. From the definition of the other thermodynamic potentials we derived in Chapter 2, further fundamental equations for a closed phase,

$$dS = \frac{1}{T}dU + \frac{p}{T}dV \tag{11}$$

(2.2.6) $$dH = TdS + Vdp \tag{12}$$

(2.2.10) $$dF = -SdT - pdV \tag{13}$$

(2.2.11) $$dG = -SdT + Vdp \tag{14}$$

Other potentials can be defined, notably the Massieu function J, and the Planck function Y. Which potential we use depends on the choice made for the independent variables.

We can generalize (14) to open phases containing several constituents $B, C, \ldots$ by recognizing that for open phases, $S(U, V, n_B, \ldots)$ where n_B is the amount of species B. Then

(9) $$dG = -SdT + Vdp + \sum_{B,C,\ldots} \mu_B dn_B \tag{15}$$

where μ_B is the chemical potential of species B. Similar equations may be written for the other thermodynamic potentials.

Exercise 12.1-2

Derive the FECP for the thermodynamic potential F. Identify the partial derivatives of $F(T, V, n_B, n_C, \ldots)$.

Answers to Exercises 12.1

Exercise 12.1-1

The FECP, Eq. (4), applies to reversible processes occurring in a single closed phase in which the only kind of work done on the system is the mechanical work involving the change in volume of the phase.

Exercise 12.1-2

By definition, the Helmholtz function $F = U - TS$. Therefore

$$dF = dU - TdS - SdT$$

$$(12.1.10) \qquad = -SdT - pdV + \sum_{B,C,\ldots} \mu_B dn_B \qquad (1)$$

It is common practice, and I shall generally follow it for the sake of brevity, to include only the one work term $-pdV$, it being understood the others may be included when necessary. The complete differential of $F(T, V, n_B, n_C, \ldots)$ is

$$dF = \left(\frac{\partial F}{\partial T}\right)_{V, n_B, n_C, \ldots} dT + \left(\frac{\partial F}{\partial V}\right)_{T, n_B, n_C, \ldots} dV$$

$$+ \sum_{B,C,\ldots} \left(\frac{\partial F}{\partial n_B}\right)_{T, V, n_C \neq n_B} dn_B \qquad (2)$$

where

$$\left(\frac{\partial F}{\partial T}\right)_{V, n_B, n_C, \ldots} = -S, \qquad (3a)$$

$$\left(\frac{\partial F}{\partial V}\right)_{T, n_B, n_C, \ldots} = -p, \qquad (3b)$$

$$\left(\frac{\partial F}{\partial n_B}\right)_{T, V, n_C \neq n_B} = \mu_B \qquad (3c)$$

12.2. Entropy Production

We consider first a simple example of a single component distributed between two closed phases α and β which are in thermal contact. The entropy of this system is

$$dS = dS^\alpha + dS^\beta \qquad (1)$$

The heat absorbed by phase α is

$$dq^\alpha = d_i q^\alpha + d_e q^\alpha \qquad (2a)$$

where $d_i q^\alpha$ is the heat received by phase α from phase β (so that $d_i q^\alpha = -d_i q^\beta$) and $d_e q^\alpha$ is the heat received by phase α from outside the system.

Similarly

$$dq^\beta = d_i q^\beta + d_e q^\beta \tag{2b}$$

The entropy change is

$$dS = \frac{dq^\alpha}{T^\alpha} + \frac{dq^\beta}{T^\beta}$$

$$= \frac{d_e q^\alpha}{T^\alpha} + \frac{d_e q^\beta}{T^\beta} + d_i q^\alpha \left[\frac{1}{T^\alpha} - \frac{1}{T^\beta} \right] \tag{3}$$

The entropy production consists of two parts. The first two terms on the RS of (3) are the entropy change due to the interaction of the system with its surroundings; the third term in (3) is the entropy change due to the irreversible flow of heat inside the system, which becomes zero when $T^\alpha = T^\beta$.

In the thermodynamics of non-equilibrium processes we are concerned with the rate of entropy production σ within a system. In this example

$$\sigma = \frac{d_i S}{dt} = \frac{d_i q^\alpha}{dt} \left[\frac{1}{T^\alpha} - \frac{1}{T^\beta} \right] \geq 0 \tag{4}$$

σ consists of the product of two terms: the rate of the irreversible process $d_i q^\alpha / dt$ and the thermodynamic force causing the irreversible process, which here is the difference between the reciprocal temperatures of the two phases α and β. This force becomes zero when $T^\alpha = T^\beta$ and $d_i q^\alpha / dt$ consequently falls to zero.

In a closed system with chemical reactions

(8.2.4)
$$dS = \frac{dQ}{T} + \frac{A d\xi}{T} \tag{5}$$

where the affinity

$$A = \sum_{B,C,\ldots} \nu_B \mu_B \tag{6}$$

and ν_B is the stoichiometric coefficient of B in the chemical equation

$$0 = \sum_{B,C,\ldots} \nu_B B \tag{7}$$

Exercise 12.2-1

Consider the simple chemical process in which an amount of B is transferred from phase α to phase β. Write down the chemical equation (7) for this process and an expression for the affinity of the process.

The rate of entropy production due to a chemical reaction is

(5)
$$\frac{d_i S}{dt} = \frac{A}{T}\frac{d\xi}{dt} > 0 \qquad (8)$$

where $d\xi/dt$ is the rate of the chemical reaction (see Chapter 8).

Exercise 12.2-2

Explain why A and $d\xi/dt$ have the same sign in (8).

Formula (8) can be extended to the case of several different concurrent reactions. The entropy produced is then

$$d_i S = \frac{1}{T}\sum_r A_r d\xi_r > 0 \qquad (9)$$

where A_r is the affinity of reaction r. The rate of entropy production is

(9)
$$\frac{d_i S}{dt} = \frac{1}{T}\sum_r A_r \frac{d\xi_r}{dt} > 0 \qquad (10)$$

Note that the Second Law requires that the *sum* of the affinity × rate of the reaction for all reactions be positive, but does not require that every term in that sum be positive. Thus it may happen that, in the case of two coupled reactions,

$$A_1\frac{d\xi_1}{dt} + A_2\frac{d\xi_2}{dt} > 0, \quad A_1\frac{d\xi_1}{dt} < 0 \qquad (11)$$

in which case the two coupled reactions may proceed. We remind the reader that the affinity may be expressed in terms of the activity of reactants and products (see Chapter 8).

In an open multicomponent system, without reactions

$$dS = \frac{1}{T}dU + \frac{p}{T}dV - \sum_{B,C,\ldots}\frac{\mu_B}{T}dn_B \qquad (12)$$

A state of non-equilibrium is one in which the thermodynamic forces are not balanced and unless there are other kinetic factors preventing this, the system will change towards the equilibrium state. The processes occurring during this change are called *flows* and the amount of a flow is expressed as the *flux* and given the symbol J_k. When the forces are sufficiently small, the fluxes (e.g. heat conduction, diffusion) are proportional to the driving forces, X_k. In general, each force can affect any of the flows so that

$$\mathbf{J}_i = \sum_{k=1}^{n} L_{ik}\mathbf{X}_k \tag{13}$$

The $n \times n$ square matrix L contains the proportionality constants L_{ik} which are known as the *phenomenological coefficients*. The fundamental theorem of the thermodynamics of non-equilibrium processes (TDNEP) proved by Onsager (1931), is that when the fluxes and forces are linearly independent and are chosen so as to satisfy the entropy production equation

$$\sigma = \frac{\mathrm{d_i}S}{\mathrm{d}t} = \sum_k \mathbf{J}_k \cdot \mathbf{X}_k \tag{14}$$

the matrix L of phenomenological coefficients is symmetric

$$L_{ik}(\mathbf{B}, \omega) = L_{ki}(-\mathbf{B}, -\omega) \tag{15}$$

The set of equations (15) is known as the *Onsager reciprocal relations* (ORR).

The product σT is called the *dissipative function*, Φ. If a magnetic field is present then the sign of the magnetic induction must be reversed, as well as the order of the subscripts. The same thing applies to the angular velocity in a rotating system, that is, the sign of the angular velocity must be reversed as well as the order of the subscripts. Onsager's proof of the ORR requires a deeper knowledge of statistical mechanics than is assumed in the rest of this book and so it will not be given here. Instead we shall accept the ORR as an additional principle, secure in the knowledge that a proof exists and that it has been verified whenever it has been tested experimentally.

Answers to Exercises 12.2

Exercise 12.2-1

$$0 = B^\beta - B^\alpha$$
$$A = \mu^\alpha - \mu^\beta$$

At equilibrium, $\mu^\alpha = \mu^\beta$ and $A = 0$.

Exercise 12.2-2

If A is positive the reaction is thermodynamically feasible and may proceed as written with a positive velocity. But if A is negative, the reaction cannot occur as written, although the reverse reaction may occur.

12.3. Assumptions Upon Which TDNEP is Based

Thermodynamic variables such as $p, T, S, \ldots$ have only been defined so far for macroscopic systems in thermodynamic equilibrium. But, because of the extremely large number of particles in such systems it is possible for systems to be close to equilibrium on a small scale, but in which gradients of temperature, pressure, chemical potential, ... exist. It is necessary to suppose that if a small cell of a macroscopic system surrounding point P were to be isolated at time t, then the contents of this cell would come to equilibrium in time δt, and the $T, p, \mu_k, \ldots$ at time $t + \delta t$ taken as the $T, p, \mu_k, \ldots$ at P, at t. This assumption is valid provided $\delta t \ll$ time taken for a measurable change in the properties of the system to occur. For example, it is valid for the linear processes described in this chapter but would not be valid for fast chemical reactions like explosions. Thus, we shall without further question apply the equations of ordinary equilibrium thermodynamics to small regions of our system, and recognize that it will often be necessary to impose gradients of temperature, pressure, and chemical potential in our non-equilibrium system. Comparatively recently in the long history of thermodynamics, the subject has been extended to include systems which are changing rapidly by the inclusion of first derivatives as well as equilibrium values as variables. We shall not discuss this development, but instead refer the interested reader to the literature (e.g. Müller and Ruggeri 1993, Jou 1996).

Superficially, it may seem that there is a considerable choice available for the fluxes and forces. However, they must be linearly independent and must be chosen so as to satisfy Eq. (12.2.14) for the rate of entropy production. If the fluxes and forces are not linearly independent, they must be made so at this stage. Commonly used fluxes and forces are shown in Table 12.1.

Exercise 12.3-1

What is the force X_i in an isothermal system?

Table 12.1. This table shows some commonly used fluxes and forces. The last line states that the amount of substance B formed (or consumed) in the rth reaction is equal to its stoichiometric coefficient multiplied by the degree of advancement of that reaction.

	Flux J	Force X	Ref. Eq.
Heat	J_q	$X_u = -\dfrac{1}{T^2}\dfrac{dT}{dx}$	(12.4.4)
Matter	J_i	$X_i = -T\dfrac{d(\mu_i/T)}{dx}$	(12.4.3)
Electric current density	$j = I/A$	$X_e = -\dfrac{1}{T}\dfrac{d\phi}{dx}$	(12.4.1)
Chemical reaction	$J_r = \dfrac{d\xi_r}{dt}$	$X_r = A_r = -\sum_{B}\nu_B\tilde{\mu}_B$	(12.2.11)
	$\nu_{B,r}d\xi_r = dn_{B,r}$		

Answer to Exercise 12.3-1

From Table 12.1

$$X_i = -T\frac{d(\mu_i/T)}{dx} = -\frac{d\mu_i}{dx}, \quad \text{in an isothermal system.}$$

12.4. Examples of Non-equilibrium Processes

12.4.1. *Electrical conduction*

When the potential difference is not too large (how large is a matter to determine experimentally) the current density j is proportional to the applied electric field E

$$j = \kappa E \quad \text{(Ohm's law)} \tag{1}$$

In general, j and E are vectors and the conductivity κ is a second rank tensor, but I have assumed a one-dimensional configuration here, as is often the case. The current density j is the total current I divided by the area A of the electrodes, and the electric field E is the applied potential difference between the electrodes $\Delta\phi$ divided by the length x between the electrodes.

Exercise 12.4-1

Find an expression for the conductivity κ in terms of the resistance R of the material between the electrodes and the geometry of the conductivity cell.

The energy dissipated as heat divided by T is the entropy produced by the current flow, so that

$$\sigma = \frac{I \, \Delta\phi}{T} \tag{2}$$

12.4.2. *Diffusion*

The flux of B, that is the amount of B crossing $1\,\mathrm{m}^2$ (or $1\,\mathrm{cm}^2$, as one chooses) of a plane normal to the direction of diffusion in one second is proportional to the concentration gradient of B. This is Fick's first law

$$J_B = -D_B \frac{\partial c_B}{\partial x} \tag{3}$$

D_B is the diffusion coefficient of B.

12.4.3. *Heat conduction*

Consider a bar of material of length Δx. The temperature at one end is T and at the other end $T + \Delta T$. The heat flux J_q is proportional to the temperature gradient

$$J_q = -k \frac{\mathrm{d}T}{\mathrm{d}x} \quad \text{(Fourier's law)} \tag{4}$$

where k is the thermal conductivity. The above three examples illustrate the fact that the phenomenological equation describing a single non-equilibrium process, is of the form

$$J = LX \tag{5}$$

J is the flux of charge, matter, heat ..., X is the force, a gradient of electrical potential, concentration, temperature... and L is a *phenomenological coefficient*. The range of validity of (5) must be determined by experiment, but (5) always holds when the forces are sufficiently small.

12.4.4. *Coupling of linear processes*

As an example of the coupling of non-equilibrium processes, consider the phenomenon of *thermal diffusion* in a mixture of two gases. A mixture of two gases B and C is subjected to a temperature gradient in the x direction.

At time $t = 0$, the concentrations of B and C are uniform. But after a finite time t (the time scale would need to be determined by experiment) the concentrations of both B and C are found to depend on the distance x. The flux equations for B and C must therefore be of the form

$$J_B = -D_{BC}\frac{\partial c_B}{\partial x} - D'_{BC}\frac{dT}{dx} \tag{6}$$

$$J_C = -D_{CB}\frac{\partial c_C}{\partial x} - D'_{CB}\frac{dT}{dx} \tag{7}$$

The rates of diffusion of B and C depend not only on their concentration gradients but also on the temperature gradient.

Answer to Exercise 12.4-1

$$\kappa = \frac{j}{E} = \frac{I \cdot x}{A\,\Delta\phi} = \frac{x}{RA}$$

12.5. Thermoelectric Effects

(Note: Equations (12.5.9) and (12.5.10) occur in the problems.)
A thermocouple consists of two different metals (e.g. Cu/Fe) joined together. Because of the difference in the electrochemical potential of electrons in the two metals, there exists a potential difference between them. If two such junctions are formed and the two junctions maintained at different temperatures, the potential difference between the two copper leads can be measured with a high impedance voltmeter. The phenomenon is known as the Seebeck effect. If the temperature at the hot junction is T and that of the cold junction T_0 is maintained at the ice-point, we have a *thermocouple*, a useful device for measuring the temperature T inside a furnace, for example.

The rate of entropy production per m^3, σ, is given by

(T12.1) $$\sigma = J_q\left(-\frac{1}{T^2}\frac{dT}{dx}\right) + j\left(-\frac{d\phi}{dx}\right)\left(\frac{1}{T}\right) \tag{1}$$

Exercise 12.5-1

σ is the rate of entropy production per unit volume. Relate σ to the amount of entropy dS produced per mole in time dt.

Exercise 12.5-2

Explain the origin of the factor $1/T$ in the second term on the RS of Eq. (1).

The phenomenological equations are

(1) $$J_q = L_{11}\left(-\frac{1}{T^2}\frac{dT}{dx}\right) + L_{12}\left(-\frac{d\phi}{dx}\right)\left(\frac{1}{T}\right) \qquad (2)$$

(1) $$j = L_{21}\left(-\frac{1}{T^2}\frac{dT}{dx}\right) + L_{22}\left(-\frac{d\phi}{dx}\right)\left(\frac{1}{T}\right) \qquad (3)$$

where

$$L_{12}(\mathbf{B}) = L_{21}(-\mathbf{B}) \qquad (4)$$

The *electrical conductivity* in an isothermal system is

(3) $$\kappa = \frac{j}{-d\phi/dx} = \frac{L_{22}}{T} \qquad (5)$$

The *thermal conductivity*

$$k = \frac{J_q}{-dT/dx} = \frac{L_{11}}{T^2} \qquad (6)$$

The *thermoelectric power*. Consider a thermocouple: when $j = 0$, the voltage measured by a potentiometer is the difference in the voltages generated at the two metallic junctions. When T increases, the *thermoelectric* power of the thermocouple ε_{AB} is the change in this voltage divided by the temperature increment. The sign of the thermopower is chosen as positive if the voltage generated drives the current from A to B at the hot junction. The *absolute thermoelectric power* of A is defined as

$$\varepsilon_A = \frac{L_{21}^A}{T\,L_{22}^A} \qquad (7)$$

If an electric current is passed through an isothermal junction between two different materials, the heat generated is called the *Peltier heat*. The source of this heat is a discontinuity in current across the junction. Because T is constant

(8), (2) $$J_q = \varepsilon j T \qquad (8)$$

(3) $$J_q^B - J_q^A = (\varepsilon^B - \varepsilon^A) j T \qquad (9)$$

The *Peltier coefficient* π_{AB} is the heat that must be supplied when unit electric current flows from A to B

$$\pi_{AB} = \left(\varepsilon^B - \varepsilon^A\right) T \tag{10}$$

which is called the second Kelvin relation.

Answers to Exercises 12-5
Exercise 12.5-1

The rate of entropy production in $1\,\mathrm{m}^3$ is

$$\left(\frac{\mathrm{d}S}{\mathrm{d}t}\right)\frac{\rho}{M} = \sigma,$$

where ρ is the density and M the molar mass.

Units check : $\dfrac{\mathrm{J\,K^{-1}\,mol^{-1}\,kg\,mol}}{\mathrm{s\,m^3\,kg}} = \dfrac{\mathrm{J\,K^{-1}}}{\mathrm{s\,m^3}}$

Exercise 12.5-2

Firstly, the equation is dimensionally correct. Since $\mathrm{d}S = (1/T)\mathrm{d}\tilde{\mu} = -(1/T)\mathrm{d}\phi, -1/T$ appears in the force X acting on an electron due to the electric field.

12.6. Diffusion and Conduction in Electrolyte Solutions

(Note that Eq. (12.6.18) occurs in the problems.)

Phenomenological equations and Hittorf transport number

Consider an aqueous solution of a single electrolyte, which ionizes according to the equation

$$M_{\nu_+}A_{\nu_-} = \nu_+ M^{z+} + \nu_- A^{z-} \tag{1}$$

The solution as a whole is electrically neutral, so that

$$\nu_+ z_+ = 0\nu_- z_- \tag{2}$$

Let J_+ denote the flux of cations and similarly J_- the flux of anions, both measured in $\mathrm{mol\,m^{-2}\,s^{-1}}$. The flux equations are

$$J_+ = L_{++}X_+ + L_{+-}X_- \tag{3}$$
$$J_- = L_{-+}X_+ + L_{--}X_- \tag{4}$$

with the ORR

$$L_{-+} = L_{+-} \tag{5}$$

The system is of uniform T and p, so that the thermodynamic forces are

$$X_+ = -\left(\frac{\partial \tilde{\mu}_+}{\partial x}\right)_{T,p} \tag{6}$$

$$X_- = -\left(\frac{\partial \tilde{\mu}_-}{\partial x}\right)_{T,p} \tag{7}$$

The cation electrochemical potential is

$$\tilde{\mu}_+ = \mu_+ + z_+ F \phi \tag{8}$$

so that the thermodynamic force acting on the cations is

$$X_+ = -\left(\frac{\partial \mu_+}{\partial x}\right)_{T,p} - z_+ F \left(\frac{\partial \phi}{\partial x}\right)_{T,p} \tag{9}$$

This force consists of two parts. The first term on the RS of (9) is the gradient of chemical potential and the second term is the gradient of electrical potential. The chemical potential may be written as

$$\mu_+ = \mu_+^{\ominus}(T,p) + RT \ln a_+(T,p,m) \tag{10}$$

$\mu_+^{\ominus}$ is the chemical potential of cations in their standard state and a_+ is the activity of cations, a function of the temperature, pressure and the molality of the solution

$$(10) \qquad -\left(\frac{\partial \mu_+}{\partial x}\right)_{T,p} = -RT\left(\frac{\partial \ln a_+}{\partial x}\right)_{T,p}$$

$$= -RT\left(\frac{\partial \ln a_+}{\partial m}\right)_{T,p}\left(\frac{\partial m}{\partial x}\right)_{T,p} \tag{11}$$

Equation (11) is the first term in the force (9). The second term is

$$-z_+ F\left(\frac{\partial \phi}{\partial x}\right)_{T,p} = z_+ F E \tag{12}$$

where the electric field strength is

$$E = -\left(\frac{\partial \phi}{\partial x}\right)_{T,p} \tag{13}$$

Our final form for the thermodynamic force is therefore

$$(11), (12) \qquad X_+ = -RT \left(\frac{\partial \ln a_+}{\partial m} \right)_{T,p} \left(\frac{\partial m}{\partial x} \right)_{T,p} + z_+ FE \qquad (14)$$

Similarly, the force on the anions is

$$X_- = -RT \left(\frac{\partial \ln a_-}{\partial m} \right)_{T,p} \left(\frac{\partial m}{\partial x} \right)_{T,p} + z_- FE \qquad (15)$$

Consider the effect of an applied electric field on a solution of uniform concentration. The forces acting on cations and anions are, respectively,

$$(14) \qquad X_+ = z_+ FE; \quad X_- = z_- FE \qquad (16)$$

The current density is

$$j = z_+ FJ_+ + z_- FJ_-$$
$$= z_+ F(L_{++}X_+ + L_{+-}X_-) + z_- F(L_{-+}X_+ + L_{--}X_-) \qquad (17)$$

The Hittorf transport number is

$$t_+ = \frac{z_+ FJ_+}{z_+ FJ_+ + z_- FJ_-}$$
$$= \frac{z_+^2 L_{++} + z_+ z_- L_{+-}}{z_+^2 L_{++} + z_+ z_- (L_{+-} + L_{-+}) + z_-^2 L_{--}} \qquad (19)$$

The four phenomenolgical coefficients $L_{++}, L_{+-}, \ L_{-+}, \ L_{--}$ cannot be determined from t_+ and κ alone so we consider *cells with transport*, and in particular the cell

$$\text{Ag}|\text{AgCl}(c)|\text{HCl}(m_1) \ | \ \text{HCl}(m_2)|\text{AgCl}(c)|\text{Ag} \qquad (20)$$

At the liquid junction, marked by a vertical line in (20), there is a region where both $(\partial m/\partial x)_{T,p}$ and $(\partial \phi/\partial x)_{T,p}$ are non-zero. We measure the cell potential when $j = 0$, so that

$$z_+ FJ_+ + z_- FJ_- = 0 \qquad (21)$$

$(21), (9)$ $z_+ \left\{ L_{++} \left[-\dfrac{\partial \mu_+}{\partial x} - z_+ F \dfrac{\partial \phi}{\partial x} \right] + L_{+-} \left[-\dfrac{\partial \mu_-}{\partial x} - z_- F \dfrac{\partial \phi}{\partial x} \right] \right\}$

$$+ z_- \left\{ L_{-+} \left[-\dfrac{\partial \mu_+}{\partial x} - z_+ F \dfrac{\partial \phi}{\partial x} \right] + L_{--} \left[-\dfrac{\partial \mu_-}{\partial x} - z_- F \dfrac{\partial \phi}{\partial x} \right] \right\} = 0 \tag{22}$$

Solve (22) for $\partial \phi / \partial x$, giving

$$\frac{\partial \phi}{\partial x} = -\frac{1}{F} \left[\frac{t'_+}{z_+} \left(\frac{\partial \mu_+}{\partial x} \right) + \frac{t'_-}{z_-} \left(\frac{\partial \mu_-}{\partial x} \right) \right] \tag{23}$$

where

$$\frac{t'_+}{z_+} = \frac{z_+ L_{++} + z_- L_{-+}}{z_+^2 L_{++} + z_{+-}(L_{+-} + L_{-+}) + z_-^2 L_{--}} \tag{24}$$

On comparing (24) with (19), we see that t'_+ is identical with t_+, provided

$$L_{+-} = L_{-+} \tag{25}$$

Equation (23) tells us how to calculate $\partial \phi / \partial x$ from the gradients of chemical potential. The liquid junction potential is

$$\phi_{\text{LJP}} = \phi_2 - \phi_1 = \int_1^2 \frac{\partial \phi}{\partial x} \, dx$$

$$= -\frac{1}{F} \int_1^2 \left[\frac{t'_+}{z_+} \left(\frac{\partial \mu_+}{\partial x} \right) + \frac{t'_-}{z_-} \left(\frac{\partial \mu_-}{\partial x} \right) \right] dx \tag{26}$$

$$= -\frac{RT}{F} \int_1^2 \left[\frac{t'_+}{z_+} d \ln a_+ + \frac{t'_-}{z_-} d \ln a_- \right] \tag{27}$$

Knowing the mean ion activities $(a_\pm)_2$ and $(a_\pm)_1$, the measurement of E of cell (20) gives t'_+, which can be compared with t_+ from Hittorf measurements. If $t'_+ = t_+$, then $L_{+-} = L_{-+}$. Measurements of t_+ and t'_+ have confirmed the ORR to within about 10%. Greater accuracy than this cannot be expected since the cross-coefficients are much smaller than L_{++}.

Diffusion of an electrolyte

In a diffusion experiment there is, to begin with, a concentration gradient, but no applied field. However, anions and cations diffuse at different rates,

Table 12.2. Diffusion coefficient of NaCl in
water at 25°C (Katchalsky and Curran 1966).

$c/\text{mol dm}^{-3}$	$10^9\,D/\text{m}^2\,\text{s}^{-1}$
10^{-4}	1.60
10^{-3}	1.59
10^{-2}	1.55
10^{-1}	1.49

soon setting up a diffusion potential. Experiment shows that the flux of electrolyte is proportional to the concentration gradient

$$J = -D\frac{\partial c}{\partial x} \quad \text{(Fick's law)} \tag{28}$$

Data for NaCl in water in Table 12.2 show that D is of the order of $10^{-9}\,\text{m}^2\text{s}^{-1}$. D decreases with increasing concentration, due to ion interactions. From TDNEP

$$J = \frac{J_+}{\nu_+} = \frac{1}{\nu_+}[L_{++}\,X_+ + L_{+-}\,X_-] \tag{29}$$

(22) $$X_+ = \left[-\frac{\partial \mu_+}{\partial x} - z_+ F\frac{\partial \phi}{\partial x}\right] \tag{30}$$

Substitute for $\partial\phi/\partial x$ from (23) and recall that

$$\mu = \nu_+\mu_+ + \nu_-\mu_- = \mu^{\ominus}(T,p) + RT\,\ln a(c,T,p) \tag{31}$$

so that

$$\left(\frac{\partial\mu}{\partial x}\right)_{T,p} = RT\left(\frac{\partial\ln a}{\partial c}\right)_{T,p}\left(\frac{\partial c}{\partial x}\right)_{T,p} \tag{32}$$

Finally, comparing the result with (28) above yields

$$D = -\frac{z_+ z_-(L_{++}L_{--} - L_{+-}L_{-+})RT\,(\partial\ln a/\partial c)_{T,p}}{\nu_+\nu_-[z_+^2 L_{++} + z_+ z_-(L_{+-} + L_{-+}) + z_-^2 L_{--}]} \tag{33}$$

Since $(\partial\ln a/\partial c)_{T,p}$ is known from equilibrium measurements, we now have D in terms of the four PCs and can proceed to evaluate the L's.

(a) By measuring t' and t, it is verified that

$$L_{+-} = L_{-+} \tag{34}$$

Table 12.3. Phenomenological coefficients for NaCl in water at 25°C (Katchalsky and Curran 1966).

Phenomenological coefficient	Concentration/mol dm^{-3}				
10^{-13} mol^2 J^{-1} m^{-1} s^{-1}	0.50	0.10	0.05	0.02	0.01
L_{++}	232.9	49.8	25.4	10.4	5.2
L_{--}	356.0	76.2	38.9	15.8	8.0
L_{+-}	42.4	5.5	2.3	0.6	0.2

(b) The three independent coefficients can be calculated from measurements of

(c) t_+, κ, and D. Results for NaCl in H$_2$O in Table 12.3 show that the ratios L_{+-}/L_{++} and L_{+-}/L_{--} decrease with c. Some general conclusions may be drawn:

(i) In very dilute solutions, that is for $I < 0.01$ mol kg^{-1}, the cross-coefficients are negligible and cations and anions diffuse essentially independently. This means that the cation flux is approximately $L_{++}X_+$.

In such dilute solutions the conductivity is

$$\kappa = F^2[z_+^2 L_{++} + z_-^2 L_{--}] \tag{18}$$

In very dilute solutions of strong electrolytes, cations and anions migrate essentially independently, so that

$$\kappa = F[u_+^0 c_+ |z_+| + u_-^0 c_- |z_-|] \tag{35}$$

(18), (35)
$$L_{++} = \frac{u_+^0 c_+}{|z_+| F}, \quad L_{--} = \frac{u_-^0 c_-}{|z_-| F} \tag{36}$$

In very dilute solutions, the cross-coefficients may be neglected and (33), (36) lead to

$$D = \frac{RT}{F} \left(\frac{u_+^0 u_-^0}{u_+^0 + u_-^0} \right) \frac{c}{z_+ \nu_+} \left(\frac{\partial \ln a}{\partial c} \right)_{T,p} \tag{37}$$

which is the *Nernst–Hartley equation*.

(ii) For concentrations at which the activity coefficient correction is not negligible, $D(c)$ agrees well with that predicted from the Nernst–Hartley equation.

Table 12.4. Phenomenological coefficients for the system NaCl + KCl + H$_2$O (Dunlop and Gostling 1959, Fujita and Gostling 1960).

$C_{\text{NaCl}}/\text{mol dm}^{-3}$	0.25	0.50	0.25	0.50
$C_{\text{KCl}}/\text{mol dm}^{-3}$	0.25	0.25	0.50	0.50
L_{12}/L_{21}	1.03	1.01	1.02	0.98

For $I > 0.01$ mol kg^{-1}, the cross-terms L_{+-}, L_{-+} cannot be neglected. They arise from cation–anion electrostatic interactions which become increasingly important as c increases. The concentration dependence of the reduced diagonal coefficients (i.e. the phenomenological coefficient divided by concentration) is due to the Onsager electrophoretic effect, while that of the reduced cross-coefficients is due to the relaxation effect. The results in Table 12.4 demonstrate the validity of the ORR for this system.

Problems 12

12.1 The experimental set-up for studying electrokinetic phenomena consists of two reversible electrodes immersed in an ionic solution and separated by a membrane, such as a glass frit or an ion exchange membrane. A pressure difference Δp is applied to the solution in one compartment and any solvent flow J_s through the membrane is measured in the other one. Either the potential difference E or the electric current I flowing between the electrodes can be measured.

The dissipation function is

$$\Phi = \sigma T = J_s \Delta p + IE \tag{1}$$

Write down the phenomenological equations describing the solvent flow and the flow of current and state the result of using Onsager's theorem. Suggest experiments that could be performed to evaluate the phenomenological coefficients.

12.2 Express the phenomenological equations (12.5.2) and (12.5.3) in terms of experimentally accessible quantities, the conductivity κ, the thermal conductivity k and the absolute thermoelectric power ε.

12.3 Substitute for X_+, X_- from (12.6.16) in (12.6.17) and then find a formula for the electrical conductivity κ. (**Hint**: Use Ohm's law (12.4.1).)

12.4 It is stated in the text that in very dilute solutions, that is for $I < 0.01 \text{ mol kg}^{-1}$, the cross-coefficients are negligible and cations and anions diffuse essentially independently, so that the cation flux is approximately $L_{++}X_+$. Does this imply that the anions play no role in the diffusion of cations?

Solutions to the Problems

Chapter 1

1.1 For $f(x,y) = x^3 - 3xy^2 + 2y^3$, $\frac{\partial f}{\partial x} = 3x^2 - 3y^2$

$$\frac{\partial f}{\partial y} = -6xy + 6y^2, \quad \frac{\partial^2 f}{\partial x^2} = 6x, \quad \frac{\partial^2 f}{\partial y^2} = -6x + 12y,$$

$$\frac{\partial^2 f}{\partial x \partial y} = -6y = \frac{\partial^2 f}{\partial y \partial x}$$

1.2 If $f(p, V, T) = 0$, then

$$df = \left(\frac{\partial f}{\partial p}\right)_{V,T} dp + \left(\frac{\partial f}{\partial V}\right)_{p,T} dV + \left(\frac{\partial f}{\partial T}\right)_{p,V} dT$$

Suppose that p, V and T undergo small changes dp, dV and dT. Then the equation of state becomes $f + df = 0$. But since $f = 0$ and $f + df = 0$, df is zero.

Since f is an implicit function of (p, V, T), that is $f(p, V, T) = 0$

A1.2
$$\left(\frac{\partial p}{\partial V}\right)_T \left(\frac{\partial V}{\partial T}\right)_p \left(\frac{\partial T}{\partial p}\right)_V = -1 \tag{1}$$

(1)
$$\left(\frac{\partial p}{\partial V}\right)_T \left(\frac{\partial V}{\partial T}\right)_{pV} = -\left(\frac{\partial p}{\partial T}\right)_V \tag{2}$$

(2)
$$\left(\frac{\partial T}{\partial p}\right)_V = \frac{k_T}{\alpha} \tag{3}$$

1.3 $du = ydx - xdy = Mdx + Ndy \cdot \frac{\partial M}{\partial y} = 1, \frac{\partial N}{\partial x} = -1$. Therefore, du is not a complete differential. Try $t = \frac{1}{y^2}$ as an integrating factor

$$tdu = \frac{1}{y}dx - \frac{x}{y^2}dy = P\,dx + Q\,dy, \quad \frac{\partial P}{\partial y} = -\frac{1}{y^2}, \quad \frac{\partial Q}{\partial x} = -\frac{1}{y^2} = \frac{\partial P}{\partial y}.$$

Therefore, $t = \frac{1}{y^2}$ is an integrating factor for du. As an exercise, show that $-1/y^2$ is also an integrating factor for du.

1.4 Along path a, the integral $\int_{x_1, y_1}^{x_2, y_2} du = (x_2 - x_1)y_1$ and along path b it is $(x_2 - x_1)y_2$. The difference in area along these two paths is $(x_2 - x_1)(y_2 - y_1)$, which is equal to the area enclosed by the two paths.

1.5 Euler's theorem states that if $f(x_i)$ is a homogeneous function of degree r, then $\sum_{i=1}^{n} x_i(\frac{\partial f}{\partial x_i}) = rf$ Here, $f(\lambda x, \lambda y, \lambda z) = \lambda^3(x^2 y - 2xyz - 2xz^2)$ so that f is a homogeneous function of degree 3

$$x\frac{\partial f}{\partial x} + y\frac{\partial f}{\partial y} + z\frac{\partial f}{\partial z} = x(2xy) + x(-2yz) + x(-2z^2)$$

$$+y(x^2) + y(-2xz) + z(-2xy) + z(-4xz)$$

$$= 3[x^2 y - 2xyz - 2xz^2] = 3f,$$

in accordance with Euler's theorem.

1.6 For an ideal gas, $pV = nRT$, $dU = C_V dT$.
   Therefore,

$$dq = dU - dw = \left(\frac{\partial U}{\partial T}\right)_V dT + \left(\frac{\partial U}{\partial V}\right)_T dV + p\,dV$$

For a monatomic ideal gas, $U = \frac{3}{2}RT$, so that

$$\left(\frac{\partial U}{\partial T}\right)_V dT = C_V dT, \quad \left(\frac{\partial U}{\partial V}\right)_T dV = 0.$$

Therefore

$$dq = \frac{3}{2}RdT + p\,dV = MdT + NdV \text{ with}$$

$$\left(\frac{\partial M}{\partial V}\right)_T = 0, \quad \left(\frac{\partial N}{\partial T}\right)_V = \left(\frac{\partial p}{\partial T}\right)_V \neq 0.$$

Therefore, for a monatomic ideal gas, dq is not a complete differential.

1.7 $dw = -pdV$ where $V = nRT/p$, so that

$$dw = -p\left[\frac{nRdT}{p} - \frac{nRTdp}{p^2}\right]$$

$$= -nRdT + \frac{nRT}{p}dp = MdT + Ndp \cdot \frac{\partial M}{\partial p} = 0, \quad \frac{\partial N}{\partial T} = \frac{nR}{p} \neq \frac{\partial M}{\partial p}.$$

Therefore dw is not a complete differential.

1.8(a) $w = -p_0 \Delta V = -p_0(V^G - V^L) = -nRT$, $V^L \ll V^G$. The vaporization occurs at the normal boiling point of benzene which is $80.1°C = 353.25\,K$, so that

$$w = -1 \times 8.3145 \times 353.25\,J = -2937\,J.$$

(b) $\Delta U = q + w = 3.08 \times 10^4 - 0.294 \times 10^4\,J = 27.9\,kJ\,mol^{-1}$

1.9 There is no change in volume during BC and DA, so that the work done during the isochoric steps is zero. During AB, $w_1 = RT_1 \ln(\frac{V_2}{V_1})$ and during CD $w_3 = RT_2 \ln(\frac{V_1}{V_2})$. The total work done around the cycle is therefore $w_1 + w_3 = R(T_1 - T_2)\ln(\frac{V_2}{V_1}) \neq 0$ Since $w \neq 0$ around the cycle, w is not a function of state.

1.10(a) At constant pressure, the work done on the gas during an adiabatic compression is

$$w = -p_0\,\Delta V = 10^4\,Pa \times 1.35\,m^3 = 1.35 \times 10^4\,J.$$

(b) For an adiabatic process, the change in internal energy is equal to the work done on the system, that is $1.35 \times 10^4\,J$.

(c) In an adiabatic process, q is zero.

1.11 The molecules have a finite volume, therefore the volume in which they are free to move is not V_m but $V_m - b$, where b is the excluded volume

$$b = \frac{1}{2}\left(\frac{4\pi}{3}\right)(2r)^3 N_A = 4\left(\frac{4\pi}{3}\right)r^3 N_A \tag{1}$$

In this formula, the factor $1/2$ prevents counting each molecule twice. The radius of the sphere is $2r$ because that is the closest possible distance between the centres of two molecules. The attractive forces between the molecules tend to hold them together and thus reduce the pressure. The pressure exerted by a gas is the product of the frequency of collision with the walls and the momentum change at each collision. Both factors are reduced by intermolecular forces and van der Waals surmised that both effects might be proportional to the density of the gas molecules, so that the reduction in pressure is proportional to V_m^{-2}. Therefore

$$p = \frac{RT}{V_m - b} - \frac{a}{V_m^2} \tag{2}$$

which is van der Waals equation of state. The first term on the RS contains a correction for the finite volume of the molecules and the second one is an approximation to simulate the attractive effect of intermolecular forces. For example, for CO_2, $a = 0.364\,\text{Pa}\,\text{m}^6\,\text{mol}^{-2}$ and $b = 0.427 \times 10^{-3}\,\text{m}^3\,\text{mol}^{-1}$. van der Waals equation does not apply in the coexistence region since the system consists of two phases, liquid as well as vapour. In this region the isotherm is a horizontal line parallel to the V_m axis. Equation (2) is a cubic equation in V_m and therefore has one or three real roots. Three real roots for V_m would be absurd since V_m is a function of state. However, there is no inconsistency, since in the coexistence region the system is univariant, $p(T)$ and p is constant along any isotherm

(2)
$$p = \frac{RT}{V_m - b} - \frac{a}{V_m^2} \tag{3}$$

(3)
$$\left(\frac{\partial p}{\partial V}\right)_{T_C} = \frac{-RT_C}{(V_{m,C} - b)^2} + \frac{2a}{V_{m,C}^3} = 0 \tag{4}$$

(4)
$$\left(\frac{\partial^2 p}{\partial V^2}\right)_{T_C} = \frac{2RT_C}{(V_{m,C} - b)^3} - \frac{6a}{V_{m,C}^4} = 0 \tag{5}$$

(4), (5)
$$\frac{2}{(V_{m,C} - b)} = \frac{3}{V_{m,C}}, \quad V_{m,C} = 3b \tag{6}$$

(4), (6)
$$T_C = \frac{8a}{27Rb} \tag{7}$$

(6), (7), (3)
$$p_C = \frac{a}{27b^2} \tag{8}$$

(6), (7), (8)
$$\frac{p_C V_C}{RT_C} = \frac{a \times 3b \times 27Rb}{27b^2 \times R \times 8a} = \frac{3}{8} \tag{9}$$

1.12 Since $V(p, T)$,

$$dV = \left(\frac{\partial V}{\partial p}\right)_T dp + \left(\frac{\partial V}{\partial T}\right)_p dT$$

$$d\ln V = -k_T\,dp + \alpha\,dT$$

$d\ln V$ is a complete differential, therefore

$$\left(\frac{\partial \alpha}{\partial p}\right)_T = \left(\frac{\partial \kappa_T}{\partial T}\right)_p$$

Chapter 2

2.1(a) In the adiabatic expansion of a gas, $q = 0$ and $\Delta U = w$. Since the system does work on its surroundings, w is negative and so, therefore, is $\Delta U = C_V dT$. The temperature of the gas therefore decreases as a result of adiabatic expansion.

(b) For the adiabatic expansion of an ideal gas,

$$C_V dT = -p\,dV \tag{1}$$

The work done

$$dw = -p\,dV = V\,dp - R\,dT \quad \text{(one mole of ideal gas)} \tag{2}$$

and the equation of state is

$$pV = RT, \tag{3}$$

$$(1),\ (2),\ (3) \qquad \frac{1}{RT}dT = \frac{-1}{C_V}d\ln V, \quad \frac{1}{R}\ln\left(\frac{T_2}{T_1}\right) = \frac{-1}{C_V}\ln\left(\frac{V_2}{V_1}\right) \tag{4}$$

$$(1),\ (3) \qquad C_V dT = V\,dp - R\,dT, \quad \frac{1}{R}d\ln T = \frac{1}{C_p}d\ln p \tag{5}$$

$$(4),\ (5) \qquad \frac{1}{C_p}\ln\frac{p_2}{p_1} = \frac{-1}{C_V}\ln\frac{V_2}{V_1} = \frac{1}{R}\ln\frac{T_2}{T_1} \tag{6}$$

Let γ denote the ratio of the heat capacities C_p/C_V

$$(6) \qquad \frac{-C_V}{C_p}\ln\frac{p_2}{p_1} = \ln\frac{V_2}{V_1}, \quad \frac{p_1}{p_2} = \left(\frac{V_2}{V_1}\right)^{\gamma}, \quad p_1V_1^{\gamma} = p_2V_2^{\gamma} \tag{7}$$

$$(6) \qquad \frac{1}{C_p - C_V}\ln\frac{T_2}{T_1} = \frac{-1}{C_V}\ln\frac{V_2}{V_1}, \quad \frac{T_2}{T_1} = \left(\frac{V_1}{V_2}\right)^{\gamma-1}, \quad T_1V_1^{\gamma-1} = T_2V_2^{\gamma-1} \tag{8}$$

$$(6) \qquad \frac{C_p - C_V}{C_p}\ln\frac{p_2}{p_1} = \ln\frac{T_2}{T_1} = \frac{\gamma-1}{\gamma}\ln\frac{p_2}{p_1}, \quad \frac{T_2}{T_1} = \left(\frac{p_2}{p_1}\right)^{\frac{\gamma-1}{\gamma}} \tag{9}$$

(c) In the isothermal expansion of an ideal gas, the slope of the pV isotherm in the p-V plane is $\partial(pV)/\partial V = p$. In an adiabatic expansion,

$$(7) \qquad pV^{\gamma} = \text{constant}, \quad \text{where } \gamma > 1.$$

The slope of the isotherm in the p, V plane is $\partial(pV^{\gamma})/\partial V = \gamma pV^{\gamma-1} > p$. Therefore the adiabatic is steeper in the p-V plane than the corresponding isothermal from the same initial state.

(2.2.2) $$dU = TdS - p\,dV \tag{1}$$

For an adiabatic change, $dS = dq/T = 0$, so that

(1) $$C_V dT = -nRT\frac{dV}{V}, \quad C_V d\ln T = -nR\,d\ln V,$$

$$\frac{1}{nR}\ln\frac{T_2}{T_1} = \left(\frac{-1}{C_V}\right)\ln\frac{V_2}{V_1} \tag{4}$$

(2.2.9) $$dH = T\,dS + V\,dp \tag{3}$$

For reversible adiabatic changes in state, $dS = dq/T = 0$, so for an ideal gas

(3) $$C_p dT = nRT\frac{dp}{p}, \quad C_p d\ln T = nR\,d\ln p,$$

$$\frac{1}{nR}\ln\frac{T_2}{T_1} = \frac{1}{C_p}\ln\frac{p_2}{p_1} \tag{4}$$

(2), (4) $$\frac{1}{nR}\ln\frac{T_2}{T_1} = \frac{1}{C_p}\ln\frac{p_2}{p_1} = \left(\frac{-1}{C_V}\right)\ln\frac{V_2}{V_1} \tag{5}$$

(4) $$\log\left(\frac{T_2}{298\,\mathrm{K}}\right) = \frac{2R}{5R}\log\frac{500}{5000}, \quad T_2 = 119\,\mathrm{K} \tag{6}$$

For a reversible adiabatic process $\Delta S = 0$.
For an irreversible adiabatic process, $q = 0$, $\Delta U = w$, so

$$C_V(T_2 - T_1) = -p_0(V_2 - V_1) = -p_0\left(\frac{RT_2}{p_0} - \frac{RT_1}{p_1}\right)$$

$$\frac{3R}{2}(T_2 - 298\,\mathrm{K}) = -RT_2 + 29.8\,\mathrm{K}$$

$$T_2 = 191\,\mathrm{K} \tag{7}$$

S is a function of state and so ΔS is independent of the path by which the system reaches the final state. Therefore ΔS may be calculated from the fundamental equation, which for an ideal gas becomes

$$dS = C_p d\ln T - (V/T)/dp \tag{8}$$

Integrating (8) yields

$$\Delta S = C_p \ln\left(\frac{T_2}{T_1}\right) - R\ln\left(\frac{p_2}{p_1}\right) = (5R2)\ln\left(\frac{190.7}{298}\right) - R\ln\left(\frac{1}{10}\right)$$

$$= 9.87\,\mathrm{J\,K^{-1}\,mol^{-1}}$$

In the reversible expansion, the system does maximum work on its surroundings, more than in an irreversible expansion, when $p < p_0$. Thus it uses up more internal energy and the final temperature is consequently lower for the reversible expansion, *cf.* (6) and (7). For a reversible adiabatic expansion $\Delta S = 0$ but in the irreversible change $\Delta S > 0$, in accordance with the Second Law $\Delta S_{\text{net}} \geq 0$.

2.2 At constant T

(2.2.11)
$$dG = V\,dp = nRT\,d\ln p$$
$$\Delta G = 2 \times 8.3145\,\mathrm{J\,K^{-1}} \times 300\,\mathrm{K} \times 2.303 \times (-1)$$
$$= -4989\,\mathrm{J}$$

G is a function of state so ΔG is the same for the same change in state, whether the path is reversible or an irreversible one.

2.3 Yes. The definition of an ideal gas is one whose internal energy U is a function of T only and which obeys the equation of state $pV = nRT$ at all pressures. Therefore, for an ideal gas, $H_m = U_m + pV_m = U_m(T) + RT$, so that $H(T)$ only

2.4(a)
$$pV_m = RT[1 + B'(T)p]$$

$$\frac{pV_m}{RT} = 1 + \frac{B'}{RT}\left[\frac{RT}{V_m} + \frac{RT}{V_m^2}B + \cdots\right] = 1 + \frac{B'}{V_m} + O\left(\frac{1}{V_m^2}\right)$$

Therefore $B' = B$, to order V_m^{-1}.

(b)
$$\left(\frac{\partial U}{\partial p}\right)_T = \left(\frac{\partial U}{\partial V}\right)_T \left(\frac{\partial V}{\partial p}\right)_T = \left[T\left(\frac{\partial p}{\partial T}\right)_V - p\right]\left(\frac{\partial V}{\partial p}\right)_T$$

$$= -T\left(\frac{\partial V}{\partial T}\right)_p - p\left(\frac{\partial V}{\partial p}\right)_T$$

(c)   If $V_m = \dfrac{RT}{p} + B$, $\left(\dfrac{\partial U}{\partial p}\right)_T$

$$= -T\left(\frac{R}{p} + \frac{dB}{dT}\right) + p\frac{RT}{p^2} = -T\frac{dB}{dT}$$

(d) (5) $\left(\dfrac{\partial U}{\partial p}\right)_T = -T\dfrac{dB}{dT} = -\dfrac{0.1515\,\mathrm{N\,m^4 mol^{-2}}}{8.314\,\mathrm{J\,K^{-1}mol^{-1}} \times 298\,\mathrm{K}}$

$$= -6.11 \times 10^{-5}\,\mathrm{m^3 mol^{-1}}$$

(e) The experimental result measured directly in a Washburn experiment is $(\partial U/\partial p)_T = -6.13 \times 10^{-5}\,\mathrm{m^3\,mol^{-1}}$. The agreement is excellent and confirms the Second Law, from which (4) is derived, via the fundamental equations.

2.5 The fundamental equation for dS, where $S(UV)$, is

$$dS = (1/T)dU + (p/T)dV.$$

Therefore:

$$\left(\frac{\partial S}{\partial U}\right)_V = \frac{1}{T}; \quad \left(\frac{\partial S}{\partial V}\right)_U = \frac{p}{T}$$

Similarly from

$$dH = TdS + Vdp, \quad \left(\frac{\partial H}{\partial S}\right)_p = T; \quad \left(\frac{\partial H}{\partial p}\right)_S = V$$

from

$$dF = -SdT - pdV, \quad \left(\frac{\partial F}{\partial V}\right)_T = -p; \quad \left(\frac{\partial F}{\partial T}\right)_V = -S \quad \text{and}$$

from

$$dG = -SdT + Vdp, \quad \left(\frac{\partial G}{\partial p}\right)_T = V; \quad \left(\frac{\partial G}{\partial T}\right)_p = -S$$

2.6 Let R be a Carnot engine that absorbs heat q_1 at T_1 and performs work $-w$ and let E be any other engine that operates in a cycle, at least one step of which is an irreversible process. Let E absorb heat q_1' at T_1 and deliver the same amount of work $-w$. Assume that

$$\eta_E > \eta_R \qquad\qquad (1)$$

(1) $$\frac{-w}{q_1'} > \frac{-w}{q_1} \qquad\qquad (2)$$

(2) $$q_1 > q_1' \qquad\qquad (3)$$

Now let E drive R in reverse. Since R operates in a reversible cycle, the heat rejected at T_1 is $\bar{q}_1$, equal to the heat q_1 that is absorbed at T_1 when R operates on the engine cycle. From the First Law

for R, $$\Delta U = 0 = q_2 - \bar{q}_1 + w \tag{4}$$

for E, $$\Delta U = 0 = q_1' - \bar{q}_2' - w \tag{5}$$

(4), (5) $$q_2 - \bar{q}_2' = \bar{q}_1 - q_1' = q_1 - q_1' \tag{6}$$

Therefore, the two engines operated in this way constitute a self-acting device which transfers a positive amount of heat $q_1 - q_1'$ (see (3)) from a colder to a hotter body, which contravenes the Second Law. Therefore the hypothesis (1) is incorrect, which proves Carnot's theorem.

2.7 For step (1) (see Exercise 1.4-1)

$$q_1 = -w = \int_{V_1}^{V_2} p\,dV = \int_{V_1}^{V_2} \frac{nRT}{V}\,dV = nRT_1 \ln(V_2/V_1) \tag{1}$$

Since $V_2 > V_1$, w is negative because the expanding gas is doing work on its surroundings. For step (3)

$$q_2 = -\bar{q}_2 = -w = \int_{V_3}^{V_4} p\,dV = \int_{V_3}^{V_4} \frac{nRT}{V}\,dV = nRT_2 \ln(V_4\backslash/V_3) \tag{2}$$

(1), (2) $$\frac{q_1}{\bar{q}_2} = \frac{T_1 \ln(V_2/V_1)}{T_2 \ln(V_3/V_4)} \tag{3}$$

Step (2) is the adiabatic expansion of an ideal gas at T_1 to T_2, for which

(1.5.2), $$C_V dT = -p\,dV = -nRT\frac{dV}{V} \tag{4}$$

(4) $$\frac{1}{nR} \int_{T_1}^{T_2} (C_V/T)dT = \ln\left(\frac{V_3}{V_2}\right) \tag{5}$$

For step (4), which is the adiabatic compression of an ideal gas at T_2 to T_1

(5) $$\frac{1}{nR} \int_{T_2}^{T_1} (C_V/T)dT = \ln\left(\frac{V_1}{V_4}\right) \tag{6}$$

(5), (6) $$\ln\left(\frac{V_3}{V_4}\right) = \ln\left(\frac{V_2}{V_1}\right) \tag{7}$$

(3), (7) $$\frac{q_1}{\bar{q}_2} = \frac{T_1}{T_2}$$ (8)

(8) $$\frac{q_1}{T_1} + \frac{q_2}{T_2} = 0$$ (9)

Equation (8) states that the ratio of the heat absorbed and rejected by an ideal gas Carnot engine, working between two heat reservoirs at temperatures $T_1 > T_2$, is equal to the ratio of these temperatures, T_1/T_2. But since the efficiency of a Carnot engine is independent of its working substance, this result is true for any heat engine operating in a Carnot cycle between these two temperatures

$$\eta = \frac{-w}{q_1} = 1 - \frac{T_2}{T_1} < 1$$ (10)

2.8 In the proof of Carnot's theorem, replace E by R_1, so that the efficiency of R_1 is not greater than that of R_2. Similarly, the efficiency of R_2 is not greater than that of R_1. Therefore all Carnot engines have the same efficiency.

2.9 Substituting $q_1 = -1\,\text{kJ}$, $T_1 = 268\,\text{K}$, $T_2 = 296\,\text{K}$ in Eq. (SP2.8.8) gives

$$q_2 = \frac{1000\,\text{J} \times 268\,\text{K}}{296\,\text{K}} = 905\,\text{J}$$
$$w = -(q_1 + q_2) = 95\,\text{J}$$

2.10 Set up a rectangular coordinate system with T as the ordinate and S as the abscissa. The first step is an isothermal expansion so draw a straight line parallel to the x-axis at T_1. Step 2 is an adiabatic expansion during which the entropy remains constant as T falls from T_1 to T_2. Represent this by a straight line parallel to the y-axis. Step 3 is represented by a straight line parallel to the x-axis of equal length to step 1. Step 4, adiabatic compression to the initial state, completes the TS rectangle. The TS diagram for the Carnot cycle is thus a rectangle in the TS plane.

2.11 To calculate ΔS, replace the spontaneous freezing of the super-cooled water at 263.15 K (d) by the reversible path (a) + (b) + (c):

(a) $H_2O(\text{liq.}, 263.15\,\text{K}) = H_2O(\text{liq.}, 273.15\,\text{K})$

(b) $H_2O(\text{liq.}, 273.15\,\text{K}) = H_2O(\text{cr}, 273.15\,\text{K})$

(c) $H_2O(\text{cr}, 273.15\,\text{K}) = H_2O(\text{cr}, 263.15\,\text{K})$

(d) $H_2O(\text{li.}, 263.15\,\text{K}) = H_2O(\text{cr}, 263.15\,\text{K})$

The entropy change during each of these steps is: $\Delta S / J\,\mathrm{K}^{-1}\,\mathrm{mol}^{-1} =$

(a) $\int_{263}^{273} C_p(\mathrm{liq})\mathrm{d}\ln T = 75.3\ln \frac{273}{263} = 2.81$

(b) $-\frac{5950}{273.16} = -21.78$

(c) $\int_{273}^{263} C_p(\mathrm{cr})\mathrm{d}\ln T = 36.9\ln \frac{263}{273} = -1.38$

(d) $2.81 - 21.78 - 1.38 = -20.35$

The entropy change in the surrounding is given by:

$$\Delta H(263.15\,\mathrm{K}) = \Delta H(273.15\,\mathrm{K}) + [C_p(\mathrm{cr}) - C_p(\mathrm{liq.})]\Delta T$$

$$= 5950 - (38,4 \times 10)\,\mathrm{J\,mol}^{-1} = 5566\,\mathrm{J\,mol}^{-1}$$

$$\Delta S_0(263.15\,\mathrm{K} = \frac{5566}{263} = 21.16\,\mathrm{J\,K}^{-1}\mathrm{mol}^{-1}$$

Thus for the freezing of the super-cooled water at $-10°\mathrm{C}$, the net entropy change is

$$\Delta S + \Delta S_0 = -20.35 + 21.16 = +0.85\,\mathrm{J\,K}^{-1}\,\mathrm{mol}^{-1}$$

The net entropy change is positive, in accordance with the Second Law.

2.12 $\quad \mathrm{d}J = \dfrac{F\mathrm{d}T}{T^2} - \dfrac{\mathrm{d}F}{T} = \dfrac{(U - TS)\mathrm{d}T}{T^2} + \dfrac{S\mathrm{d}T + p\mathrm{d}V}{T} = \dfrac{U\mathrm{d}T}{T^2} + \dfrac{p\mathrm{d}V}{T}$

$\quad\;\; \mathrm{d}Y = \dfrac{G\mathrm{d}T}{T^2} - \dfrac{\mathrm{d}G}{T} = \dfrac{(H - TS)\mathrm{d}T}{T^2} + \dfrac{S\mathrm{d}T - V\mathrm{d}p}{T} = \dfrac{H\mathrm{d}T}{T^2} - \dfrac{V\mathrm{d}p}{T}$

Chapter 3

3.1

(3.3.9)
$$E = \frac{\hbar^2 k^2}{2m} = \frac{n^2 h^2}{8ma^2} \qquad (1)$$

But the linear momentum p of the particle is related to its kinetic energy E by

$$E = \frac{p^2}{2m} \qquad (2)$$

k is the wave number $= 2\pi/\Lambda$, so that

(1), (2)
$$p = \hbar k = h/\Lambda \qquad (3)$$

which is the de Broglie relation, Λ being the thermal de Broglie wavelength. In three dimensions, the same formulae apply, except that p and k are vectors. k is the wave vector, of magnitude $2\pi/\Lambda$ and orientation along the direction of propagation of the wave (or direction of motion of the particle).

3.2

$$p^2 = 2mE = 2 \times 1.675 \times 10^{-27} \text{ kg} \times 0.1 \times 1.602 \times 10^{-19} \text{ J}$$
$$= 0.537 \times 10^{-46} \text{ kg}^2 \text{m}^2 \text{s}^{-2}$$

Therefore, $p = 0.733 \times 10^{-23} \text{ kg m s}^{-1}$, and so

$$\Lambda = h/p = 6.626 \times 10^{-34} \text{ J s}/0.733 \times 10^{-23} \text{ kg m s}^{-1}$$
$$= 9.04 \times 10^{-11} \text{ m} = 0.9 \times 10^{-10} \text{ m}$$

i.e. about 1 Å. Slow neutrons should thus be useful in studying diffraction by crystals, as is indeed the case.

3.3 The Maxwell distribution function for molecular speeds is

(3.7.6)
$$f(v)dv = \left(\frac{m}{2\pi kT}\right)^{\frac{3}{2}} \exp(-mv^2/2kT)4\pi v^2 dv \qquad (1)$$

Differentiating $f(v)$ with respect to v and setting the derivative equal to zero gives for the most probable speed

$$v_p = \left(\frac{2kT}{m}\right)^{\frac{1}{2}} \qquad (2)$$

3.4 See Figure S3.1

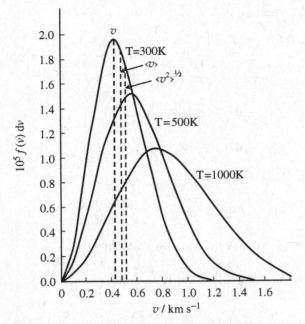

Figure S3.1. Maxwell distribution of molecular speeds in nitrogen at 300 K, 500 K and 1000 K.

3.5 Consider a small amount of a volatile solid on a pan that can be weighed in an oven maintained at a constant temperature. If one wall of the oven has a small hole in it of radius r, then a molecule striking this hole will leave the container. Provided r is much less than the average distance between collisions, there will be no collective movement of gas molecules near the hole and the Maxwell distribution of speeds will be maintained. The rate of *effusion*, that is the loss of mass of the solid per time per area of the hole, is

(3.7.15)
$$\Delta w = m Z_{\mathrm{w}} = p \left(\frac{M}{2\pi RT} \right)^{1/2}$$
(1)

Δw is the mass of the solid lost by effusion, m is the mass of one molecule of effusing vapour and M is the molar mass of the vapour.

Units check: LS: kg m^{-2} s^{-1}, RS = N m^{-2}(kg mol^{-1}/J K^{-1}mol^{-1} K)$^{1/2}$ = kg m s^{-2} m^{-2}(m^{-1} s) = kg s^{-1} m^{-2}

Thus, if Δw is measured and the molar mass is known, p can be calculated.

3.6 For a closed phase

(3.9.27) $\mathrm{d}J = -\mathrm{d}(F/T)$

$$= \frac{F}{T^2}\mathrm{d}T - \frac{1}{T}\left[\left(\frac{\partial F}{\partial T}\right)_V \mathrm{d}T + \left(\frac{\partial F}{\partial V}\right)_T \mathrm{d}V\right] \qquad \text{(SP3.6.1)}$$

(2.2.10) $$= \frac{F}{T^2}\mathrm{d}T + \frac{S}{T}\mathrm{d}T + \frac{p}{T}\mathrm{d}V$$

$$= \frac{U}{T^2}\mathrm{d}T + \frac{p}{T}\mathrm{d}V \qquad \text{(SP3.6.2)}$$

For an open phase, $J(T, V, n_B)$

(P3.6.2), (2.7.6) $$\mathrm{d}J = \frac{U}{T^2}\mathrm{d}T + \frac{p}{T}\mathrm{d}V - \frac{1}{T}\sum_B \mu_B \mathrm{d}n_B \qquad \text{(P3.6.3)}$$

Similarly, for the Planck function for a closed phase

(3.9.28) $$\mathrm{d}Y = -\mathrm{d}\left(\frac{G}{T}\right) = \frac{G}{T^2}\mathrm{d}T - \frac{1}{T}\mathrm{d}G$$

$$= \left(\frac{G}{T^2}\right)\mathrm{d}T - \frac{1}{T}\left[\left(\frac{\partial G}{\partial T}\right)_p \mathrm{d}T + \left(\frac{\partial G}{\partial p}\right)_T \mathrm{d}p\right] \qquad \text{(P3.6.4)}$$

(2.2.11) $$= \left(\frac{G}{T^2}\right)\mathrm{d}T - \frac{1}{T}[-S\mathrm{d}T + V\mathrm{d}p]$$

$$= \frac{H}{T^2}\mathrm{d}T - \frac{V}{T}\mathrm{d}p \qquad \text{(P3.6.5)}$$

For an open phase

(P3.6.5), (2.7.7) $$\mathrm{d}Y = \frac{H}{T^2}\mathrm{d}T - \frac{V}{T}\mathrm{d}p - \frac{1}{T}\sum_B \mu_B \mathrm{d}n_B$$

3.7 Since y and z are independent variables, they may always be chosen to make $By + Cz = 0$, whatever the values of B and C. This makes $Ax = 0$ for all values of x, so that $A = 0$. Similarly, $B = 0, C = 0$. The condition (P3.6.1) together with the independence of x, y and z, thus ensures that A, B and C are all zero. This is the argument used in the derivation of the Boltzmann distribution.

3.8 The Helmholtz energy of this system is

(3.9.27) $F = -kT\ln Q$ \hfill (1) where the partition function Q is

(3.8.11)
$$Q = \frac{q^N}{N!} \tag{2}$$

(1), (2)
$$F = -NkT \ln q + N \ln N - N \tag{3}$$

The chemical potential μ is therefore

(3)
$$\mu = \left(\frac{\partial F}{\partial N}\right)_{T,V} = -kT\ln\left(\frac{q(T,V)}{N}\right) \tag{4}$$

3.9

(3.9.2)
$$\langle E \rangle = NkT^2 \frac{\partial \ln q}{\partial T} = U \tag{1}$$

(3.9.6)
$$p = NkT\frac{\partial \ln q}{\partial V} \tag{2}$$

With c set $= 0$

(3.9.19)
$$S = kT\frac{\partial \ln Q}{\partial T} + k\ln Q$$

$$= Nk\left[T\frac{\partial \ln q}{\partial T} + \ln q + 1 - \ln N\right]$$

$$= Nk\left[T\frac{\partial \ln q_t}{\partial T} + \ln\left(\frac{q_t}{N}\right) + 1 + T\frac{\partial \ln q_r}{\partial T}\right.$$

$$\left. + \ln q_r + T\frac{\partial \ln q_v}{\partial T} + \ln q_v\right] \tag{3}$$

The first three terms are the translational contribution, the next two terms are the rotational contribution and the last two terms the vibrational contribution.

(3.6.12)
$$q_t = V(2\pi mkT/h^2)^{3/2} \tag{4}$$

(3.10.26)
$$q_r = \frac{T}{\sigma\Theta_r} \tag{5}$$

(3.10.45)
$$q_v = \frac{\exp(-\Theta_v/2T)}{1 - \exp(-\Theta_v/T)} \tag{6}$$

For nitrogen, $\Theta_v = 3374\,\mathrm{K}$, $\Theta_r = 2.88\,\mathrm{K}$ and $\sigma = 2$

$$S_t/R = \ln[(V/N)(2\pi mkT/h^2)^{3/2}] + 1 + T(3/2T)$$

$$= (5/2) + \ln(2\pi mkT/h^2)^{3/2}(kT/p)$$

$$= (5/2) + (3/2) \ln \frac{2\pi(0.0280134 \, \text{kg mol}^{-1})(1.38066 \times 10^{-23} \, \text{J K}^{-1})(T)}{(6.0220 \times 10^{23} \, \text{mol}^{-1})(6.6262 \times 10^{-34} \, \text{J s})^2}$$

$$+ \ln\left(\frac{(1.38066 \times 10^{-23} \, \text{J K}^{-1})T}{101.325 \, \text{kPa}}\right) + \left(\frac{5}{2}\right) \ln\left(\frac{T}{K}\right)$$

$$= 2.5 + 1.33415 + (5/2)\ln(T/K) \tag{7}$$

(5)
$$\frac{\mathrm{d}\ln q_{\mathrm{r}}}{\mathrm{d}T} = \frac{1}{T} \tag{8}$$

(6)
$$\frac{\mathrm{d}\ln q_{\mathrm{v}}}{\mathrm{d}T} = \frac{\Theta_{\mathrm{v}}}{2T^2} + \frac{(\Theta_{\mathrm{v}}/T^2)\exp(-\Theta_{\mathrm{v}}/T)}{(1 - \exp(-\Theta_{\mathrm{v}}/T))}$$

$$= \frac{\Theta_{\mathrm{v}}}{T^2}\left[\frac{1}{2} + \frac{\exp(-\Theta_{\mathrm{v}}/T)}{(1 - \exp(-\Theta_{\mathrm{v}}/T))}\right] \tag{9}$$

(1), (4), (5), (6)
$$U = RT\left[\frac{3}{2} + 1 + \frac{\Theta_{\mathrm{v}}}{T}\left(\frac{1}{2} + \frac{\exp(-\Theta_{\mathrm{v}}/T)}{1 - \exp(-\Theta_{\mathrm{v}}/T)}\right)\right] \tag{10}$$

T/K	U_{t}/kJ mol^{-1}	U_{r}/kJ mol^{-1}	U_{v}/kJ mol^{-1}	U/kJ mol^{-1}
300	3.742	2.805	14.026	20.573
500	6.235	4.157	14.595	24.987
1000	12.472	8.315	14.987	35.774

$$H_{\mathrm{t}} = U_{\mathrm{t}} + pV = U_{\mathrm{t}} + RT, \quad H_{\mathrm{r}} = U_{\mathrm{r}}, \quad H_{\mathrm{v}} = U_{\mathrm{v}} \tag{11}$$

T/K	H_{t}/kJ mol^{-1}	H/kJ mol^{-1}
300	6.236	23.067
500	10.392	29.144
1000	20.786	44.088

(3)
$$S = kT\left(\frac{\partial \ln Q}{\partial T}\right)_V + k\ln Q$$

$$= Nk\ln q - Nk\ln N + Nk + NkT\left(\frac{\partial \ln q}{\partial T}\right)_V \tag{12}$$

For one mole of substance, $S = S_{\mathrm{t}} + S_{\mathrm{r}} + S_{\mathrm{v}}$, where

$$\frac{S_{\mathrm{t}}}{R} = \ln\frac{q_{\mathrm{t}}}{N} + 1 + T\left(\frac{\partial \ln q_{\mathrm{t}}}{\partial T}\right)_V, \quad \frac{S_{\mathrm{r}}}{R} = \ln q_{\mathrm{r}} + T\left(\frac{\partial \ln q_{\mathrm{r}}}{\partial T}\right)_V,$$

$$\frac{S_{\mathrm{v}}}{R} = \ln q_{\mathrm{v}} + T\left(\frac{\partial \ln q_{\mathrm{v}}}{\partial T}\right)_V \tag{13}$$

(8) $$\frac{S_t}{R} = \frac{5}{2} + \ln\left(\frac{2\pi mkT}{h^2}\right)^{3/2}\frac{kT}{p} = \frac{5}{2} + 1.33415 + \frac{5}{2}\ln\frac{T}{K} \quad (14)$$

(5), (13) $$S_r = R - R\ln(\sigma\Theta_r) + R\ln(T/K)$$
$$= 8.3145[1 - 1.7509 + \ln(T/K)] \quad (15)$$

(6), (9) $$\frac{S_v}{R} = -\frac{\Theta_v}{2T} - \ln[1 - \exp(-\Theta_v/T)]$$
$$+ \frac{\Theta_v}{T}\left[\frac{1}{2} + \frac{\exp(-\Theta_v/T)}{(1 - \exp(-\Theta_v/T))}\right]$$

At $T = 300\,\mathrm{K}$, $\Theta_v/T = 3374/300 = 11.2466$, $\exp(-\Theta_v/T) = 1.305 \times 10^{-5}$
$-\ln[1 - \exp(-\Theta_v/T)] \approx \exp(-\Theta_v/T)$

$$\frac{S_v}{R} = \exp(-\Theta_v/T)] + \frac{\Theta_v}{T}\left[\frac{\exp(-\Theta_v/T)}{(1 - \exp(-\Theta_v/T))}\right] = [1 + \frac{\Theta_v}{T}]\exp(-\Theta_v/T)$$

T/K	Θ_v/T	$\dot{S}_t/\mathrm{J\,K^{-1}}$ $\mathrm{mol^{-1}}$	$S_r/\mathrm{J\,K^{-1}}$ $\mathrm{mol^{-1}}$	$S_v/\mathrm{J\,K^{-1}}$ $\mathrm{mol^{-1}}$	$S/\mathrm{J\,K^{-1}}$ $\mathrm{mol^{-1}}$
300	11.2467	150.4	41.1	0.0013	191.5
500	6.748	161.1	45.4	0.0091	206.5
1000	3.374	175.5	51.2	1.2457	228.0

T/K	U/kJ $\mathrm{mol^{-1}}$	$S/\mathrm{J\,K^{-1}}$ $\mathrm{mol^{-1}}$	H/kJ $\mathrm{mol^{-1}}$	F/kJ $\mathrm{mol^{-1}}$	G/kJ $\mathrm{mol^{-1}}$
300	20.573	191.5	23.067	−41.377	−34.383
500	24.987	206.5	29.144	−78.263	−74.106
1000	35.774	228.0	44.088	−192.223	−183.912

Chapter 4

4.1 In Section 2.8 it is stated that a temperature of zero is unattainable in systems in equilibrium by using a device operating in a reversible cycle. But this statement does not include processes that are not cyclic in nature. The distinction is important, because in Chapter 4 we proved the equivalence of the unattainability of a temperature of zero and the Planck statement of the Third Law.

Thus we have not proved that the Third Law is deducible from the Second, as would have been the case, but for the limitation of the proof of inaccessibility in Chapter 2 to cyclic processes. Adiabatic demagnetization of systems with either electronic or nuclear magnetic moments can result in extremely low temperatures, but zero K has not and cannot be obtained in systems in equilibrium.

4.2 A quartet spin state means that the spin degeneracy w_e is 4 and consequently the contribution to the entropy from electron spin is

$$S_e = R \ln w_e = 8.3145 \ln 4 \, \mathrm{J \, K^{-1} \, mol^{-1}} = 11.53 \, \mathrm{J \, K^{-1} \, mol^{-1}}.$$

4.3 The entropy of isotopic mixing is

$$S = -R \sum_B x_B \ln x_B, \quad \text{where } x_B \text{ is the mole fraction isotope B}$$

For lead

$$S = -R[0.015 \ln 0.015 + 0.236 \ln 0.236 + 0.226 \ln 0.226 + 0.523 \ln 0.523]$$

$$= 8.3145 \, \mathrm{J \, K^{-1} \, mol^{-1}} \times 1.079 = 8.97 \, \mathrm{J \, K^{-1} \, mol^{-1}}$$

This contribution to the total entropy is not part of the Third Law entropy and is by convention ignored on the grounds that it remains unchanged during all terrestrial physical and chemical changes (except those designed to change the isotopic constitution).

4.4 Adopting the usual convention that $c = 0$, in accordance with the Third Law that the entropy of a pure, perfect, crystalline substance at zero K is zero (see Section 4.3). The entropy of silver at $T_1 = 298.15 \, \mathrm{K}$ is

$$S(T_1 = 298.15 \, \mathrm{K}) = \int_{T_2}^{T_1} (C_p/T) \, \mathrm{d}T + \int_0^{T_2} (C_p/T) \mathrm{d}T$$

where $T_2 = 15\,\mathrm{K}$ and C_p below $15\,\mathrm{k}$ is approximately equal to C_V. The value of the first integral is $44.3\,\mathrm{J\,K^{-1}mol^{-1}}$ and that of the second is $0.22\,\mathrm{J\,K^{-1}mol^{-1}}$. Therefore the entropy of silver at T_1 is $44.5\,\mathrm{J\ K^{-1}mol^{-1}}$.

4.5 The two hydrogen atoms of the central molecule are oriented towards two of the four adjacent O atoms. There are, therefore, $4!/2!2! = 6$ possible orientations of the central molecule. Each adjacent site has 2 out of 4 directions available for forming H bonds. The chance of both H atoms bonding is therefore $(1/2)(1/2)$. The ground state degeneracy is therefore $\Omega_0 = 6/4 = 1.5$. Therefore, the configurational entropy of ice is $S_0 = R\ln 1.5 = 3.37\,\mathrm{J\ K^{-1}\ mol^{-1}}$.

4.6 For deuterium, the sum of the entropies from nuclear spin degeneracy (the first two terms on the RS of (1)), and ortho-para mixing (the other two terms) is

$$S_0 = \frac{2}{3}R\ln 3 + \frac{1}{3}R\ln 1 - R\left[\frac{2}{3}\ln\frac{2}{3} + \frac{1}{3}\ln\frac{1}{3}\right] = 11.4\,\mathrm{J\,K^{-1}\,mol^{-1}} \quad (1)$$

If ordinary deuterium were to be converted to pure $p\text{-}D_2$ the entropy extrapolated to $0\,\mathrm{K}$ would be zero. However, if the usual custom of neglecting the entropy from nuclear spin degeneracy and ortho-para mixing had been followed, then S_0 for ordinary deuterium would have been zero and that for pure $p\text{-}D_2$ would be $-11.4\,\mathrm{J\ K^{-1}\ mol^{-1}}$.

4.7 From data supplied, the entropy of vaporization of water at $298.15\,\mathrm{K}$ is

$$\Delta S = 188.83 - 69.91\,\mathrm{J\,K^{-1}\,mol^{-1}} = 118.92\,\mathrm{J\,K^{-1}mol^{-1}}$$

The enthalpy of vaporization under these conditions, is

$$\Delta H = T\Delta S = 118.92 \times 298.15\,\mathrm{J\,mol^{-1}} = 35.46\,\mathrm{kJ\,mol^{-1}}$$

The entropy of vaporization is expected to be positive because the gaseous state is more disordered than the liquid state. For normal liquids, that is with no features (such as hydrogen bonding or association) tending to create a more ordered environment, the entropy of vaporization is usually about $87\,\mathrm{J\ K^{-1}\ mol^{-1}}$ (Trouton's rule). For example, the entropy of vaporization at $298.16\,\mathrm{K}$ of CCl_4 is $85.8\,\mathrm{J\ K^{-1}\ mol^{-1}}$ and that of H_2S is $87.9\,\mathrm{J\ K^{-1}\ mol^{-1}}$. The higher value for water is due to H-bonding in water.

4.8 Since Θ_r is relatively high, the rotational partition function will be evaluated term-by-term

$$(3.9.19) \qquad S = kT\frac{\partial \ln Q}{\partial T} + k\ln Q \quad (c \text{ set} = 0) \qquad (1)$$

where

$$Q = \frac{q^N}{N!} \quad \text{so that} \quad \ln Q = N\ln q - N\ln N + N \qquad (2)$$

For one mole

$$(1), (2) \qquad S_r = R\ln q_r + RT\frac{d\ln q_r}{dT} \qquad (3)$$

$$(3) \qquad C_{V,r} = T\left(\frac{\partial S_r}{\partial T}\right)_V = 2RT\frac{d\ln q_r}{dT} + RT^2\frac{d^2\ln q_r}{dT^2} \qquad (4)$$

Since the (complete) state function is antisymmetric with respect to the interchange of indistinguishable particles, there are two kinds of hydrogen:

ortho-hydrogen, with parallel nuclear spins, $S = 1$, and $J = 1, 3, 5, \ldots$.
para-hydrogen, with anti-parallel nuclear spins, $S = 0$, and $J = 0, 2, 4, \ldots$.

The rotational partition function of o—hydrogen is

$$(3.40) \qquad q_o = (2S+1)\sum_{J\,odd}(2J+1)\exp[-J(J+1)\Theta_r/T] \qquad (5)$$

where S is the total nuclear spin quantum number, equal to 1 for *ortho*-hydrogen. For *para*-hydrogen (with $S = 0$)

$$q_p = (2S+1)\sum_{J\,even}(2J+1)\exp[-J(J+1)\Theta_r/T] \qquad (6)$$

$$(5) \qquad \frac{d\ln q_o}{dT} = -\frac{\Theta_r}{T^2 q_o}(2S+1)\sum_{J\,odd}-J(J+1)(2J+1)\exp(-J(J+1)\Theta_r/T)$$

$$= \frac{\Theta_r q_o'}{T^2 q_o} \qquad (7)$$

which defines q_o'

$$\frac{d^2 q_o'}{dT} = \sum_{J\,odd}\frac{\Theta_r}{T^2}J^2(J+1)^2(2J+1)\exp(-J(J+1)\Theta_r/T) \qquad (8)$$

$$\frac{d^2 q_o}{dT} = \frac{\Theta_r}{T^2}\left[-\frac{2}{T}\left(\frac{q_o'}{q_o}\right) - \frac{\Theta_r}{T^2}\left(\frac{q_o'}{q_o}\right)^2 + \frac{\Theta_r}{T^2}\left(\frac{q_o''}{q_o}\right)\right] \qquad (9)$$

where

$$q_o'' = (2S+1) \sum_{J \, odd} J^2 (J+1)^2 (2J+1) \exp(-J(J+1)\Theta_r/T) \qquad (10)$$

Finally, we have

(3), (7) $$S_r/R = \ln q_r + \frac{\Theta_r q_o'}{T^2 q_o} \qquad (11)$$

(8), (11) $$C_{V,r}/R = 2\left(\frac{\Theta_r q_o'}{T q_o}\right) + \left[-2\frac{\Theta_r}{T}\frac{q_o'}{q_o} - \left(\frac{\Theta_r}{T}\frac{q_o'}{q_o}\right)^2 + \left(\frac{\Theta_r}{T}\right)^2 \frac{q_o''}{q_o}\right]$$

$$= \left(\frac{\Theta_r}{T}\right)^2 \left[\frac{q_o''}{q_o} - \left(\frac{q_o'}{q_o}\right)^2\right] \qquad (12)$$

The same equations hold for p-H_2 except that $(2S+1) = 1$ and the sums run over $J = 0, 2, 4, \ldots$. We first calculate the values of q_o, q_o' and q_o'' for *ortho*-H_2

(5) $$q_o = 3 \sum_{J \, odd} (2J+1) \exp[-J(J+1)\Theta_r/T]$$

For $T = 140\,\mathrm{K}$

$$\frac{\Theta_r}{T} = \frac{85.4}{140} = 0.610$$

$$q_o = 3[3\exp(-1.22) + 7\exp(-7.32) + \cdots]$$
$$= 3[0.8857 + 0.00046] = 2.6564$$

$$q_o' = (2S+1) \sum_{J \, odd} J(J+1)(2J+1) \exp(-J(J+1)\Theta_r/T)$$

$$= 3[6\exp(-1.22) + 84\exp(-7.32]$$
$$= 3(1.77138 + 0.05562) = 5.481$$

$$\frac{q_o'}{q_o} = \frac{5.481}{2.6564} = 2.0633$$

(10) $$q_0'' = 3[12 \times 0.2952 + 1008 \times 0.00046]$$
$$= 12.018$$

$$\frac{q_o''}{q_o} = \frac{12.018}{2.6564} = 4.5242$$

$$(14) \qquad \frac{C_{V,o}}{R} = \left(\frac{\Theta_{\rm r}}{T}\right)^2 \left[\frac{q_o''}{q_o} - \left(\frac{q_o'}{q_o}\right)^2\right] = 0.3721 \times 0.2670 = 0.0994$$

$$C_{V,o} = 0.826 \,\mathrm{J\,K^{-1}\,mol^{-1}}$$

For *para*-H_2

$$q_p = \omega_p \sum_{J=0,2,4,\ldots} (2J+1)\exp[-J(J+1)\Theta_{\rm r}/T]$$

$$= 1[\exp(-0) + 5\exp(-3.66) + 9\exp(-12.2) + \cdots]$$

$$= 1.0 + 0.1287 + 0.000045$$

$$= 1.1287$$

$$q_p' = \sum_{J=0.2.4.} J(J+1)(2J+1)\exp(-J(J+1)\Theta_{\rm r}/T)$$

$$q_p' = 0 + 0.7720 + 0.0009 = 0.7729$$

$$\frac{q_p'}{q_p} = \frac{0.7229}{1.1287} = 0.6405$$

$$q_p'' = \sum_{J=0,2,4,} J^2(J+1)^2(2J+1)\exp(-J(J+1)\Theta_{\rm r}/T)$$

$$q_p'' = 0 + 180\exp(-3.66) + 3600\exp(-20 \times 0.610)$$

$$= 4.6318 + 0.0181 = 4.6500$$

$$\frac{q_p''}{q_p} = \frac{4.6500}{1.1287} = 4.1198$$

$$(3) \qquad S_p = R\ln q_p + RT\frac{\mathrm{d}\ln q_p}{\mathrm{d}T}$$

$$(7) \qquad \frac{\mathrm{d}\ln q_p}{\mathrm{d}T} = \frac{\Theta_{\rm r}q_p'}{T^2 q_p} = \frac{85.4 \times 0.6405}{140^2} = 0.0028$$

$$S_p = R(0.1211 + 0.0028) = 1.02\,\mathrm{J\,K^{-1}\,mol^{-1}}$$

$$(3) \qquad C_{V,p} = T\left(\frac{\partial S_p}{\partial T}\right)_V = 2RT\frac{\mathrm{d}\ln q_p}{\mathrm{d}T} + RT^2\frac{\mathrm{d}^2\ln q_p}{\mathrm{d}T^2}$$

where

$$RT^2\frac{\mathrm{d}^2\ln q_p}{\mathrm{d}T^2} = R\left[-\frac{2\Theta_{\rm r}}{T}\left(\frac{q_p'}{q_p}\right) - \frac{\Theta_{\rm r}^2}{T^2}\left(\frac{q_p'}{q_p}\right)^2 + \frac{\Theta_{\rm r}^2}{T^2}\left(\frac{q_p''}{q_p}\right)\right]$$

and

$$2T\frac{d \ln q_p}{dT} = \frac{2\Theta_r}{T}\left(\frac{q_p'}{q_p}\right)$$

$$\frac{C_{V,p}}{R} = -\frac{\Theta_r^2}{T^2}\left(\frac{q_p'}{q_p}\right)^2 + \frac{\Theta_r^2}{T^2}\left(\frac{q_p''}{q_p}\right)$$

$$= 0.610^2[4.1198 - 0.6405^2] = 1.380$$

$$C_{V,p} = 11.474\,\mathrm{J\,K^{-1}\,mol^{-1}}$$

4.9 The Maxwell relation

$$\left(\frac{\partial S}{\partial p}\right)_T = \left(\frac{\partial V}{\partial T}\right)_p = \alpha V$$

holds at any temperature. Therefore, since $\alpha \to 0$ as $T \to 0\,\mathrm{K}$, S is independent of p in the limit $T \to 0\,\mathrm{K}$. Thus no specification of pressure is required in the statement of the Third Law.

Chapter 5

5.1 Plot $(Z-1)/p$ against p and calculate the area under the curve between 0 and p, which is $\ln(\tilde{p}/p)$ (Eq. (5.5.6)). The result is $\tilde{p} = 3500\,\text{kPa}$ at $p = 4000\,\text{kPa}$. Fit the data in Table P5.4 to the virial equation (1.3.11) and hence calculate the fugacity of CO_2 at 333 K. Repeat, using Eq. (1.3.12).

5.2 The virial equation to order B_2/V_m is

$$\frac{pV_m}{RT} = 1 + \frac{B_2}{V_m} \tag{1}$$

whence

$$V_m = \frac{RT}{p} + B_2 \tag{2}$$

Therefore, to the same order,

(2)
$$\frac{-1}{V_m}\frac{\partial V_m}{\partial p} = \frac{RT}{p^2 V_m} \tag{3}$$

(3)
$$\kappa^{-1} = \frac{p^2 V_m}{RT} = p\left(1 + \frac{B}{V}\right) \tag{4}$$

The virial equation in terms of p, to the same order, is

$$pV_m = RT + B_2 p \tag{5}$$

so that

$$V_m = \frac{RT}{p} + B_2 \tag{6}$$

and

$$\kappa = \frac{RT}{pRT + B_2 p^2}, \quad \kappa^{-1} = p + \frac{B_2 p^2}{RT}$$

$$V_m = \frac{RT}{p} + B_2 = \frac{8.314\,\text{J K}^{-1}\text{mol}^{-1}250\,K}{10\,\text{MPa}} - 16.2\,\text{cm}^3\,\text{mol}^{-1}$$

$$= (207.85 - 16.2)\text{cm}^3\,\text{mol}^{-1}$$

$$V_m = 191.65\,\text{cm}^3\,\text{mol}^{-1}$$

(4)
$$\kappa^{-1} = p\left(1 + \frac{B}{V}\right) = 10\left(1 - \frac{16.2}{191.65}\right) = 9.155\,\text{MPa}$$

$$\kappa = 0.1092\,\text{MPa}^{-1}$$

$$V_{\mathrm{m}} = \frac{RT}{p} + B_2 = \frac{8.314\,\mathrm{J\,K^{-1}mol^{-1}}500\,K}{10\,\mathrm{MPa}} + 16.9\,\mathrm{cm^3\,mol^{-1}}$$

$$= 432.6\,\mathrm{cm^3\,mol^{-1}}$$

$$\kappa^{-1} = p\left(1 + \frac{B}{V}\right) = 10\left(1 + \frac{16.9}{432.6}\right) = 10.391\,\mathrm{MPa}$$

$$\kappa = 0.0962\,\mathrm{MPa^{-1}}$$

5.3

(6), (9)
$$\Delta\mu_{\mathrm{real}}/RT = \ln\left(\frac{\tilde{p}}{p}\right) = \int_0^p \left(\frac{Z-1}{p}\right)\mathrm{d}p$$

$$= \int_0^p (B_2' + B_3'p)\mathrm{d}p = B_2'p + \frac{1}{2}B_3'p^2$$

For N_2 at 273 K and 4 MPa,

$$\Delta\mu_{\mathrm{real}}/RT = -(4 \times 4.53 \times 10^{-3}) + (8 \times 2.25 \times 10^{-4}) = -16.32 \times 10^{-3}$$

$$\Delta\mu_{\mathrm{real}} = -8.3145\,\mathrm{J\,K^{-1}\,mol^{-1}} \times 273\,K \times 16.3 \times 10^{-3} = -37.0\,\mathrm{J\,mol^{-1}}$$

5.4 The ratio of the adiabatic to isothermal compressibilities is

$$\frac{k_S}{k_T} = \frac{(\partial V/\partial p)_S}{(\partial V/\partial p)_T}.$$

Since $S(p, V)$,
$$\left(\frac{\partial S}{\partial p}\right)_V \left(\frac{\partial V}{\partial S}\right)_p = -\left(\frac{\partial V}{\partial p}\right)_S$$

and since $V(T, p)$,

$$\left(\frac{\partial V}{\partial T}\right)_p \left(\frac{\partial T}{\partial p}\right)_V = -\left(\frac{\partial V}{\partial p}\right)_T$$

Therefore,

$$\frac{k_S}{k_T} = \frac{\left(\frac{\partial S}{\partial p}\right)_V \left(\frac{\partial V}{\partial S}\right)_p}{\left(\frac{\partial V}{\partial T}\right)_p \left(\frac{\partial T}{\partial p}\right)_V} = \frac{\left(\frac{\partial S}{\partial T}\right)_V}{\left(\frac{\partial S}{\partial T}\right)_p} = \frac{C_V}{C_p}$$

5.5 The Berthelot equation of state for 1 mole of gas is

$$p = \frac{RT}{V-b} - \frac{a}{TV^2} \tag{1}$$

At $T = T_C$,

$$\left(\frac{\partial p}{\partial V}\right)_T = -\frac{RT}{(V-b)^2} + \frac{2a}{TV^3} = 0 \tag{2}$$

and

$$\left(\frac{\partial^2 p}{\partial V^2}\right)_T = \frac{2RT}{(V-b)^3} - \frac{6a}{TV^4} = 0 \tag{3}$$

(2), (3)
$$\frac{2}{V-b} = \frac{3}{V}, \quad V = 3b \tag{4}$$

(2), (4)
$$\frac{RT^2}{4b^2} = \frac{2a}{27b^3}, T_c = \left(\frac{8a}{27Rb}\right)^{1/2} \tag{5}$$

(1), (4), (5)
$$p = \frac{RT}{V-b} - \frac{a}{TV^2} = \frac{RT^2V^2 - 2ab}{(18b^3)T}$$

$$= \frac{R(9b^2)(8a/27Rb) - 2ab}{(18b^3)T}$$

$$p = \frac{(8ab/3) - 2ab}{(18b^3)T} = \frac{2ab}{(18b^3)}\left(\frac{3Rb}{8a}\right)^{1/2}$$

$$= \frac{1}{18b}\left(\frac{3aR}{2b}\right)^{1/2}$$

$$p_C = \frac{1}{6b}\left(\frac{aR}{6b}\right)^{1/2} \tag{6}$$

Units of a: Pa K m^6 mol^{-2} Units of b: m^3 mol^{-1}

　　Check: units of $T_c = \left(\frac{8a}{27Rb}\right)^{1/2}$ are $\left(\frac{\mathrm{Pa\,K\,m^6mol^{-2}}}{\mathrm{J\,K^{-1}mol^{-1}m^3mol^{-1}}}\right)^{1/2} = K$

(6)　　　Units of p_c are $\left(\frac{\mathrm{Pa\,K\,m^6mol^{-2}J\,K^{-1}\,mol^{-1}}}{\mathrm{m^6\,mol^{-2}\,m^3mol^{-1}}}\right)^{1/2} = \mathrm{Pa}$

To find the reduced equation of state, set $p = p_r\, p_c$, $V = V_r V_c$, $T = T_r\, T_c$. Then

(6)
$$p_r\, p_c = \frac{1}{V_r V_c}\left(\frac{aR}{24b}\right)^{1/2}$$

so that

$$\frac{p_r V_r}{T_r} \left(\frac{p_c V_c}{RT_c}\right) = \frac{3}{8}$$

(4) $\qquad b = V_c/3 = 94.2\,\text{cm}^3\,\text{mol}^{-1}/3 = 31.4\,\text{cm}^3\,\text{mol}^{-1}$

(5) $\quad T_c = \left(\frac{8a}{27Rb}\right)^{1/2}, \quad a^{1/2} = T_c\left(\frac{9RV_c}{8}\right)^{1/2}, \quad a = \left(\frac{9RV_c}{8}\right)T_c^2 \quad$ (7)

(7) $\qquad a = 81.4\,\text{m}^6\,\text{Pa}\,\text{K}\,\text{mol}^{-2}$

(1) $\qquad p = \frac{RT}{V-b} - \frac{a}{TV^2}$

$$p = \frac{8.3145 \times 300 \times 1000}{(10 - 0.0942)} - \frac{81.4}{300 \times 0.01} = 0.251779\,\text{MPa}$$

The van der Waals equation is

$$p = \frac{RT}{V-b} - \frac{a}{V^2}$$

with, for CO_2, $a = 0.364\,\text{m}^6\,\text{Pa}\,\text{mol}^{-2}$ and $b = 42.7\,\text{cm}^3\,\text{mol}^{-1}$. Therefore,

$$p = \frac{8.3145 \times 300}{0.001(10 - 0.0427)} - \frac{0.364}{0.1} = 250627.4 = 0.250627\,\text{MPa}$$

The ideal value

$$p = \frac{8.3145 \times 300}{0.01}\,\text{MPa} = 0.249435\,\text{MPa}$$

5.6 The first thermodynamic equation of state is

(2.3.8) $\qquad \left(\frac{\partial U}{\partial V}\right)_T = T\left(\frac{\partial p}{\partial T}\right)_V - p$ $\qquad$ (1)

From van der Waals equation of state,

$$p = \frac{RT}{V-b} - \frac{a}{V^2} \qquad (2)$$

(2) $\qquad T\left(\frac{\partial p}{\partial T}\right) = \frac{RT}{V-b}$ $\qquad$ (3)

(1), (2), (3) $\qquad \left(\frac{\partial U}{\partial V}\right)_T = \frac{a}{V^2}$ $\qquad$ (4)

Similarly, from the Berthelot equation of state,

$$p = \frac{RT}{V - b} - \frac{a}{TV^2} \tag{5}$$

(1), (5)
$$\left(\frac{\partial U}{\partial V}\right)_T = T\left(\frac{\partial p}{\partial T}\right)_V - p = \frac{2a}{TV^2} \tag{6}$$

(c) (2.2.14)
$$\left(\frac{\partial S}{\partial V}\right)_T = \left(\frac{\partial p}{\partial T}\right)_V \tag{7}$$

For the isothermal change in volume of a gas that obeys van der Waals equation

(7), (2)
$$\Delta S = \int_{V_1}^{V_2} \frac{R}{V - b} dV = R \ln \frac{V_2 - b}{V_1 - b} \tag{8}$$

For the Berthelot equation,

(7), (5)
$$\Delta S = \int_{V_1}^{V_2} \left(\frac{R}{V - b} + \frac{a}{T^2 V^2}\right) dV$$

$$= R \ln \frac{V_2 - b}{V_1 - b} - \frac{2a}{T^2}\left(\frac{1}{V_2} - \frac{1}{V_1}\right) \tag{9}$$

5.7 The square-well potential (Figure 5.3) is defined by:

$$\varphi(r) = \infty, \quad \text{for } r < \sigma_1$$
$$\varphi(r) = -\varepsilon, \quad \text{for } \sigma_1 < r < \sigma_2$$
$$\varphi(r) = 0, \quad \text{for } r > \sigma_2$$
$$B(T) = \frac{N_A}{2} \int_0^\infty [1 - \exp\{-\phi(r)/kT\}]4\pi r^2 dr$$

We break the integral into three ranges:

$$0 \to \sigma_1, \quad \phi(r) = \infty, \quad I_1 = \frac{N_A}{2} \int_0^{\sigma_1} 4\pi r^2 dr = \frac{2}{3}\pi N_A \sigma_1^3$$

$$\sigma_1 \to \sigma_2, \quad \phi(r) = -\varepsilon, \quad I_2 = \frac{N_A}{2} \int_{\sigma_1}^{\sigma_2} 4\pi r^2 [1 - \exp\{-\varepsilon/kT\}]dr$$

$$= \frac{2}{3}\pi N_A(\sigma_2^3 - \sigma_1^3)[1 - \exp\{-\varepsilon/kT\}]$$

$$\sigma_2 \to \infty, \quad \phi(r) = 0, \quad I_3 = 0 \ \ I_1 + I_2 + I_3$$

$$= \frac{2}{3}\pi N_A[\sigma_1^3 + (\sigma_2^3 - \sigma_1^3)[1 - \exp\{-\varepsilon/kT\}]$$

If $\sigma_1 = \sigma$, $\sigma_2 = 1.611\sigma$, $\frac{2}{3}\pi N_A \sigma^3 = 0.440 V_c$, $\varepsilon = 0.75\,kT_C$, then
$$B(T)/V_C = 0.440 + 1.40\{1 - \exp(0.75 T_c/T)\}$$
which provides a good fit to the second virial coefficients of several gases (Guggenheim 1977). The square well is a rather rough approximation but it has the two essential features of intermolecular potentials, a hard core and an attractive region. This exercise illustrates an important feature of the experimental data — and that is how difficult it is to deduce information about intermolecular potentials from experimental data.

5.8 Using your computer, fit $\log(p/\text{kPa})$ to Eq. (P5.9.1). The result is

$$(5.6.30) \qquad \ln p = -\frac{3348.6}{T} - 0.8403 \log T + 9.6753 \qquad (3)$$

It is shown in Section 5.6 that for low vapour pressure over a restricted temperature range, integration of the Clausius–Clapeyron equation yields the Equation (3)

$$(5.6.30) \qquad \ln p = -\frac{\Delta_{\text{vap}} H(0\,\text{K})}{RT} + \frac{\Delta C_p \ln T}{R} + a \qquad (3)$$

The assumptions made in the derivation of (3) are that the vapour behaves as an ideal gas and that the heat capacities of liquid and gas are constant over the temperature range employed. Both these assumptions are favoured by restricting the temperature range employed to lower temperatures. Then c may be associated with an approximate value for the heat of vaporization and b with an approximate value for the difference between the heat capacity of the vapour and that of the liquid.

Chapter 6

6.1

(18) $\quad \mu_B = \mu_B^*(p^{\ominus}) + RT\ln(y_B p/p^{\ominus})$

$$+ p\{B_{BB} - y_A^2(B_{AA} - 2B_{AB} + B_{BB})\} \tag{1}$$

(18) $\quad \mu_A = \mu_A^*(p^{\ominus}) + RT\ln(y_A p/p^{\ominus})$

$$+ p\{B_{AA} - y_B^2(B_{AA} - 2B_{AB} + B_{BB})\} \tag{2}$$

(1), (2) $\quad G = n_A\mu_A^*(p^{\ominus}) + n_B\mu_B^*(p^{\ominus}) + (n_A + n_B)RT\ln(p/p^{\ominus})$

$$+ n_A RT\ln y_A + n_B RT\ln y_B + p[n_{AA}B_{AA}$$

$$+ n_{BB}B_{BB} - y_A n_B(B_{AA} - 2B_{AB} + B_{BB})] \tag{3}$$

$$S = -\left(\frac{\partial G}{\partial T}\right)_p = -n_A\left(\frac{\mathrm{d}\mu_A^*(p^{\ominus})}{\mathrm{d}T}\right)_p - n_A\left(\frac{\mathrm{d}\mu_A^*(p^{\ominus})}{\mathrm{d}T}\right)_p$$

$$+ (n_A + n_B)R\ln(p/p^{\ominus}) - n_A R\ln y_A - n_B R\ln y_B$$

$$- p\left[n_A\frac{\mathrm{d}B_{AA}}{\mathrm{d}T} + n_B\frac{\mathrm{d}B_{BB}}{\mathrm{d}T} - y_A n_B\frac{\mathrm{d}}{\mathrm{d}T}(B_{AA} - 2B_{AB} + B_{BB})\right] \tag{4}$$

$$H = G + TS = n_A\left[\mu_A^*(p^{\ominus}) - T\frac{\mathrm{d}\mu_A^*(p^{\ominus})}{\mathrm{d}T}\right]$$

$$+ n_B\left[\mu_B^*(p^{\ominus}) - T\frac{\mathrm{d}\mu_B^*(p^{\ominus})}{\mathrm{d}T}\right]$$

$$+ pn_A\left(B_{AA} - T\frac{\mathrm{d}B_{AA}}{\mathrm{d}T}\right) + pn_B\left(B_{BB} - T\frac{\mathrm{d}B_{BB}}{\mathrm{d}T}\right)$$

$$- py_A n_B\left(1 - T\frac{\mathrm{d}}{\mathrm{d}T}\right)(B_{AA} - 2B_{AB} + B_{BB}) \tag{5}$$

6.2 The reduced temperature is $T_r = 673/134.4 = 5.02$ and the reduced pressure is $35/3.46 = 10.12$. By interpolation in published graphs, at $p_r = 10.12$ and $T_r = 5.02$, $\tilde{p}/p = 1.16$. Therefore, the fugacity of CO is estimated to be about

$$\tilde{p} = 0.25 \times 1.16 \times 35 = 10.15\,\text{MPa}$$

This is only an estimate because of the use of the Principle of Corresponding States and the Lewis and Randall rule, but nevertheless a better estimate than assuming ideal behaviour.

6.3

$$x_B = \frac{y_B p_A^*}{p_B^* - y_B(p_B^* - p_A^*)} \tag{6.4.14}$$

$$p = p_A^* + \frac{(p_B^* - p_A^*)y_B p_A^*}{p_B^* + (p_B^* - p_A^*)y_B} \tag{6.4.15}$$

$$= \frac{p_A^* p_B^*}{p_B^* - y_B(p_B^* - p_A^*)} \tag{6.4.16}$$

6.4 From Section 6.4, Eq. (6.4.32)–(6.4.34), we learn that for a regular mixture

$$\ln f_A = \frac{w}{RT}x_B^2 \quad \ln f_B = \frac{w}{RT}x_A^2 \tag{P6.4.1}$$

Table S6.5 has the values of f_A, f_B for the various solutions. The required values of $\ln f_A$, $\ln f_B$ as well as x_A^2, x_B^2 are given in Table S6.5.

$$\text{(P6.4.1)} \qquad \ln f_A = \frac{w}{RT}x_B^2 \quad \ln f_B = \frac{w}{RT}x_A^2 \tag{P6.4.2}$$

The plot of $\ln f_A$ versus x_B^2 is linear, showing that these mixtures are regular. The slope of this plot is 1.063. Therefore

$$w = 1.06\,RT = 1.06 \times 8.3145 \times 308.4\,\text{J mol}^{-1} = 2720\,\text{J mol}^{-1}$$

$$\text{(P6.4.2)} \qquad \frac{\partial \ln f_A}{\partial T} = -\frac{w}{RT^2}x_B^2 \quad \frac{\partial \ln f_B}{\partial T} = -\frac{w}{RT^2}x_A^2 \tag{P6.4.3}$$

so that w is the relative partial molar enthalpy

$$\text{(6.4.40)} \qquad w = H_B - H_B^{\ominus} \tag{P6.4.4}$$

Its physical origin is that when a small amount of B is added to a large amount of A, each B molecule is now surrounded by A molecules. The B–A interactions have a different energy to the A–A interactions and therefore w is not, in general, zero.

$$\text{(6.4.40)} \qquad w = H_B - H_B^{\ominus} \tag{P6.4.4}$$

However, if the molecules A and B are chemically similar, this difference may be very small, so that w is undetectably small and an ideal mixture results.

Table S6.5. Test to See if chloroform (A) + acetone (B) form a regular mixture. Using data from Table 6.2.

x_A	x_B	x_A^2	x_B^2	$\ln f_A$	$\ln f_B$	$\ln f_A / x_B^2$
0	1.000	0	1	—	0	
0.059	0.941	0.00348	0.8885	−0.6311	−0.0030	
0.123	0.877	0.01513	0.7691	−0.5692	−0.00904	
0.185	0.815	0.0342	0.6422	−0.5310	−0.01207	
0.297	0.703	0.0882	0.4942	−0.4526	−0.0502	
0.423	0.577	0.1789	0.3329	−0.3327	−0.1312	0.999
0.514	0.486	0.2642	0.2362	−0.2459	−0.2157	1.041
0.663	0.337	0.4396	0.1136	−0.1301	−0.3857	1.145
0.800	0.200	0.6400	0.0400	−0.0440	−0.6088	1.10
0.9175	0.0825	0.8418	0.0068	−0.0070	−0.7831	1.03
1.000	0.000	1	0	1	—	

6.5 The mole fraction of B in the liquid and vapour phases is given by

$$x_B = \frac{\lambda_A - f_A}{\lambda_A \lambda_B f_B - f_A} \tag{6.4.62}$$

(6.4.62), (6.4.63)

$$\frac{y_B}{f_B x_B} = \frac{(\lambda_A - f_A)\lambda_B f_B}{f_B(\lambda_A - f_A)}$$

$$= \lambda_B = \exp(\Lambda_{B,b})$$

$$\ln\left(\frac{y_B}{f_B x_B}\right) = \Lambda_{B, b} = \frac{\Delta H^\ominus_{vap,B}}{R}\left[\frac{1}{T_{B,b}} - \frac{1}{T}\right]$$

6.6 Using data in Table P6.3 the plot of $p_{Zn}(x_{Zn})$ shows that this system has a miscibility gap in the range $0.29 < x_{Zn} < 0.94$. In this range, p_{Zn} is constant, so that $v = 1 = 2 - \phi + 2$, and $\phi = 3$, that is, the vapour phase and two liquid phases. Since $p_{Zn}(x_{Zn})$ is approximately linear for $x_{Zn} < 0.143$, f_{Zn} at ∞ dilution is given by

$$\frac{p_{Zn}}{x_{Zn} p^*_{Zn}} = \frac{11.4}{0.143 \times 17.7} = 4.50$$

$$\mu_{Zn} - \mu^*_{Zn} = RT \ln a_{Zn} = RT \ln \frac{p_{Zn}}{p^*_{Zn}}$$

$$= 8.3145 \times 1027 \, J \times \ln \frac{11.5}{17.7} = -3680 \, J$$

6.7 (a) In mixture I, for water $L_A = \partial \Delta H / \partial n_A = -20.92 \, kJ \, mol^{-1}$ and for sulphuric acid $L_B = \partial \Delta H / \partial n_B = -7.91 \, kJ \, mol^{-1}$. Similarly, for mixture II, $L_A = -6.07 \, kJ \, mol^{-1}$, and $L_B = -31.46 \, kJ \, mol^{-1}$.

(b) From Eq. (6.4.38), $-\Delta H = -(n_A L_A + n_B L_B) = 1(20.92) + 1(7.91) = 28.83\,\mathrm{kJ}$. Similarly, for II, $-\Delta H = 3(6.07) + 1(31.46) = 49.67\,\mathrm{kJ}$.

(c) The heat evolved on dilution of I with 2 mol water is $49.67 - 28.83 = 20.84\,\mathrm{kJ}$.

6.8 The necessary calculations are given in Table S6.4.

The ideal reference state is defined by the state in which the activity coefficient $\to 1$. The standard state is the state of unit activity. Therefore, for the Raoult's law activity coefficient f_A the ideal reference state is pure ethanol, but for the Henry's law activity coefficient $\gamma_{x,A}$ it is the infinitely dilute solution of ethanol in chloroform. The standard state in which $a_A = 1$, that is both f_A and x_A are 1, is again pure ethanol for Raoult's law, but for Henry's law it is a hypothetical state of pure ethanol with $\gamma_{x,A} = 1$.

6.9 When a solution is separated from pure solvent by a semi-permeable membrane that allows solvent molecules to pass through the membrane but not the solute, the equilibrium state (equality of chemical potentials of solvent) is one in which the pressure on the solution side is higher than that on the solvent side. To cause water to flow from the solution into pure water, the chemical potential of water on the solution side must be increased by increasing the pressure over the solution. Maintaining this pressure while continually removing water from the solvent side will result in the production of potentially drinkable water from seawater. The water from these pressure units is generally filtered through activated carbon and subjected to further purification before it is considered fit to drink.

6.10 When $x \ll 1$, (6.4.79) becomes

$$p_B = p_B^* x \exp(w/RT)$$

which is Henry's law with activity coefficient

$$\gamma_{x,B} = \exp(w/RT)$$

Table S6.6. Calculation of Raoult's law f_A and Henry's law $\gamma_{x,A}$ coefficients for ethanol in the ethanol + chloroform mixtures. k_H is the limiting slope of the plot of p_A against x_A.

x_A	0	0.05	0.2	0.4	0.6	0.8	1.0
$x_A p_A^*/\mathrm{kPa}$	0	0.685	2.741	5.481	8.222	10.96	13.70
$p_A = y_A p/\mathrm{kPa}$	0	1.568	5.605	7.211	8.757	11.00	13.70
$f_A = p_A/x_A p_A^*$	—	2.288	2.945	1.316	1.065	1.004	1.0
$k_H x_A/\mathrm{kPa}$	0	1.568	6.271	12.54	18.81	25.09	31.36
$\gamma_{x,A}$	1.0	1.0	0.893	0.575	0.466	0.438	0.436

6.11 Let $AB = BC = CA = a$, and consider the triangles PAB, PAC, PBC. The sum of their areas is S, the area of triangle ABC. Therefore

$$a(PQ) + a(PR) + a(PS) = 2S = a(AT)$$

so that

$$PQ + PR + PS = AT$$

6.12 As water evaporates, the state point moves along the line Pd. At d, $(NH_4)_2SO_4$ starts to precipitate and the point describing the solution moves along db. At e the solution is saturated with respect to both solutes, and on further evaporation deposits both NH_4Cl and $(NH_4)_2SO_4$ until point f is reached when H_2O has completely evaporated.

Chapter 7

7.1
$$dU = TdS - p\,dV + \gamma dA + \sum_B \mu_B dn_B \tag{1}$$

U, S, V, A and n_B are extensive properties and therefore homogeneous functions of degree one, whereas T, p, γ and μ_B are intensive properties and therefore homogeneous functions of degree zero. On using Euler's theorem (Appendix A1),

(1)
$$U = TS - pV + \gamma A + \sum_B \mu_B n_B \tag{2}$$

(2)
$$U - TS + pV - \gamma A = G = \sum_B \mu_B n_B \tag{3}$$

(3)
$$G = -TS + pV + \sum_B \mu_B n_B + \gamma A \tag{4}$$

Differentiate (4) and subtract (1), giving

(4), (1)
$$0 = S\,dT - V dp - A d\gamma + \sum_B n_B d\mu_B \tag{5}$$

which is the Gibbs–Duhem equation for a surface phase.

7.2 From Table 7.1, $\gamma = 72.94\,\text{mJ m}^{-2}$. Therefore

(7.1.3)
$$\delta p = \frac{2\gamma}{r} = \frac{2 \times 72.94 \times 10^{-3}\,\text{J m}^{-2}}{2.0 \times 10^{-7}\,\text{m}} = 0.729\,\text{MPa}$$

The pressure inside the bubble is $p + \delta p = 0.83\,\text{MPa} = 8.2\,\text{atm}$.

7.3 Neglecting the density of air in comparison with that of mercury

(7.2.8)
$$h = \frac{2\gamma \cos\theta}{\rho g_0 r} - \frac{r}{3}$$

$$= \frac{2(0.4865\,\text{Nm}^{-1})(-0.766)}{(13.59 \times 10^{-3}\,\text{kg m}^{-3})(9.806\,\text{m s}^{-2})(0.375 \times 10^{-3}\text{m})}$$

$$- \frac{0.375\,\text{mm}}{3}$$

$$= -(14.8 - 0.125)\,\text{mm} = -14.7\,\text{mm}$$

7.4
$$W = \gamma^{l,\text{air}} \cos\theta + \gamma^{s,l} - \gamma^{s,\text{air}} = -\Delta G$$

The equilibrium condition is $W = 0$, or $\gamma^{l,\text{air}} \cos\theta + \gamma^{s,l} = \gamma^{s,\text{air}}$. However, most systems display hysteresis, in that the contact angle depends on whether the liquid is advancing over a dry surface or receding from one that was previously wetted. For further information, see Adam (1941).

7.5

(7.4.9)
$$\Gamma = \frac{w_o \Delta c}{MA}$$
$$= \frac{2.3\,\text{g} \times 1.30 \times 10^{-5}}{150\,\text{g mol}^{-1} \times 310\,\text{cm}^2}$$
$$= 6.43 \times 10^{-10}\,\text{mol cm}^{-2}$$
$$= 3.87 \times 10^{14}\,\text{molecules cm}^{-2}$$

(7.3.13)
$$\Gamma = -\frac{1}{RT}\frac{\mathrm{d}\gamma}{\mathrm{d}\ln m}$$
$$= \frac{0.00400\,\text{N m}^{-1} \times 4.00}{8{:}314\,\text{J K}^{-1}\,\text{mol}^{-1} \times 298.15\,\text{K}}$$
$$= 6.45 \times 10^{-10}\,\text{mol cm}^{-2}$$
$$= 3.88 \times 10^{14}\,\text{molecules cm}^{-2}$$

7.6

(7.6.19)
$$\frac{\theta}{1-\theta} = \frac{(A_\mathrm{m}/N_\mathrm{A})p}{\nu_\mathrm{a}(2\pi mkT)^{1/2}}\exp(-\Delta_\mathrm{ads}U_\mathrm{m}/RT) \qquad (1)$$

$A_\mathrm{m}/N_\mathrm{A}$ is the area occupied by a N_2 molecule $= 0.162\,\text{nm}^2$ so that

$$\frac{\theta}{1-\theta} = \frac{0.162\,\text{nm}^2 \times 0.1\,\text{MPa} \times \exp(-1500/8.3145 \times 78)}{5 \times 10^{12}\,\text{s}^{-1}(2\pi \times 28.014 \times 1.6605 \times 10^{-27}\,\text{kg} \times 1.3806 \times 10^{-23}\,\text{J K}^{-1} \times 78\,\text{K})^{1/2}}$$
$$\theta = 0.099$$

7.7 For a reversible isochoric (at constant volume) process in a closed system, the First Law of Thermodynamics is

(1.5.3)	$\mathrm{d}U = \mathrm{d}q + \mathrm{d}w$	(1)
(1.4.9), (2.1.3)	$= T\mathrm{d}S + \gamma\mathrm{d}A$	(2)
(2)	$\mathrm{d}S = \dfrac{1}{T}\mathrm{d}U - \dfrac{\gamma}{T}\mathrm{d}A$	(3)
(2), (2.2.7)	$\mathrm{d}F = -S\mathrm{d}T + \gamma\mathrm{d}A$	(4)
(4), (A2.30)	$-\left(\dfrac{\partial S}{\partial A}\right)_T = \left(\dfrac{\partial \gamma}{\partial T}\right)_A$	(5)

Dividing each side of (2) by $\mathrm{d}A$ at constant T gives

(5)
$$\left(\frac{\partial U}{\partial A}\right)_T = T\left(\frac{\partial S}{\partial A}\right)_T + \gamma = -T\left(\frac{\partial \gamma}{\partial T}\right)_A + \gamma \qquad (6)$$

which is a thermodynamic equation of state for a surface. It describes the dependence of internal energy on the area of the surface at constant temperature.

7.8

(7.7.10), (7.7.4)
$$d\mathcal{U}^\sigma = TdS^\sigma - pdV^\sigma + \sum_B \mu_B^\sigma dn_B^\sigma - A^\sigma d\pi$$

$$d\mathcal{H}^\sigma = TdS^\sigma + V^\sigma dp + \sum_B \mu_B^\sigma dn_B^\sigma - A^\sigma d\pi$$

$$d\mathcal{F}^\sigma = -S^\sigma dT - pdV^\sigma + \sum_B \mu_B^\sigma dn_B^\sigma - A^\sigma d\pi$$

$$d\mathcal{G}^\sigma = -S^\sigma dT + V^\sigma dp + \sum_B \mu_B^\sigma dn_B^\sigma - A^\sigma d\pi$$

If dq is the heat absorbed by σ during the adsorption of dn_B^σ moles of B, then the differential heat of adsorption is

$$\frac{dq}{dn_B^\sigma} = \left(\frac{\partial \mathcal{U}^\sigma}{\partial n_B^\sigma}\right)_{T,V^\sigma,\pi}$$

$$\frac{dq}{dn_B^\sigma} = \left(\frac{\partial \mathcal{H}^\sigma}{\partial n_B^\sigma}\right)_{T,p,\pi}$$

The last was called by Hill (1949) the *equilibrium heat of adsorption*.

$$\mu_B^\sigma = \sum_B \left(\frac{\partial \mathcal{U}^\sigma}{\partial n_B^\sigma}\right)_{S^\sigma,V^\sigma,n_{A\neq B}^\sigma,A^\sigma} = \sum_B \left(\frac{\partial \mathcal{H}^\sigma}{\partial n_B^\sigma}\right)_{S^\sigma,p,n_{A\neq B}^\sigma,A^\sigma}$$

$$= \sum_B \left(\frac{\partial \mathcal{F}^\sigma}{\partial n_B^\sigma}\right)_{V^\sigma,p,n_{A\neq B}^\sigma,A^\sigma} = \sum_B \left(\frac{\partial \mathcal{G}^\sigma}{\partial n_B^\sigma}\right)_{T,p,n_{A\neq B}^\sigma,A^\sigma}$$

7.9 The first term on the RS of (1) in Section 7.8 is

$$n(\mu^l - \mu^g) = nkT\ln(p^*/p) = nkT\ln(1/2) \tag{1}$$

The volume of a spherical nucleus containing n molecules of H_2O is

$$\frac{nV_m}{N_A} = \frac{4}{3}\pi r^3 \tag{2}$$

(1), (P7.8.1), (P7.8.2)
$$\Delta G = -(n \times 1.381 \times 10^{-23} \times 293 \times 0.693)\,\text{J}$$
$$+ (4\pi)^{1/3}(3n \times 18.06$$
$$\times (10^{-6}\,\text{m}^3\,\text{mol}^{-1}/6.022$$

$$\times 10^{23}\, mol^{-1})^{2/3}\gamma n^{2/3}$$

$$= -(5.6082 \times 10^{-21} \times n)J$$

$$+ (89.97 \times 10^{-30})^{2/3} \times 0.7294\, J \times n^{2/3}$$

$$= -(5.6082 \times 10^{-21} \times n)J + 146.5$$

$$\times 10^{-21}\, J \times n^{2/3} \tag{3}$$

(i) $\Delta G/J = -2.804 \times 10^{-18} + 1.465 \times 10^{-17} > 0$

(ii) $\Delta G/J = -2.804 \times 10^{-15} + 1.465 \times 10^{-15} < 0$. This is an instructive exercise, since it shows that nuclei of water are unlikely to form at this super-saturation.

These considerations show that the formation of a new phase β is not necessarily spontaneous, even when ΔG for the bulk process $\alpha \rightarrow \beta$ is favourable (negative). The above calculations are for *homogeneous* nucleation. In fact, it is found that for nucleation to occur spontaneously in water vapour cooled rapidly to 25°C, the partial pressure of water vapour should be four times the equilibrium vapour pressure. Consequently, nucleation is often *heterogeneous,* as in cloud formation initiated by dust particles or crystallization from a super-saturated solution initiated by scratching the side of the glass container. Pure water may be cooled to temperatures as low as −40°C before nucleation occurs spontaneously.

7.10 See Figure 7.6.

7.11 A cloud chamber is a sealed environment containing super-saturated water vapour. When a chamber is exposed to ionizing radiation (α-particles, for example) the tracks of these particles become visible due to condensation of water nucleated by the ions formed by the radiation.

Chapter 8

8.1 The heat absorbed by a system during a chemical reaction at constant pressure is the reaction enthalpy

(8.2.8) $$q_p = \Delta_r H = \Delta_r U + p\Delta_r V$$

(8.2.9) $$= q_V + nRT \quad \text{(ideal behaviour)}$$

8.2 The standard reaction enthalpy is, from (8.2.12) and (8.2.17)

$$\Delta_r H^\circ(298.15\,\text{K}) = \sum_B \nu_B H_B^\circ = \sum_B \nu_B \Delta_f H_B^\circ$$

For the reduction of ferric oxide by H_2

$$\text{Fe}_2\text{O}_3(\text{cr}) + 3\text{H}_2(\text{g}) = 2\text{Fe}(\text{cr}) + 3\text{H}_2\text{O}(\text{liq}) \tag{1}$$

the standard reaction enthalpy is

(1) $\quad \Delta_r H^\circ(298.15\,\text{K})/\text{kJmol}^{-1} = 3(-285.830) - (-824.2) = 33.3\,\text{kJ/mole}$

8.3 $\quad \Delta C_p/\text{J K}^{-1}\text{mol}^{-1} = 29.14 + 33.58 - 37.13 - 20.79 = 4.8$

$$\Delta H^\circ(T)/\text{kJmol}^{-1} = \Delta H^\circ(T^\circ = 298.15\,\text{K})/\text{kJ mol}^{-1} + \Delta C_p(T - T^\circ)$$

$$= 41.16 + 4.8 \times 10^{-3}(T - 298.15\,\text{K})\text{kJ mol}^{-1}$$

8.4

(8.1.4) $$\xi = \frac{n_{\text{NH}_3}(\xi) - 0}{1}, \quad n_{\text{NH}_3} = \xi$$

$$\xi = \frac{n_{\text{N}_2}(\xi) - \frac{1}{2}}{-\frac{1}{2}}, \quad n_{\text{N}_2}(\xi) = \frac{1}{2}(1 - \xi)$$

$$\xi = \frac{n_{\text{H}_2}(\xi) - \frac{3}{2}}{-\frac{3}{2}}, \quad n_{\text{H}_2}(\xi) = \frac{3}{2}(1 - \xi)$$

$$\sum_B n_B = 2 - \xi, \quad y_{\text{NH}_3} = \frac{\xi}{2 - \xi} = 0.355, \quad \xi = 0.5240$$

$$y_{\text{N}_2} = \frac{0.238}{1.476} = 0.1612$$

$$y_{\text{H}_2} = \frac{0.714}{1.476} = 0.4837$$

$$\text{Check}: \sum_B y_B = 0.9999 \approx 1$$

$$K_f = \prod_B f_B^{\nu_B} = K_p \prod_B \phi_B^{\nu_B} = p^{\sum_B \nu_B} K_y \prod_B \phi_B^{\nu_B}$$

$$= (30.3\,\mathrm{MPa})^{-1} \frac{0.355}{(0.4837)^{3/2}(0.1612)^{1/2}} \times \frac{0.908}{(1.094)^{3/2}(1.14)^{1/2}}$$

$$= 8.64 \times 10^{-2} \times 0.743\,\mathrm{MPa}^{-1}$$

$$= 6.42 \times 10^{-2}\,\mathrm{MPa}^{-1}$$

The difference between K_f and K_p is due to non-ideal behaviour of this gas mixture, which is responsible for the factor 0.743. In the absence of knowledge of the partial fugacities of the three gases, the Lewis and Randall rule was used in the estimation of K_f.

8.5

(8.3.4) $$\Delta a/\mathrm{J\,K^{-1}mol^{-1}} = 26.36 - \frac{1}{2}(26.88) - \frac{1}{2}(37.25) = -1.45$$

$$\Delta b \times 10^3/\mathrm{J\,K^{-1}mol^{-1}} = 3.83 - \frac{1}{2}(3.59) - \frac{1}{2}(0.78) = +1.645$$

$$\Delta c \times 10^6/\mathrm{J\,K^{-1}mol^{-1}} = 0.17 - \frac{1}{2}(0.11) + \frac{1}{2}(0.05) = +0.14$$

The change in $\Delta_f H^{\ominus}$ between 298.15 K and 500 K is, therefore, from (8.3.6)

$$-1.45(500 - 298.15) + 0.8225 \times 10^{-3}(500^2 - 298.15^2) - 0.14$$

$$\times 10^6 \left(\frac{1}{298.15} - \frac{1}{500} \right)$$

$$\Delta_f H^{\ominus}(500\,\mathrm{K})/\mathrm{J\,K^{-1}mol^{-1}} = -27000 - 292.7 + 132.5 - 189.6 = -26970$$

$$\Delta_f H^{\ominus}(500\,\mathrm{K}) = -27.35 \pm 0.2\,\mathrm{kJ\,mol^{-1}}$$

We may only assert that $\Delta_f H^{\ominus}$ (500 K) for the formation of HI from $H_2(g)$ and $I_2(g)$ lies between 27.15 and 27.55 kJ mol^{-1} because of the rather large uncertainty in the value of the enthalpy change for reaction (a) at 298.15 K.

8.6 For the reaction, $2Cu + \frac{1}{2}O_2 = Cu_2O$

$$\Delta G/\mathrm{kJ\,mol^{-1}} = -195.4 - 0.0164(873)\log(873) + 0.1427 \times 873 = -112.9$$

$$= -(8.3145 \times 873/\mathrm{J\,mol^{-1}})\ln K$$

$$K = 5.66 \times 10^6 = \left(\frac{p^{\ominus}}{p} \right)^{1/2}$$

Substituting 10^5 Pa for the standard pressure gives $p = 3.1 \times 10^{-9}$ Pa. The proposed method is effective.

8.7 When the extent of reaction is ξ, the partial pressures of the three gases are

$$y(PCl_5(g))p = \frac{(1-\xi)p}{1+\xi}, \quad y(PCl_3(g))p = \frac{\xi p}{1+\xi}, \quad y(Cl_2(g))p = \frac{\xi p}{1+\xi}$$

The equilibrium constant is, therefore, $K = \frac{\xi^2}{1-\xi^2}$

T/K	$10^3 K/T$	ξ	ξ^2	K	$\log K$
600	1.666	0.982	0.964	26.78	1.428
520	1.923	0.800	0.640	1.778	0.259

$$\log 26.78 - \log 1.7 = \frac{-\Delta H^\oplus}{2.303\,R}[1.666 - 1.923]\frac{1}{1000\,K}$$
$$\Delta H^\oplus/kJ\,mol^{-1} = 86.43$$

$\Delta G^\oplus = -RT\ln K$, so at 520 K

$$\Delta G^\oplus/J\,mol^{-1} = -8.3145 \times 520 \times \ln 1.778 = -2488$$
$$\Delta S^\oplus/JK^{-1}mol^{-1} = \frac{\Delta H^\oplus - \Delta G^\oplus}{T} = \frac{86430 + 2488}{520} = 171$$

Since

$$K_p = \frac{\xi^2 p}{1-\xi^2}$$

may be taken as constant at low pressures, say below 1 MPa, an increase in pressure will reduce the extent of reaction, in accordance with Le Chatelier's principle.

8.8 The chemical reaction is

$$2\,NO(g) + O_2(g) = 2\,NO_2(g)$$

When the extent of reaction is ξ, the amounts of the three participants, in moles, are: oxygen $1-\xi$, nitric oxide $2(1-\xi)$, and nitrogen dioxide 2ξ. The total number of moles is $3 - \xi$ and the mole fractions, therefore, are

NO(g)	O_2(g)	NO_2(g)
$\dfrac{2(1-\xi)}{3-\xi}$	$\dfrac{1-\xi}{3-\xi}$	$\dfrac{2\xi}{3-\xi}$

The amount of oxygen present is $1 - \xi = 0.71$ mol, so that $\xi = 0.29$ mol.

Substituting the mole fractions in K gives for the equilibrium constant

$$K = \frac{[y(NO_2)]^2}{[y(NO)]^2 y(O_2)} = \frac{(3 - \xi)(2\xi)^2}{[2(1 - \xi)]^2(1 - \xi)}$$

$$= \frac{(3 - 0.29)(0.29)^2}{[(1 - 0.29)]^2(1 - 0.29)} = 0.637$$

$$\Delta G = -RT \ln K, \quad \Delta G / \text{J mol}^{-1} = -8.3145 \times 800 \times \ln 0.637 = 3003$$

Chapter 9

9.1 $$m_\pm = (\nu_+^{\nu_+}\nu_-^{\nu_-})^{1/\nu}m, \quad \nu = \nu_+ + \nu_-, \quad a_\pm = \gamma_\pm m_\pm/m^\circ,$$

$$m^\circ = 1\,\mathrm{mol\,kg^{-1}}, \quad a = (a_\pm)^\nu, \quad I_m = \sum_B \frac{1}{2}m_B z_B^2$$

Solute	ν_+	ν_-	ν	m/m°	$\gamma_\pm$	$m_\pm/m^\circ$	$a_\pm$	a	$I_m/$ mol kg^{-1}
AgNO$_3$	1	1	2	0.100	0.72	0.100	0.072	0.00518	0.100
CuSO$_4$	1	1	2	0.050	0.21	0.050	0.0105	0.00010	0.200
H$_2$SO$_4$	2	1	3	0.010	0.544	0.0159	0.00865	6.47×10^{-7}	0.030
In$_2$(SO$_4$)$_3$	2	3	5	0.100	0.035	0.255	0.0089	5.67×10^{-11}	1.50

9.2

(i) $E = \dfrac{\Delta\phi}{L} = \dfrac{6.943 \times 10^{-2}\mathrm{V}}{8.908 \times 10^{-2}\mathrm{m}} = 0.7794\,\mathrm{V\,m^{-1}}$

(ii) $j = \dfrac{I}{A} = \dfrac{2.850 \times 10^{-4}\mathrm{A}}{3.142 \times 10^{-4}\mathrm{m^2}} = 9.071\,\mathrm{A\,m^{-2}}$

(iii) $\kappa = \dfrac{j}{E} = \dfrac{0.9071\,\mathrm{A\,m^{-2}}}{0.7794\,\mathrm{V\,m^{-1}}} = 1.1638\,\mathrm{S\,m^{-1}}$

(iv) $\Lambda = \dfrac{\kappa}{c} = \dfrac{1.1638\,\mathrm{S\,m^{-1}}}{0.1000\,\mathrm{mol\,dm^{-3}}} = 1.1638 \times 10^{-2}\mathrm{S\,m^2\,mol^{-1}}$

(v) $j_+ = jt_+ = J_+ z_+ F$

$$J_+ = \frac{0.9071\,\mathrm{Am^{-2}} \times 0.4898}{1 \times 96485\,\mathrm{A\,s\,mol^{-1}}} = 4.605 \times 10^{-6}\mathrm{mol\,m^{-2}s^{-1}}$$

(vi) $v_+ = \dfrac{J_+}{c_+} = \dfrac{4.605 \times 10^{-6}\mathrm{mol\,m^{-2}s^{-1}}}{0.100 \times 10^3\mathrm{mol\,m^{-3}}} = 4.605 \times 10^{-8}\,\mathrm{m\,s^{-1}}$

(vii) $u_+ = \dfrac{|v_+|}{E} = \dfrac{4.605 \times 10^{-8}\,\mathrm{m\,s^{-1}}}{0.7794\,\mathrm{V\,m^{-1}}} = 5.908 \times 10^{-8}\,\mathrm{m^2s^{-1}V^{-1}}$

9.3 The volume swept out by the moving boundary

$$V = \frac{\text{amount of Na}^+ \text{ transported}}{\text{amount of Na}^+/\text{volume}} = \frac{t_+ Q/Fz_+}{\nu_+ c} = xA$$

Hence

$$t_+ = \frac{FxA|z_+|\nu_+c}{Q}$$

$$= \frac{96485\,\text{C}\,\text{mol}^{-1} \times 1.00 \times 10^{-2}\,\text{m} \times 0.1115 \times 10^{-4}\,\text{m}^2 \times 20.0\,\text{mol}\,\text{m}^{-3}}{344.8\,\text{s} \times 1.6001 \times 10^{-3}\,\text{A}}$$

$$= 0.390$$

9.4

$\dfrac{10^3 m_1}{\text{mol}\,\text{kg}^{-1}}$	$\dfrac{10^4 m}{\text{mol}\,\text{kg}^{-1}}$	$\dfrac{10^3 I_\text{m}}{\text{mol}\,\text{kg}^{-1}}$	$\dfrac{I_\text{m}^{1/2}}{\text{mol}^{1/2}\text{kg}^{-1/2}}$	$\log$ $(m/m^\ominus)$	$\gamma_\pm$ (expt.)	$\gamma_\pm$ (DH)
0	1.761	0.1761	0.01327	-3.7542	0.986	0.985
1.301	1.813	1.482	0.03850	-3.7416	0.957	0.956
3.252	1.863	3.438	0.05683	-3.7298	0.932	0.934
6.503	1.908	6.694	0.08182	-3.7194	0.910	0.910
14.10	1.991	14.30	0.1196	-3.7009	0.872	0.869

A plot of $\log\,(m/m^\ominus)$ against $(m/m^\ominus)^{1/2}$ extrapolated to $I_\text{m} = 0$ yields $\frac{1}{2}\log K_\text{sp} = -3.7605$, whence $K_\text{sp} = 3.01 \times 10^{-8}$.

9.5

RS $\qquad\qquad$ $\text{AgCl(cr)} + \text{e}^-\,(\text{Ag}) = \text{Ag} + \text{Cl}^-\,(\text{soln})$

LS $\qquad\qquad$ $\frac{1}{2}\text{H}_2(\text{g}) = \text{H}^+ + \text{e}^-\,(\text{Pt})$

R + L $\qquad\qquad$ $\text{AgCl(cr)} + \frac{1}{2}\text{H}_2(\text{g}) = \text{Ag} + \text{H}^+ + \text{Cl}^-$

$$E^\ominus = E^{\ominus,\text{R}} - E^{\ominus,\text{L}} = 0.2225\,V$$

$$E = E^\ominus - \frac{RT}{nF}\ln\left(\frac{a_{\text{H}^+}a_{\text{Cl}^-}a_{\text{Ag}}}{a_{\text{H}_2}^{1/2}a_{\text{AgCl}}}\right)$$

$$= E^\ominus - \frac{2RT}{F}\ln\left(\frac{\gamma_\pm m_\pm}{m^\ominus}\right)$$

$$0.2333 = 0.2225 - 2 \times 0.05916\log\gamma_\pm$$

$$\log\gamma_\pm = \frac{0.0108}{0.11832}, \quad \gamma_\pm = 0.810$$

9.6

RS $\qquad\qquad Ag^+(aq) + e^-(Ag) = Ag(s)$

LS $\qquad\qquad Ag(s) + I^-(aq) = AgI(cr) + e^-(Ag)$

RS + LS $\qquad\qquad Ag^+(aq) + I^-(aq) = AgI(cr)$

$$E^\ominus = E^{\ominus,R} - E^{\ominus,L} = 0.7991 - (-0.1518)\,V = 0.9509\,V$$

$$-nFE^\ominus = \Delta G^\ominus = -1 \times 96485\,C\,mol^{-1} \times 0.9509\,V = -91\,750\,J\,mol^{-1}$$

$$= -\Delta G^\ominus \quad \text{for the reaction } AgI(cr) = Ag^+(aq) + I^-(aq)$$

$$K_{sp} = \exp(-\Delta G^\ominus/RT) = \exp\left(\frac{-91\,750\,J\,mol^{-1}}{8.3145\,J\,K^{-1}\,mol^{-1} \times 298.16\,K}\right)$$

$$= 8.426 \times 10^{-17}$$

9.7 $\qquad\qquad E_j = E_a - E_b = 0.02802 - 0.01696\,V = 0.01106\,V$

(9.9.5), (9.7.7) $\qquad E_a = 2t_+ E_b$

$$t_+ = \frac{0.02802}{2 \times 0.01696} = 0.826$$

The relatively high value of t_+ for hydrogen is because hydrogen ions in water exist as H_3O^+ and the facile transfer of a proton from one water molecule to the next provides a mechanism for the rapid transport of H^+. Similarly, OH^- is a "proton hole" and is thus expected to have a higher mobility than other anions.

9.8 If interionic effects are neglected $\xi = \Lambda/\Lambda^0$

$$K = \frac{\xi^2 c}{1 - \xi} = \frac{\Lambda^2 c}{\Lambda^0(\Lambda^0 - \Lambda)} \tag{1}$$

Rearrange (1)

$$\frac{1}{\Lambda} = \frac{1}{\Lambda^0} - \frac{\Lambda c}{K(\Lambda^0)^2} \tag{2}$$

Plot Λ^{-1} against Λc (either manually, or using Mathcad or similar software) which gives $\Lambda^0 = 200\,S\,cm^2\,mol^{-1}$ and $K = 7.10 \times 10^{-4}\,mol\,dm^{-3}$.

9.9 The physically acceptable solution of (4) is

$$S^{1/2} = \frac{1}{2}\left[Z + (Z^2 + 4)^{1/2}\right] = \frac{Z}{2} + \left[1 + \frac{Z^2}{4}\right]^{1/2} \tag{6}$$

(6) $$S = \left\{\frac{Z}{2} + \left[1 + \frac{Z^2}{4}\right]^{1/2}\right\}^2 = 1 + Z + \frac{Z^2}{2} + \frac{Z^3}{8} + \cdots \tag{7}$$

Usually, only the first two terms will be required in (7)

$$\xi = \frac{\Lambda S}{\Lambda^0} = \frac{\Lambda}{\Lambda^0}\left[1 + Z + \frac{Z^2}{2} + \frac{Z^3}{8} + \cdots\right] \tag{8}$$

The dissociation constant

$$K_a = \frac{\xi^2 c(\gamma^\pm)^2}{1 - \xi} = \frac{\Lambda^2 S^2 c(\gamma^\pm)^2}{\Lambda^0(\Lambda^0 - \Lambda S)} \tag{9}$$

where

$$-\log(\gamma^\pm)^2 = 2\Lambda(\xi c)^{1/2} \tag{10}$$

Rearranging (9) gives

$$\frac{1}{\Lambda S} = \frac{1}{\Lambda^0} + \frac{\Lambda S c(\gamma^\pm)^2}{K(\Lambda^0)^2} \tag{11}$$

Note the similarity of (11) to the Ostwald dilution law (Eq. (9.2.10)) in which interionic effects are neglected. Equation (11) holds at ionic strengths up to about $0.01\,\text{mol kg}^{-1}$.

9.10 The conductivity of acetic acid is

$$\kappa = \text{cell constant}/R = \frac{0.7096\,\text{cm}^{-1}}{4372\,\text{ohm}} = 1.623 \times 10^{-4}\,\text{S cm}^{-1}$$

If interionic effects are ignored

(9.7.30) $$\xi = \frac{\Lambda}{\Lambda^0 - (A + B\Lambda^0)(\xi c)^{1/2}}$$

$$= \frac{16.23}{390.7 - (60.2 + 0.229 \times 390.7)(0.01\xi)^{1/2}}$$

$$= \frac{16.23}{390.7 - (149.67)(0.01\xi)^{1/2}}$$

Solve for ξ by successive approximations, for example $\xi = 0.042,\ 0.04187,\ 0.041869$

$$\log \gamma^{\pm} = -0.510(0.01 \times 0.041869)^{1/2} = -0.104356$$

$$\gamma^{\pm} = 0.97625, \quad (\gamma^{\pm})^2 = 0.95308$$

$$K_a = \frac{\xi^2 c (\gamma^{\pm})^2}{1 - \xi} = \frac{(0.041869)^2 \times 0.01 \times 0.95308}{0.95813}$$

$$= 1.744 \times 10^{-5} \, \text{mol dm}^{-3}$$

9.11 The force

$$f = -\frac{dV(r)}{dr} = \frac{Q_+ Q_-}{4\pi\varepsilon_0 \varepsilon_r r^2} \tag{1}$$

where Q_+, Q_- are the charges on the ions, ε_0 is the permittivity of a vacuum, ε_r is the relative permeability of the medium, and r the distance apart between the ions. $V(r)$ is the potential energy of interaction between the ions.

$$V(r_0) - V(\infty) = -\int_{\infty}^{r_0} f(r) dr = \int_{\infty}^{r_0} \frac{e^2}{4\pi\varepsilon_0 \varepsilon_r r^2} dr$$

$$= -\frac{e^2}{4\pi\varepsilon_0 \varepsilon_r r_0} = V(r_0) \tag{2}$$

(2) $$V(r_0) = -\frac{(1.602 \times 10^{-19} C)^2}{4 \times 3.142 \times 8.8542 \times 10^{-12} \, \text{F m}^{-1} \times 3 \times 10^{-10} \, \text{m}}$$

$$= -7.688 \times 10^{-19} \, \text{J}$$

$$= -0.7688 \, \text{aJ} = -4.8 \, \text{eV}$$

In a medium of relative permittivity 78.42, this energy is reduced to $-(0.7688/78.42) \, \text{aJ} = -0.0098 \, \text{aJ} = -0.061 \, \text{eV}$. This calculation illustrates the reason why ionic crystals dissolve to form an electrolyte solution in solvents of high relative permittivity.

Chapter 10

10.1

(10.4.1)
$$\Omega = \frac{N_e!}{\prod_i n_i!}$$

$$\ln \Omega = N_e \ln N_e - N_e - \sum_i (n_i \ln n_i - n_i)$$

$$\ln \Omega = N_e \ln N_e - \sum_i n_i \ln n_i$$

The most probable distribution, which occurs n_i^* times, is determined by the maximum value of Ω, and therefore of $\ln \Omega$, subject to the constraints

$$\sum_i n_i = N_e$$

$$\sum_i n_i E_i = E_{\text{tot}}$$

$$\sum_i n_i N_i = N_{\text{tot}}$$

N_i is the number of particles in the ith distribution and E_i is the energy of the ith distribution.

We find the most probable configuration, allowing for the constraints by Lagrange's method of undetermined multipliers. The condition that Ω be a maximum is

$$d\Omega = \sum_i (-\ln n_i - \alpha - \beta E_i - \gamma N_i) dn_i = 0$$

We can choose α, β and γ so as to make the first three coefficients with $i = 1, 2, 3$ zero; the remaining dn_i may then be varied independently to make all the coefficients of the dn_i zero. The maximum value of n_i is therefore given by

$$-\ln n_i^* - \alpha - \beta E_i - \gamma N_i = 0$$
$$n_i^* = \exp(-\alpha) \exp(-\beta E_i) \exp(-\gamma N_i) \qquad (10.4.5)$$

10.2 The Fermi energy μ is given by

(10.7.6)
$$\mu = \frac{h^2}{8m} \left(\frac{3\langle N \rangle}{\pi V} \right)^{\frac{2}{3}} \qquad (6)$$

For silver

$$\mu = \frac{(6.626 \times 10^{-34}\,\text{J sec})^2}{8 \times 9.109 \times 10^{-31}\,\text{kg}} \left(\frac{3 \times 5.85 \times 10^{22}\,\text{cm}^{-3}}{\pi} \right)^{\frac{2}{3}} \times \frac{10^{19}\,\text{eV}}{1.602\,\text{J}}$$

$$= 5.49\,\text{eV}$$

For Cu, the formula is the same except for the electron concentration, giving

$$\mu = 7.01\,\text{eV}$$

10.3 An atom of He^3 contains two electrons, two protons and one neutron, a total of five particles each of spin $1/2$. It is therefore a fermion. One mole contains $4.0026\,\text{g}$ of He and occupies a volume of $4.0026/0.81 = 4.94\,\text{cm}^3$ mol^{-1}. The electron concentration is therefore

$$\frac{2 \times 6.022 \times 10^{23}\,\text{mol}^{-1}}{4.94 \times 10^{-6}\,\text{m}^3\,\text{mol}^{-1}} = 2.44 \times 10^{29}\,\text{m}^{-3}$$

The Fermi energy of He^3 is therefore

$$\mu = \frac{(6.626 \times 10^{-34}\,\text{J sec})^2}{8 \times 9.109 \times 10^{-31}\,\text{kg}} \left(\frac{3 \times 2.44 \times 10^{29}\,\text{m}^{-3}}{\pi} \right)^{\frac{2}{3}} \times \frac{10^{19}\,\text{eV}}{1.602\,\text{J}}$$

$$= 0.3760 \times 0.3287 \times 10^2\,\text{eV} = 12.36\,\text{eV}$$

10.4

(10.6.33) $$C_V = \frac{3}{2} Nk \left[1 - \frac{z}{2^{7/2}} + \cdots \right]$$

$$S = \frac{U - F}{T} = \frac{U - \mu N}{T} + k \ln \Xi = Nk \left[\frac{5}{2} - \ln z - \frac{z}{2^{7/2}} + \cdots \right]$$

10.5 In the limit $T \to 0$

(5.6.6) $$\lim_{T \to 0} \frac{1}{1 - (\mu/kT) - 1} = N \tag{11}$$

since at $T = 0\,\text{K}$ the number of particles in excited states is zero. Therefore

$$N = -kT/\mu, \quad \mu = -kT/N \tag{12}$$

10.6

(10.8.15) $$\mu = \mu(0) \left[1 + \frac{\pi^2}{8} \left(\frac{kT}{\mu} \right)^{-2/3} + \cdots \right]$$

$$= \mu(0) \left[1 - \frac{\pi^2}{12} \left(\frac{kT}{\mu(0)} \right)^2 + \cdots \right] \tag{17}$$

Thus the Fermi energy decreases slightly with increasing T.

10.7 The RS of (10.8.19) is the internal energy U. Therefore

(10.8.19)
$$U = \frac{3N}{5}\mu(0)\left[1 + \frac{5\pi^2}{12}\left(\frac{kT}{\mu(0)}\right)^2 + \cdots\right]$$

$$C_V = \frac{\pi^2 Nk}{2}\left(\frac{kT}{\mu(0)}\right) \tag{20}$$

$$S = \int_0^T \frac{C_V}{T}dT = \frac{\pi^2 Nk}{2}\left(\frac{kT}{\mu(0)}\right) \tag{21}$$

Chapter 11

11.1 Suppose that $d\mathbf{r}$ is along the x-axis and of length dl so that $d\mathbf{r} = dl\,\mathbf{e_1}$. Then

$$d\mathbf{r'} = dl\,\mathbf{e_1} + dl\,\mathbf{e_1} \cdot \Sigma$$
$$= dl\,(1 + \Sigma_{xx})\,\mathbf{e_1} + dl\,\Sigma_{xy}\,\mathbf{e_2} + dl\,\Sigma_{xz}\,\mathbf{e_3}$$

The length of $d\mathbf{r'}$ is $dl(1 + \Sigma_{xx})$, to first order in the small quantities Σ_{xx}, Σ_{xy}, Σ_{xz} (as may be verified easily by finding the square root of $d\mathbf{r'} \cdot d\mathbf{r'}$). Thus the component Σ_{xx} is the fractional change in length of an element initially along the x direction and similarly. This is the reason why Σ_{xx}, Σ_{yy} and Σ_{zz} are called the linear dilations.

11.2 The volume of the parallelopiped, to first order in small quantities, is

$$V = V_0[1 + \Sigma_1 + \Sigma_2 + \Sigma_3]$$

where V_0 is the volume of the unstrained crystal. The volume dilation is the fractional change in volume due to the strain, that is $\Sigma_1 + \Sigma_2 + \Sigma_3$.

11.3 Since the system is invariant under the change $x \to y \to z \to x$

$$c_{11} = \frac{\partial T_{xx}}{\partial \Sigma_{xx}} = \frac{\partial T_{yy}}{\partial \Sigma_{yy}} = c_{22}$$

$$c_{11} = \frac{\partial T_{xx}}{\partial \Sigma_{xx}} = \frac{\partial T_{zz}}{\partial \Sigma_{zz}} = c_{33}$$

11.4 The generators of T_d are S_4 and C_3. The transforms of $x_1 x_2 x_3$ are shown Table S11.4(a) where the notation xyz means the axis with $x = y = z$.

11.5 The formation of an interstitial is facilitated by the deformation of the ion concerned. The Ag^+ has a $3d^{10}$ configuration which is more easily deformed than K^+ or Na^+ with their filled p^6 shells. (Check their polarizabilities.) In the anion Cl^- the outer electrons are less firmly held than in Na^+ or K^+ because of the positive nuclear charge and we therefore expect Cl^- to have a higher polarizability and be more easily deformed than Na^+ or K^+. Thus anion interstitials occur in both NaCl and KCl but not cation interstitials.

Table S11.4.

(a)	x'_1	x'_2	x'_3
S_{4z}	x_2	$-x_1$	$-x_3$
C_{3xyz}	x_2	x_3	x_1

(b) S_{4z}	p,q	ij	$i'j'$	p',q'	S_{4z}	x_2	$-x_1$	$-x_3$		
	1	11	22	2	22	12	23	25	−24	−26
	2	22	33	3		11	13	15	−14	−16
	3	33	11	1			33	35	−34	−36
	4	23	13	5				55	−45	−56
	5	13	−12	−4					44	46
	6	12	−23	−6						66

(c) C_{3xyz}	p,q	ij	$i'j'$	p',q'	C_{3xyz}	x_2	x_3	x_1		
	1	11	22	2	11	12	13	14	−15	−16
	2	22	33	3		22	23	24	−25	−26
	3	33	11	1			33	34	−35	−36
	4	23	13	5				44	−45	−46
	5	13	12	6					55	56
	6	12	23	4						66

(d) $[c_{pq}]$					
11	12	13	0	0	0
	11	13	0	0	0
		33	0	0	0
			44	0	0
				44	0
					66

11.6 The conductivity of KCl:Ca^{2+} is greater than that of pure KCl. This shows that divalent Ca^{2+} ions in solid solution introduce mobile cation vacancies. Since the mass of one Ca^{2+} is less than that of the two K^+ ions that it replaces, the density of KCl crystals should decrease on doping with $CaCl_2$.

Chapter 12

12.1

$$J_s = L_{11}\Delta p + L_{12}E \qquad (2)$$

$$I = L_{21}\Delta p + L_{22}E \qquad (3)$$

There is one ORR, namely

$$L_{12} = L_{21} \qquad (4)$$

If J_s is measured when the applied potential difference E is zero, then

$$L_{11} = J_s/\Delta p \quad (E = 0) \qquad (5)$$

When $E = 0$, there is still a current flowing, the streaming current and measurement of this current

$$I = L_{21}\Delta p \qquad (6)$$

yields

$$L_{21} = I/\Delta p \quad (E = 0) \qquad (7)$$

When the applied pressure is zero

$$L_{22} = I/E = \kappa \qquad (8)$$

where κ is the conductivity of the membrane in the solution. The ratio of the solvent flow to the current I when Δp is zero is called the electro-osmotic permeability

$$\beta = \left(\frac{J_s}{I}\right)_{\Delta p=0} = \frac{L_{12}}{L_{22}} \qquad (9)$$

Thus all four of the phenomenological coefficients can be determined experimentally.

12.2

$$(12.5.2), (12.5.6) \qquad J_q = k\left(-\frac{1}{T^2}\frac{dT}{dx}\right) + \kappa\varepsilon\left(-\frac{d\phi}{dx}\right)\left(\frac{1}{T}\right) \qquad (12.5.9)$$

$$(12.5.3), (12.5.5) \qquad j = \kappa\varepsilon\left(-\frac{1}{T^2}\frac{dT}{dx}\right) + \kappa\left(-\frac{d\phi}{dx}\right)\left(\frac{1}{T}\right) \qquad (12.5.10)$$

In (9), units of LS are J m^{-2} s^{-1} and of RS are J m^{-1} s^{-1} K K^{-1} m^{-1} and J m^{-2} s^{-1} = LS. In (10), units of LS are A m^{-2} and of RS are S m^{-1} V K^{-1} K m^{-1} = A m^{-2} and A m^{-1} K K^{-1} m^{-1} = A m^{-2}. J_q is the heat flow and the

1, 1 diagonal term is Fourier's law of heat conduction, where k is the heat conductivity. κ is the electrical conductivity and the 2,2 diagonal term is a statement of Ohm's law. The cross terms show that the potential gradient contributes to the heat flow and that the thermal gradient contributes to the flow of electric current. However, these contributions from the off-diagonal terms are seen to be much smaller than those from the diagonal terms, because they contain ε, which is small.

12.3 Using $j = \kappa E$ and Eqs. (12.6.17) and (12.6.16)

$$\kappa = F^2 [z_+^2 L_{++} + z_+ z_- (L_{+-} + L_{-+}) + z_-^2 L_{--}] \qquad (12.6.18)$$

12.4 No. The anions diffuse at a different rate to the cations, so that the slower-moving ion lags behind the faster ion, thus setting up a diffuse double layer. The drag exerted by this double layer slows down the rate of diffusion of cations.

Appendices

A1 Partial Differentiation

A1.1. Definitions

For a function $f(x, y, \ldots)$ of more than one variable, the *partial derivative* of f with respect to x is defined by

$$\lim_{\delta x \to 0} \frac{f(x + \delta x, y, \ldots) - f(x, y)}{\delta x} = \left(\frac{\partial f}{\partial x} \right)_{y,z} \tag{1}$$

The subscripts $y, z \ldots$ denote that all other variables but x remain constant during the limiting process denoted by the operator $\frac{\partial}{\partial x}$. The subscripts may be omitted when there is no possibility of confusion. The *complete differential* of f is defined by

$$df = \frac{\partial f}{\partial x} dx + \frac{\partial f}{\partial y} dy + \frac{\partial f}{\partial z} dz \tag{2}$$

For only two variables, $f(x, y)$

(2)
$$df = \frac{\partial f}{\partial x} dx + \frac{\partial f}{\partial y} dy \tag{3}$$

Equation (3) holds whether x and y are independent or not. For example, if both x and y are functions of t, then dividing each side of (3) by dt gives

(3)
$$\frac{df}{dt} = \frac{\partial f}{\partial x} \frac{dx}{dt} + \frac{\partial f}{\partial y} \frac{dy}{dt} \tag{4}$$

df/dt is the complete derivative of f with respect to t.

The partial derivatives are themselves functions of $x, y, \ldots$. Therefore, they may be differentiated partially to give second- and higher-order partial derivatives

$$\frac{\partial}{\partial x} \left(\frac{\partial f}{\partial x} \right) = \frac{\partial^2 f}{\partial x^2}, \quad \frac{\partial}{\partial y} \left(\frac{\partial f}{\partial x} \right) = \frac{\partial^2 f}{\partial y \partial x} \tag{5}$$

$$\frac{\partial}{\partial x} \left(\frac{\partial f}{\partial y} \right) = \frac{\partial^2 f}{\partial x \partial y}, \quad \frac{\partial}{\partial y} \left(\frac{\partial f}{\partial y} \right) = \frac{\partial^2 f}{\partial y^2} \tag{6}$$

417

For continuous functions with continuous first- and second-order partial derivatives, the second-order partial derivatives commute

$$\frac{\partial^2 f}{\partial x \partial y} = \frac{\partial^2 f}{\partial y \partial x} \tag{7}$$

A1.2. Implicit Functions

If $f(x, y, z) \equiv 0$, then

$$df = \frac{\partial f}{\partial x}dx + \frac{\partial f}{\partial y}dy + \frac{\partial f}{\partial z}dz = 0 \tag{8}$$

If x remains constant, $dx = 0$, then

$$(8) \qquad \left(\frac{dy}{dz}\right)_x = \frac{\partial y}{\partial z} = -\frac{(\partial f/\partial z)}{(\partial f/\partial y)} \tag{9}$$

Similarly, setting $dy = 0$ and $dz = 0$, in turn

$$(8) \qquad \left(\frac{dz}{dx}\right)_y = \frac{\partial z}{\partial x} = -\frac{(\partial f/\partial x)}{(\partial f/\partial z)} \tag{10}$$

$$(8) \qquad \left(\frac{dx}{dy}\right)_z = \frac{\partial x}{\partial y} = -\frac{(\partial f/\partial y)}{(\partial f/\partial x)} \tag{11}$$

$$(11), (9), (10) \qquad \left(\frac{\partial x}{\partial y}\right)_z \left(\frac{\partial y}{\partial z}\right)_x \left(\frac{\partial z}{\partial x}\right)_y = -1 \tag{12}$$

Note that by rearranging (8) with $dx = 0$

$$(8) \qquad \left(\frac{dz}{dy}\right)_x = \frac{\partial z}{\partial y} = -\frac{(\partial f/\partial y)}{(\partial f/\partial z)} = \frac{1}{(\partial y/\partial z)} \tag{13}$$

Thus the partial differential coefficients may be inverted. Consequently, for implicit functions

$$(12) \qquad \left(\frac{\partial x}{\partial y}\right)_z \left(\frac{\partial y}{\partial z}\right)_x = -\left(\frac{\partial x}{\partial z}\right)_y, \quad f(x, y, z) = 0 \tag{14}$$

A1.3. Function of a Function

If f is a function of x and y and x,y are functions of w and z, then

$$\left(\frac{\partial f}{\partial z}\right)_w = \left(\frac{\partial f}{\partial x}\right)_y \left(\frac{\partial x}{\partial z}\right)_w + \left(\frac{\partial f}{\partial y}\right)_x \left(\frac{\partial y}{\partial z}\right)_w \tag{15}$$

The proof of (15) is straightforward. Write down the differentials of $f(x, y), x(z, w)$ and $y(z, w)$. Substitute in $\mathrm{d}f$ for $\mathrm{d}x$ and $\mathrm{d}y$. But f can be expressed through x, y as a function of z and w, so that

$$\mathrm{d}f = \left(\frac{\partial f}{\partial z}\right)_w \mathrm{d}z + \left(\frac{\partial f}{\partial w}\right)_z \mathrm{d}w \tag{16}$$

Comparing coefficients of dz in the two expressions for $\mathrm{d}f$ yields (15).

An important special case in thermodynamics arises when $z = x$, that is f is a function of x and y, and y is a function of x and w. Then

(15)
$$\left(\frac{\partial f}{\partial x}\right)_w = \left(\frac{\partial f}{\partial x}\right)_y + \left(\frac{\partial f}{\partial y}\right)_x \left(\frac{\partial y}{\partial x}\right)_w \tag{17}$$

If $x(y, w)$ with $y(z, w)$

(15)
$$\left(\frac{\partial x}{\partial z}\right)_w = \left(\frac{\partial x}{\partial y}\right)_w \left(\frac{\partial y}{\partial z}\right)_w \tag{18}$$

which is called the *chain rule*. Note the difference between (14) and (18).

A1.4. Euler's Theorem on Homogeneous Functions

If

$$f(\lambda x_1, \lambda x_2, \ldots, \lambda x_n) = \lambda^r f(x_1, x_2, \ldots, x_n) \tag{19}$$

then f is said to be *a homogeneous function of degree r* in $x_1, x_2, \ldots, x_n$.

Differentiate (19) with respect to x_i

$$\left(\frac{\partial f}{\partial \lambda x_i}\right) \frac{\mathrm{d}\lambda x_i}{\mathrm{d}x_i} = \left(\frac{\partial f}{\partial \lambda x_i}\right) \lambda = \lambda^r \left(\frac{\partial f}{\partial x_i}\right) \tag{20}$$

Differentiate (19) with respect to λ

$$\sum_i \left(\frac{\partial f}{\partial \lambda x_i}\right) x_i = r\lambda^{r-1} f \tag{21}$$

(20), (21)
$$\sum_{i=1}^{n} x_i \left(\frac{\partial f}{\partial x_i}\right) = rf \tag{22}$$

which is Euler's theorem.

There are two cases of special interest. If $r = 1$, so that

(19)
$$f(\lambda x_1, \lambda x_2, \ldots, \lambda x_n) = \lambda f(x_1, x_2, \ldots, x_n) \tag{23}$$

then

$$(22) \qquad \sum_{i=1}^{n} x_i \left(\frac{\partial f}{\partial x_i} \right) = f \qquad (24)$$

If $r = 0$, that is

$$(19) \qquad f(\lambda x_1, \lambda x_2, \ldots, \lambda x_n) = f(x_1, x_2, \ldots, x_n) \qquad (25)$$

then

$$(22) \qquad \sum_{i=1}^{n} x_i \left(\frac{\partial f}{\partial x_i} \right) = 0 \qquad (26)$$

Extensive functions of the state of a system obey Eq. (23). Intensive properties such as temperature, pressure and density are independent of the mass of a system and they obey Eq. (25). Examples are given in Chapter 2. The converse to Euler's theorem also holds, so that, for example, if (26) is true, it implies that f is a homogeneous function of degree zero.

A1.5. Complete Differentials

In thermodynamics, we frequently encounter expressions of the form

$$\mathrm{d}u = M\mathrm{d}x + N\mathrm{d}y \qquad (27)$$

where M and N are functions of x and y. These are called *Pfaff expressions* or *linear differential forms*. Now (27) does not necessarily imply that a function $u = f(x, y)$ exists, but if such a function does exist then its complete differential is

$$\mathrm{d}f = \frac{\partial f}{\partial x}\mathrm{d}x + \frac{\partial f}{\partial y}\mathrm{d}y \qquad (3)$$

Consequently, the sufficient condition that $M\,\mathrm{d}x + N\,\mathrm{d}y$ is the complete differential of a function $f(x, y)$ is that

$$M = \frac{\partial f}{\partial x}, \quad N = \frac{\partial f}{\partial y} \qquad (28)$$

It may be shown that this condition is also necessary. But if (28) is true

$$\frac{\partial M}{\partial y} = \frac{\partial^2 f}{\partial y \partial x}, \quad \frac{\partial N}{\partial x} = \frac{\partial^2 f}{\partial x \, \partial y} \qquad (29)$$

It follows from the commutative property (28) that $M\,\mathrm{d}x + N\,\mathrm{d}y$ is a complete differential if

(29)
$$\frac{\partial M}{\partial y} = \frac{\partial N}{\partial x}$$
(30)

which is known as Euler's criterion.

A1.6. Line Integrals

A line integral is one which is carried out along a specified geometrical path

(27)
$$\int_{x_1,y_1}^{x_2,y_2} \mathrm{d}u = \int_{x_1,y_1}^{x_2,y_2} M(x,y)\mathrm{d}x + \int_{x_1,y_1}^{x_2,y_2} N(x,y)\mathrm{d}y$$
(31)

These integrations on the RS of (31) cannot be carried out unless y is given as a function of x, that is, the path of integration is specified in the x,y plane. If several different paths are specified, then several different values of $\int \mathrm{d}u$ will, in general, be obtained. The integration is particularly simple, however, when $\mathrm{d}u$ is a complete differential, that is, when

$$\mathrm{d}u = M\mathrm{d}x + N\mathrm{d}y = \frac{\partial f}{\partial x}\mathrm{d}x + \frac{\partial f}{\partial y}\mathrm{d}y = \mathrm{d}f(x,y)$$
(32)

For then

$$\int_{x_1,y_1}^{x_2,y_2} \mathrm{d}u = \int_{x_1,y_1}^{x_2,y_2} \mathrm{d}f = [f(x,y)]_{x_1,y_1}^{x_2,y_2} = f(x_2,y_2) - f(x_1,y_1)$$
(33)

No relation between x and y is required and $\int \mathrm{d}u$ is *independent of the path of integration*, provided $\mathrm{d}u$ is a complete differential.

If $\mathrm{d}u = M\mathrm{d}x + N\mathrm{d}y$ is not a complete differential, but a function $t(x,y)$ can be found, such that

$$t\mathrm{d}u = (tM)\mathrm{d}x + (tN)\mathrm{d}y$$
(34)

is complete, then t is called an *integrating factor*. If t is an integrating factor for the Pfaffian $\mathrm{d}u$, then it is evident from (34) that any multiple of t is also an integrating factor for $\mathrm{d}u$.

A2 The Classical Electromagnetic Field

The reasons for this appendix are to supply some background in electromagnetic theory for readers who feel that they need this and, specifically, to derive an expression for the work done on a system by electric and magnetic fields. It may be skipped by readers who prefer to accept the results in Eqs. (17), (21) and (23).

A2.1. The Fundamental Equations

The essence of electromagnetic theory is contained in 11 equations: these are Maxwell's four field equations, the Lorentz equation for the force acting on a charged particle in an electromagnetic field, three relations which describe the response of matter to electric and magnetic fields, and the defining relations for the scalar and vector potentials. The force F acting on a charged particle in an electromagnetic field is

$$F = Q[E + (\mathbf{v} \times B)] \tag{1}$$

Q is the charge on the particle in C, $\mathbf{v}$ its velocity (in the observer's frame of reference) in m s^{-1}, E is the electric field strength in V m^{-1} and B is the magnetic induction in Tesla ($1\,\mathrm{T} = 1\,\mathrm{V\,s\,m^{-2}}$). Maxwell's equations describing the electromagnetic field are

$$\nabla \cdot D = \rho \tag{2}$$

$$\nabla \times E = -\frac{\partial B}{\partial t} \tag{3}$$

$$\nabla \cdot B = 0 \tag{4}$$

$$\nabla \times H = J + \frac{\partial D}{\partial t} \tag{5}$$

The fields are macroscopic (space averaged) fields. H is the magnetic field strength (A m^{-1}), J is the electric current density (A m^{-2}), D is the electric displacement (C m^{-2}) and ρ is the charge density (C m^{-3}). The magnetic induction and the magnetic field intensity are connected by the relation

$$B = \mu H = \mu_0 (H + M) \tag{6}$$

μ is the magnetic permeability and in vacuum it has the value $\mu_0 = 4\pi \times 10^{-7}$ H m^{-1} (or V s A^{-1} m^{-1}). μ_0 is called the magnetic constant. M is the magnetization, that is the magnetic dipole moment per unit volume in A m^{-1} (see also (1.4.29)). The electric displacement and the electric field strength are connected by the relation

$$D = \varepsilon E = \varepsilon_0 E + P \tag{7}$$

ε is the permittivity of the medium and in vacuum it has the value $\varepsilon_0 = 8.8542 \times 10^{-12}$ F m^{-1}, and is called the electric constant. P is the dielectric polarization that is the electric dipole moment per unit volume in C m^{-2}. The current density J and the electric field strength are connected by the relation

$$J = \sigma E \tag{8}$$

where σ is the conductivity (S m^{-1}). The relations (6), (7) and (8) describe the response of material systems to electric and magnetic fields and, in the above forms, are valid for isotropic media when the fields are not too large. For anisotropic materials μ, ε and σ are second rank tensors. Nevertheless, we shall continue to use the forms (6), (7) and (8), making the generalizations to cope with non-linearity or anisotropy only when it is necessary to do so. We should remark that Maxwell's equations (2)–(5) are exact and contain no approximations.

The physical significance of Maxwell's equations is:

(i) when the charges are at rest, Eq. (2) becomes Coulomb's law of electrostatics;

(ii) Eq. (3) represents Faraday's law of electromagnetic induction and shows that a time dependent magnetic induction B gives rise to an electric field normal to the direction of B;

(iii) Eq. (4) shows that there is no magnetic analogue of electric charge (*cf.* (2));

(iv) Eq. (5) shows the existence of a magnetic field near a current density and also that a time dependent electric field will produce a magnetic field.

Equations (3) and (5) together show how an electromagnetic field can be propagated through empty space where $\boldsymbol{J} = 0$. The charge density ρ in Eq. (2) is *the net extra charge density* supplied by an external source. It does not include the polarization charge density which involves the separation of existing charges but does not lead to the movement of charge into or out of the system. Similarly, $\boldsymbol{J}$ does not include polarization or magnetization currents but involves the motion of charge carriers in an applied electric field, which give rise to the longitudinal component of $\boldsymbol{J}$ and also the effects of photon absorption, which contribute to the transverse current density. This contribution is called the optical conductivity, and in an isotropic material the dc electrical conductivity is the zero-frequency limit of this optical conductivity. The scalar and vector potentials are defined by

$$\boldsymbol{E} = -\nabla\varphi - \frac{\partial \boldsymbol{A}}{\partial t} \tag{9}$$

$$\boldsymbol{B} = \nabla \times \boldsymbol{A} \tag{10}$$

together with

$$\nabla \cdot \boldsymbol{A} + \varepsilon\mu\frac{\partial\varphi}{\partial t} = 0 \tag{11}$$

The potentials are not fixed by (9) and (10) alone, but may be made to satisfy any further condition that leaves the electric and magnetic fields invariant. The process of selecting the potentials in a particular way is called *choosing the gauge*. One possibility is the *Lorentz gauge*, Eq. (11). Another frequently used choice is *the Coulomb gauge*

$$\nabla \cdot \boldsymbol{A} = 0 \tag{11'}$$

Neither (9), (10) and (11), nor (9), (10) and (11') determine the potentials uniquely. This further arbitrariness may be removed by setting $\varphi(\boldsymbol{r}, t) = 0$, which may be done provided there is no net charge density, $\rho = 0$. These 11 equations are the foundations of classical electricity and magnetism, but it is useful to have, in addition, the following equation, which is not an independent postulate and which states that a charged particle moving through an electric field with velocity $\mathbf{v}$ is subject to a magnetic field of induction

$$\boldsymbol{B} = -\frac{1}{c^2}(\mathbf{v} \times \boldsymbol{E}) \tag{12}$$

A2.2. Energy Accumulation in the Electromagnetic Field

$$(5) \qquad \boldsymbol{E} \cdot (\nabla \times \boldsymbol{H}) = \boldsymbol{E} \cdot \boldsymbol{J} + \boldsymbol{E} \cdot \frac{\partial \boldsymbol{D}}{\partial t} \qquad (13)$$

$$(3) \qquad \boldsymbol{H} \cdot (\nabla \times \boldsymbol{E}) = -\boldsymbol{H} \cdot \frac{\partial \boldsymbol{B}}{\partial t} \qquad (14)$$

$$(13),\ (14) \qquad -\nabla \cdot (\boldsymbol{E} \times \boldsymbol{H}) = \boldsymbol{E} \cdot \frac{\partial \boldsymbol{D}}{\partial t} + \boldsymbol{H} \cdot \frac{\partial \boldsymbol{D}}{\partial t} + \boldsymbol{E} \cdot \boldsymbol{J} \qquad (15)$$

$\boldsymbol{E} \times \boldsymbol{H}$ is called the Poynting vector and is the outward power flow per unit area over the bounding surface. The LS of (15) is therefore the energy gained by the system per unit time per unit surface area. $\boldsymbol{E} \cdot \boldsymbol{J}$ is the energy dissipated as heat per unit volume per unit time and the first two terms on the RS are the work done on the system per unit time per unit volume. On multiplying by $\mathrm{d}V$ and integrating over the volume of the system, the work done during a quasistatic (reversible) change in state is

$$(15) \qquad \mathrm{d}w = \int (\mathbf{E} \cdot \mathrm{d}\boldsymbol{D} + \boldsymbol{H} \cdot \mathrm{d}\boldsymbol{B})\mathrm{d}V \qquad (16)$$

We now consider the magnetic and electrical terms separately. For the magnetic case, if the magnetic field $\boldsymbol{H}$ is uniform, as it would be at the centre of a long solenoid, the work done on the system (per unit volume) during an infinitesimal change in state is

$$(16) \qquad \mathrm{d}w = \boldsymbol{H} \cdot \mathrm{d}\boldsymbol{B} \qquad (17)^{\mathrm{a}}$$

The work done comprises two parts: the work done in creating the field within the empty solenoid and the work done in magnetizing the system of interest. To see this

$$(6),\ (16) \qquad \mathrm{d}w = \mu_0 \int (\mathbf{H} \cdot \mathrm{d}\mathbf{H} + \mathbf{H} \cdot \mathrm{d}\mathbf{M})\mathrm{d}V$$

$$= \int \frac{1}{2}\mu_0 \mathbf{H}^2 \mathrm{d}V + \mu_0 V \mathbf{H} \cdot \mathrm{d}\mathbf{M} \qquad (18)$$

in a uniform external magnetic field. The first term on the RS of (18) is the work required to create the field in vacuum and since this tells us nothing

---

[a]Some care is required here; the field H is not the local field within the body, but the field within the empty solenoid. It is usually called the *external field* (Pippard 1966).

about the system, it is commonly included with the internal energy change dU by replacing dU with

$$dU' = dU - d\{\int \frac{1}{2}\mu_0 \mathbf{H}^2 dV\} \tag{19}$$

The fundamental equation is then

(18), (19) $\qquad dU' = TdS - pdV + \mu_0 V \boldsymbol{H} \cdot d\boldsymbol{M} \tag{20}$

The case of an electric field is similar. For a capacitor which has plates whose area is much larger than the cross-sectional area presented by the system, edge effects are negligible and the work done (per unit volume) during an infinitesimal change in state is

(16) $\qquad dw = \boldsymbol{E} \cdot d\boldsymbol{D} \tag{21}$

This work consists of two parts: the work done in creating the electric field in vacuum and the work done in polarizing the system between the plates

(7), (16) $\qquad dw = \int \frac{1}{2}\varepsilon_0 \boldsymbol{E}^2 dV + V \boldsymbol{E} \cdot d\boldsymbol{P} \tag{22}$

$\boldsymbol{E}$ is the electric field with vacuum between the plates of a capacitor, and the first term on the RS of (20) is the work required to create the electric field in vacuum. The second term is the work done in polarizing the material between the plates of the capacitor in a constant external field, that is, the electric field between the plates of the capacitor in vacuum. Again, the first term may be combined with dU, giving

(20), (22) $\qquad dU' = TdS - pdV + \mu_0 V \boldsymbol{H} \cdot d\boldsymbol{M} + V \boldsymbol{E} \cdot d\boldsymbol{P} \tag{23}$

Usually, the prime is dropped, giving

(23) $\qquad dU = TdS - pdV + \mu_0 V \boldsymbol{H} \cdot d\boldsymbol{M} + V \boldsymbol{E} \cdot d\boldsymbol{P} \tag{23a}$

Thus dU has four possible meanings, but which one is intended will be obvious from the fundamental equation employed.

A3 Sources of Thermodynamic Data

The (US) National Institute of Standards and Technology (NIST) was set up to provide access to reference data. The data provided includes physical properties, vibrational, spectroscopic and thermodynamic data, with references. The data are updated continually. Data for compounds can be searched for by the name of the compound, registry number, molecular formula or structure. Species can be searched for by structure classification, such as bond type, carbon rings and specific elements. See:

NIST Chemistry *WebBook*
webbook.nist.gov/chemistry

For those who prefer the printed page, there are:
Barin 1995, Binnewies and Milke 2002, Chase 1998, Chase *et al.* 1985, Cox *et al.* 1989, Gamsjäger *et al.* 2008, Knacke *et al.* 1991, and Wagman *et al.* 1982.

References

Adam, N.K. (1941) *The Physics and Chemistry of Surfaces* 3rd edn. Oxford: Oxford University Press.

Adamson, A.W. and Gast, A.P. (1997) *Physical Chemistry of Surfaces*. New York: John Wiley.

Allnatt, A.R. and Jacobs, P.W.M. (1968) *Can. J. Chem.* **46**, 111.

Atkins, P.W. (1983) *Molecular Quantum Mechanics*, 2nd edn. Oxford: Oxford University Press.

Atkins, P.W. (1986) *Physical Chemistry* 3rd edn. Oxford: Oxford University Press and New York: W.H. Freeman.

Bard, A.J., Parsons, R. and Jordan, J. (1985) *Standard Potentials in Aqueous Solution*. New York: Dekker.

Barin, I. (ed.) (1995) *Thermochemical Data of Pure Substances* 3rd edn. Weinheim: Wiley-VCH.

Barnes, P., Finney, J.L., Nicholas, J.D. and Quinn, J.E. (1979) *Nature* **282**, 459.

Bates, R.G. and Guggenheim, E.A. (1960) *Pure and Appl. Chem.* **1**, 163.

Benson, G.C. and Shuttleworth, R. (1951) *J. Chem. Physics* **19**, 130.

Benton, A.F. and Drake, L.C. (1932) *J. Amer. Chem. Soc.* **54**, 2186.

Bernal, J.D. and Fowler, R.H. (1933) *J. Chem. Physics* **1**, 515.

Berry, R.S., Rice, S.A. and Ross, J. (2002) *Matter in Equilibrium* 2nd edn. Oxford: Oxford University Press.

Binnewies, M. and Milke, M. (2002) *Thermochemical Data of Elements and Compounds*. Weinheim: Wiley-VCH.

Born, M. and Huang, K. (1954) *Dynamical Theory of Crystal Lattices*. Oxford: Clarendon Press.

Buck, R.P., Rondinini, S., Covington, A.K., Baucke, F.G.K., Brett, C.M.A., Camoes, M.F.M., Milton, J.T., Mussini, T.R., Naumann, T., Pratt, K.W., Spitzer, P. and Wilson, G.S. (2002) *Pure Appl. Chem.* **74**, 2169.

Cailletet, L. and Mathias, E.C. (1886) *Compt. rend.* **102**, 1202.

Callen, H.B. (1960) *Thermodynamics*. New York: John Wiley.

Carnahan, N.F. and Starling, K.E. (1969) *J. Chem. Physics* **51**, 635.

Castellan, G.W. (1983) *Physical Chemistry* 3rd edn. Reading, MA: Addison Wesley.

Catlow, C.R.A., Corish, J. and Jacobs, P.W.M. (1979) *J. Phys. C*, **12**, 3433.

Catlow, C.R.A. and Mackrodt, W.C. (eds) (1982) *Lecture Notes in Physics*: *Computer Simulation of Solids*. Berlin: Springer.

Chase, M.W. (1998) *NIST JANAF Thermochemical Tables* 4th edn. *J. Phys. Chem. Ref. Data* Monograph 9.

Chase, M.W., Davies, C.A., Downey, J.R., Frurip, D.J., McDonald, R.A. and Syverud, A.N. (1985) *JANAF Thermochemical Tables* 3rd edn. *J. Phys. Chem. Ref. Data* **14**, Suppl. 1.

Cohen, E.R., Cvitaš, T., Frey, J.G., Holmström, B., Kuchitsu, K., Marquardt, R., Mills, I., Pavese, F., Quack, M., Stohner, J., Strauss, H.L., Takami, M. and Thor, A.J. (2007) *Quantities, Units and Symbols in Physical Chemistry* 3rd edn. Cambridge: The Royal Society of Chemistry.

Corish, J. (1989) *J. Chem. Soc. Faraday Trans.* **85**, 237.

Corish, J. (1990) *J. Imaging Science* **34**, 84.

Corish, J. and Jacobs, P.W.M. (1972) *J. Phys. Chem. Solids* **33**, 1799.

Cotton, F.A. (1963) *Chemical Applications of Group Theory* 3rd ed. New York: John Wiley.

Cox, J.D., Wagman, D.D. and Medvedev, V.A. (1989) *CODATA Key Values for Thermodynamics*. New York: Hemisphere Publishing.

Darwin, C.G. and Fowler, R.H. (1922) *Phil. Mag.* **44**, 450.

Davidson, N. (1962) *Statistical Mechanics*. New York: McGraw Hill.

Davis, H.T. (1996) *Statistical Mechanics of Phases, Interfaces and Thin Films*. NewYork: Wiley-VCH.

Debye, P. and Hückel, E. (1923) *Physikal. Zeit.* **24**, 185.

Denbigh, K.G. (1955) *The Principles of Chemical Equilibrium*. Cambridge: Cambridge University Press.

Drain, L.E. and Morrison, J.A. (1953) *Trans. Faraday Soc.* **49**, 654.

Dunlop P.J. and Gostling L.J. (1959) *J. Phys. Chem.* **63**, 86.

Dymond, J.H. and Smith, E.B. (1980) *The Virial Coefficients of Pure Gases and Gas Mixtures*. Oxford: Clarendon Press.

Eston, R.E. (1959) *J. Chem. Physics* **31**, 892.

Everett, D.H. (1950a) *Trans. Faraday Soc.* **46**, 453.

Everett, D.H. (1950b) *Trans. Faraday Soc.* **46**, 942.

Everett, D.H. (1950c) *Trans. Faraday Soc.* **46**, 957.

Everett, D.H. and Young, D.M. (1952) *Trans. Faraday Soc.* **48**, 1164.

Eyring, H., Henderson, D., Stover, B.J. and Eyring, E.M. (1964) *Statistical Mechanics and Dynamics*. New York: John Wiley & Sons Inc.

Falkenhagen, H. (1934) *Electrolytes*. Oxford: Oxford University Press.

Fernandez, D.P., Mulev, Y., Goodwin, A.R.H. and Levelt Sengers, J.M.H. (1997) *J. Phys. Chem. Ref. Data* **26**, 1125.

Fowler, R.H. and Guggenheim, E.A. (1939) *Statistical Thermodynamics*. Cambridge: Cambridge University Press.

Fujita, H. and Gostling, L.J. (1960) *J. Phys. Chem.* **64**, 1256.

Fuller, R.G. (1966) *Phys. Rev.* **142**, 524.

Fuoss, R. and Onsager, L. (1957) *J. Physic. Chem.* **61**, 668.

Fuoss, R. and Onsager, L. (1958) *J. Physic. Chem.* **62**, 1339.

Gamsjäger, H., Lorimer, J.W., Scharlin, P. and Shaw, D.G. (2008) *Pure Appl. Chem.* **80**, 233.

Gibbs, J.W. (1875) *Collected Works*. London: Longmans.

Gibbs, J.W. (1906) *The Scientific Papers of J. Willard Gibbs,* Vol. 1. London: Longmans, Green and Company.

Gibbs, J.W. (1928a) *Collected Works,* Vol. 1, *Thermodynamics.* London: Longmans.

Gibbs, J.W. (1928b) *Collected Works,* Vol. 2, *Elementary Principles in Statistical Mechanics.* London: Longmans.

Gomer, R. and Tryson, S. (1977) *J. Chem. Physics* **66**, 4413.

Gouq-Jen, S. (1946) *Ind. Eng. Chem.* **38**, 803.

Grosse, A.V. (1962) *J. Inorg. Nucl. Chem.* **24**, 147.

Guggenheim, E.A. (1930) *J. Physic. Chem.* **34**, 1758.

Guggenheim, E.A. (1935) *Phil. Mag.* **19**, 588.

Guggenheim, E.A. (1936) *Proc. Roy. Soc. London A* **155**, 63.

Guggenheim, E.A. (1945) *J. Chem. Phys.* **13**, 253.

Guggenheim, E.A. (1966a) *Applications of Statistical Mechanics.* Oxford: Clarendon Press.

Guggenheim, E.A. (1966b) *Thermodynamics* 5th edn. Amsterdam: North-Holland.

Guggenheim, E.A. (1967) *Thermodynamics* 6th edn. Amsterdam: North-Holland.

Haar, L. (1968) *J. Res. Nat. Bur. Std.* **72A**, 207.

Habgood, H.W. and Schneider, W.G. (1954) *Can. J. Chem.* **32**, 98.

Harding, J.H. (1990) *Rep. Prog. Phys.* **53**, 1403.

Hill, T.L. (1962) *An Introduction to Statistical Thermodynamics* 2nd edn. Reading, MA: Addison-Wesley.

Hooton, I.E. and Jacobs, P.W.M. (1990) *J. Phys. Chem. Solids* **51**, 1207.

Hougen, O.A. and Watson, K.M. (1947) *Chemical Process Principles II.* New York: John Wiley.

Jacobs, P.W.M. (1992) *Rev. Solid State Sci.* **5**, 507.

Jacobs, P.W.M. (2005) *Group Theory with Applications in Chemical Physics.* Cambridge: Cambridge University Press.

Jacobs, P.W.M., Corish, J. and Catlow, C.R.A. (1980) *J. Phys. C.* **13**, 1977.

Jacobs, P.W.M and Ong, S.H. (1980) *Crystal Lattice Defects* **8**, 177.

Jacobs, P.W.M. and Rycerz, Z.A. (1997) "Molecular Dynamics Methods in Computer Modelling", in C.R.A. Catlow (ed.) *Inorganic Crystallography.* London: Academic Press, pp. 83–115.

Jacobs, P.W.M. and Vernon, M.L. (1997) *J. Phys. Chem. Solids* **58**, 1007.

Jacobs, P.W.M. and Vernon, M.L. (1998) *Can. J. Chem.* **76**, 1540.

Jahnke, E. and Emde, F. (1945) *Tables of Functions* 4th edn. New York: Dover.

Jou, D. (1996) *Extended Irreversible Thermodynamics.* New York and Berlin: Springer-Verlag.

Katchalsky, A. and Curran, P.F. (1966) *Non-equilibrium Thermodynamics in Biophysics.* Cambridge, MA: Harvard University Press.

Kittel, C. (1969) *Thermal Physics.* New York: John Wiley.

Kittel, C. (1986) *Introduction to Solid State Physics* 6th edn. New York: John Wiley.

Knacke, O., Kubachewski, O. and Hesselmann, K. (eds) (1991) *Thermochemical Properties of Inorganic Substances* 2nd edn. Berlin: Springer-Verlag.

Knuth, E.L. (1966) *Introduction to Statistical Thermodynamics.* New York: McGraw Hill.

Larson, A.T. and Dodge, R.L. (1923) *J. Amer. Chem. Soc.* **45**, 2918.

Leslie, M. (1985) *Pyysica* **131B**, 145.

Longsworth, L.G. (1932) *J. Amer. Chem. Soc.* **54**, 2741.

Lorimer, J.W. (1966) *J. Chem. Educ.* **43**, 39.

Margenau, H. and Murphy, G.M. (1962) *The Mathematics of Physics and Chemistry* 2nd edn. New York: D. Van Nostrand.

Maxwell, J.C. (1860) *Phil. Mag.* **19**, 19.

Maxwell, J.C. (1868) *Phil. Mag.* **35**, 129, 185.

McGlashan, M.L. (1979) *Chemical Thermodynamics.* London: Academic Press.

McNaught, A.D. and Wilkinson, A. (1997) *IUPAC. Compendium of Chemical Terminology* 2nd edn. ("Gold Book"). Oxford: Blackwell Scientific Publications, Oxford. *Cf.* Nic, M., Jirat, J. and Kosata, B. On-line corrected version: http://goldbook.iupac.org with updates compiled by A. Jenkins (2006).

Mendoza, E. (ed.) (1977) *Reflections on the Motive Force of Fire and other Papers on the Second Law of Thermodynamics by E. Clapeyron and R. Clausius.* Gloucester, MA: Peter Smith.

Michels, A., Blaisse, B. and Michels, C. (1937) *Proc. Roy. Soc. London* **A160**, 367.

Mills, I., Cvitas, T., Homann, K., Kallay, N. and Kuchitsu, K. (1988) *Quantities, Units and Symbols in Physical Chemistry.* Oxford: Blackwell Scientific Publications.

Moelwyn-Hughes, E.A. (1961) *Physical Chemistry* 2nd edn. Oxford: Pergamon Press.

Moore, W.J. (1983) *Basic Physical Chemistry.* London: Prentice-Hall.

Müller, I. and Ruggeri, T. (1993) *Extended Thermodynamics.* New York: Springer-Verlag.

Nernst, W. (1906) *Ges. Wiss. Gottingen Klasse Math. Phys.* **1**.

Newton, R.H. (1935) *Ind. Eng. Chem.* **27**, 302.

Nye, J.F. (1957) *Physical Properties of Crystals.* Oxford: Oxford University Press.

Okamoto, H. (1991) *J. of Phase Equilibria* **12**, 504.

Onsager, L. (1931a) *Phys. Rev.* **37**, 405.

Onsager, L. (1931b) *Phys. Rev.* **38**, 2265.

Ott, J.B., Coates, J.R. and Hall, H.T. (1971) *J. Chem. Educ.* **48**, 515.

Pauling, L. (1935) *J. Amer. Chem. Soc.* **57**, 2680.

Pippard, A.B. (1966) *Elements of Classical Thermodynamics.* Cambridge: Cambrige University Press.

Planck, M. (1945) *Treatise on Thermodynamics* 3rd edn. New York: Dover.

Ries, H.E. (1984) *Colloids and Surfaces* **10**, 283.

Rigby, M. (1970) *Quart. Rev.* **24**, 416.

Robinson, R.A. (1945) *Proc. Roy. Soc. N.Z.* **75**, 203.

Robinson, R.A. and Stokes, R.H. (1968) *Electrolyte Solutions.* London: Butterworths.

Roebuck, J.R. and Osterberg, H. (1935) *Phys. Rev.* **48**, 450.

Rossini, F.D. and Frandsen, M. (1932) *J. Res. Nat. Bur. Std.* **9**, 733.

Rowlinson, J.S. (1969) *Liquids and Liquid Mixtures* 2nd edn. London: Butterworths.

Sackur, O. (1911) *Ann. Physik* **36**, 958.

Sackur, O. (1912) *Ann. Physik* **40**, 67.

Shedlovsky, T. (1950) *J. Amer. Chem. Soc.* **72**, 3680.

Simon, F.E. and Swenson, C.A. (1950) *Nature* **165**, 829.

Stephenson, G. (1961) *Mathematical Methods for Science Students*. London: Longmans.

Taylor, P.B. (1927) *J. Physic. Chem.* **31**, 1478.

Tetrode, H. (1912a) *Ann. Physik* **38**, 434.

Tetrode, H. (1912b) *Ann. Physik* **39**, 255 (corrections).

Tetrode, H. (1915) *Proc. Kon. Ned. Akad. Wet* **17**, 1167.

Thomsen, J.S. (1962) *Am. J. Phys.* **30**, 294.

Tilley, D.R. and Tilley, J. (1990) *Superfluidity and Superconductivity* 3rd edn. London: Taylor and Francis.

Verwey, E.J.W. (1941) *Rec. Trav. Chim. Pays-Bas* **69**, 887.

Vielstich, W., Lamm, A. and Gasteiger, H.A. (2003) *Handbook of Fuel Cells*. New York: John Wiley and Sons.

von Zawidski, J. (1900) *Z. Phys. Chem.* **35**, 129.

Wagman, D.D., Evans, W.H., Parker, V.B., Schumm, R.H., Halow, I., Bailey, S.M., Churney, K.L. and Nuttall, R.L. (1982) *The NBS Tables of Chemical Thermodynamic Properties. J. Phys. Chem. Ref. Data* **11**, Suppl. 2.

Waldram, J.R. (1985) *The Theory of Thermodynamics*. Cambridge: Cambridge University Press.

Zemansky, M.W. and Dittman, R.H. (1997) *Heat and Thermodynamics: An Intermediate Textbook* 7th edn. New York: McGraw Hill.

Zhukovskii, Y.F., Jacobs, P.W.M. and Causa, M. (2003) *J. Phys. Chem. Solids* **64**, 1317.

Zhukovskii, Y.F., Kotomin, E.A., Jacobs, P.W.M. and Stoneham A.M. (2000) *Phys. Rev. Lett.* **84**, 1256.

Index

1953 Stockholm convention, 270
2-hydroxytetracosanoic, 199

absolute activity, 78
acetic acid, 259
activated state, 249
activity, 152
activity coefficient, 153
adiabatic, 27
adiabatic demagnetization, 49, 105, 108
adiabatically isolated, 4
adsorbent, 203
adsorption, 195, 201, 206
adsorption isotherm, 202
affinity, 220, 221, 347
AgCl, 330
ammonia, 237
ammonia synthesis, 236
amorphous materials, 327
amount of substance, 219
anharmonicity corrections, 82
anion interstitials, 331
anion vacancies, 331
anisotropic stresses, 321
antisymmetric, 84
apparent molar properties, 147
atomic masses, 248
attraction, 117
attractive forces, 255
average speed, 69
Avogadro constant, 6
azeotrope, 161, 162

barium fluoride, 329
batteries, 290
boiling point constant, 181
Boltzmann distribution, 62
Boltzmann entropy equation, 97, 98
Boltzmann tombstone, 98
Boltzmann's constant, 7
Bose–Einstein statistics, 65, 302
bosons, 302
boundary conditions, 54
Boyle temperature, 7, 114, 115
Brunauer, Emmett, Teller (BET) isotherm, 203
bubble-point curve, 155, 167

calorimetric entropy, 77, 99, 100
canonical ensemble, 296
Carnot, Sadi, 46
Carnot cycle, 49
cation vacancies, 329
cell constant, 257
cell potential, 270
cells with transport, 287
charge conservation, 258
chemical potential, 124, 151
chemical reactions, 347
chemisorption, 201
Clapeyron equation, 101, 133
Clausius, 46, 49
Clausius–Clapeyron equation, 133
coexistence region, 8, 9
colligative properties, 182
collinear interstitialcy, 332

commutantive property, 30
compilations of thermodynamic data, 240
complete differential, 12, 44
component, 2
compressibility, 128
compression factor, 6, 114
condition for equilibrium, 266
conduction band, 338
contact angle, 193
continuum, 62
critical mixing, 169
critical point, 114
critical solution temperature, 168
critical state, 113
critical temperature (T_c), 9, 113
critical volume (V_c), 113
cross-coefficients, 358
curved interface, 191
cyclic process, 12, 46

de Broglie, Louis, 59
Debye, Peter, 88
Debye length, 280
Debye T^3 law, 89, 99, 136
Debye–Hückel limiting law, 277
Debye–Hückel model, 277
Debye–Hückel–Onsager equation, 279
degeneracy, 59, 60
degrees of freedom, 3, 61, 71, 73, 75
    classical, 73
    unexcited, 72
density of states, 312
depression of freezing point, 183
deuterium, 102
dew-point curve, 155, 167
diamagnetism, 106
diatomic molecule, 78, 80, 84, 88
differential heat of adsorption, 208
differential isosteric heat of
    adsorption, 208
diffusion, 352
diffusion coefficient, 331
dipole moment, 117
displacement, 343
dissociation constant, 259

dissociation pressure of silver oxide, 241
distance of closest approach, 274
distillation, 160
distribution coefficient, 185
distribution of
    velocity components, 70
doping, 339
Duhem–Margules equation, 160

efficiency (of a heat engine), 45
eigenfunctions, 57
eigenvalues, 57
Einstein condensation, 309
Einstein function, 88
Einstein summation convention, 319
electrical conduction, 351
electrical conductivity, 354
electrochemical cell, 270
electrochemical potential, 266
electrode potential E, 267
electrolysis, 255
electrolyte solution, 255
electrolytes, 258
elevation of boiling point, 180
ensemble, 295
ensemble average, 296, 298, 300
enthalpy, 35
enthalpy change, 125
enthalpy of activation, 252
enthalpy of mixing, 150
enthalpy of vaporization, 36
entropy, 75, 76, 95, 230, 298, 344
entropy change on isotopic mixing, 41
entropy of activation, 252
entropy of mixing, 150
entropy of vaporization, 101
equation of state, 6
equations of phase equilibrium, 145
equilibrium, 10, 249
    thermodynamic, 3
equilibrium condition, 40, 78
equilibrium constant K_c, 232, 236, 237, 245, 259, 279
equilibrium state, 95
equipartition principle, 75, 88

Euler's theorem, 3, 43, 45, 195, 325
eutectic, 175
excess chemical potential, 163
exp, 6, 118
extent of dissociation, 259
extent of reaction, 219, 244
extrinsic semiconductivity, 339

Fermi level, 303
Fermi–Dirac statistics, 65, 302
fermions, 302
Fick's law, 352, 359
First Law of Thermodynamics, 20,
  21, 343, 344
fixed point, 6
fluctuations, 66
fluid, 8
flux, 348, 350
flux equations, 355
flux of species k, 260
forces, 350
Fourier's law, 352
free rotation, 139
freezing point constant, 184
Freundlich isotherm, 204
fuel cell, 291
fugacity, 151
fugacity coefficient, 127
fugacity coefficient product, 237
full band, 338
function of state, 29
fundamental equation, 30, 207, 298,
  301, 324
fundamental equation for a closed
  phase (FECP), 29, 344, 345
fundamental equation
  thermodynamic potential, 29
fundamental equations, 207
fundamental equations for a closed
  phase, 345
fundamental theorem, 349

gas constant, 6
gas mixture, 149
gas scale, 5
general chemical reaction, 219

generalized force, 11, 40, 343
geometric progression, 90, 304
Giaque function, 244, 246
Gibbs adsorption equation, 196
Gibbs convention, 194, 196
Gibbs energy, 30, 36, 40
Gibbs energy of activation, 252
Gibbs energy of formation, 237
Gibbs energy of mixing, 150, 165
Gibbs monolayer, 198
Gibbs–Duhem equation, 43–45, 159,
  195, 325
Gibbs–Helmholtz equation, 236
grand canonical ensemble, 296,
  299
grand partition function, 300
group generators, 327

Hamiltonian operator
  separable, 55
hard-sphere, 118
hard-sphere model, 128
hard-sphere pressure
  Carnahan–Starling
    approximation, 124
harmonic oscillator, 86
heat capacity, 129, 228, 237
  constant pressure, 125, 229
heat engine, 45
heat flux, 352
heat of adsorption, 201, 208
heat of mixing, 163
heat reservoir, 28
Helmholtz energy, 35, 40, 41
Helmholtz function, 30, 75
Henry's law, 158
Hertz–Knudsen, 71
Hess' law, 222
Hooke's law, 327
hydrogen, 102
hydrophilic, 197

ice, 102
ideal Bose–Einstein gas, 307
ideal Fermi–Dirac gas, 312
ideal gas, 66

ideal gas equation, 7
ideal liquid mixture, 153
impurities, 340
incongruent melting points, 175
independent variables, 30, 345
inner potential, 266
integrating factor, 28
interfacial tension, 189
interior of the phase, 266
intermolecular forces, 117
intermolecular potential, 131
internal energy, 131
interstitial, 332
inversion curve
        calculated, 123
        experimental, 123
ion atmosphere, 275
ion–ion interactions, 258
ionic solution, 255
ionic strength, 276
irreversible, 27, 28
isothermal absorption of heat, 28

JANAF tables, 240
Joule–Thomson (J–T) effect, 120
Joule–Thomson (J–T) inversion
    curve, 121
Joule's experiment, 27

Kelvin, 46, 49
Kelvin relation, 355
Kirchhoff equation, 227

Lagrange, Joseph Louis, 63
Lagrange's method, 63, 299
Lamé constants, 327
λ-transitions, 140
λ-type transition, 308
Landé splitting factor, 107
Langmuir isotherm, 197, 203
Langmuir trough, 199
laws of electrolysis, 255
Le Chatelier's Principle, 243
lead acid accumulator, 290
Lennard-Jones, John, 117, 118, 120
Lewis and Randall rule, 151, 237

linear dilation, 323
linearized Boltzmann distribution,
    274
liquid, 128, 129
liquid mixtures, 152
liquid phase extraction, 185
liquid–solid equilibrium, 172
liquidus, 176
lithium ion batteries, 291
lithium oxide, 329
localized, 204
London dispersion force, 191
Lorimer, J.W., 63

magnetic moment, 106
Massieu function, 77, 345
Maxwell distribution function, 69
Maxwell relations, 30, 325
Maxwell–Boltzmann distribution, 53
McBain's method, 198
mean ionic activity, 276
mean molar properties, 148
mechanical equilibrium, 40, 343
mechanical properties, 296
microcanonical ensemble, 296, 298
molar conductivities, 258
molar enthalpy of vaporization, 180
molecular dynamics, 131
molecular speeds, 68
molecules, 78
moments of inertia, 83, 84
monolayer, 204
most probable speed, 69
moving boundary method, 265

natural process, 11, 27, 28, 343
Nernst equation, 283
Nernst–Hartley equation, 360
Nernst heat theorem, 98
net entropy change, 27, 38
Neumann's principle, 319
neutron scattering, 131
new notation, 229
non-equilibrium processes, 344
non-ideal mixtures, 150
normal stress, 328

normalization, 57
n-type, 339
nuclear spin states, 102
nucleation, 213, 215

Ohm's law, 351
Onsager, Lars, 349
Onsager electrophoretic effect, 361
open phase, 42, 345
ORR, 349, 356, 358, 361
ortho-hydrogen, 86, 103
orthogonal, 57
osmotic coefficient, 182
osmotic pressure, 181
Ostwald dilution law, 259

para-hydrogen, 86, 103
paramagnetism, 105, 106
partial miscibility, 174
partial molar, 43, 44
partial molar properties, 146, 147
partition chromatography, 185
partition function, 66, 71, 297
    system, 74
path (of process), 3
Pauli exclusion principle, 84, 302
Peltier heat, 354
peritectic point, 176
permittivity, 258
phase, 1
phase factor, 85
phase rule, 146
phase transitions
    first order, 138
    in solids, 138
    second order, 140
phenomenological coefficients, 349
phenomenological equations, 354, 355
Planck function, 77, 345
Poisson's equation, 273
polar vector, 320
polyatomic molecule, 74, 82
positive hole, 339
postulates of statistical
    thermodynamics, 296
potential difference, 255

pressure, 243
principle of corresponding states
    (PCS), 115, 132, 137, 189
probability, 95, 297
process
    thermodynamic, 3
properties
    macroscopic, 1
property, 3
    extensive, 3
    intensive, 3
p-type doping, 339

quantization, 56
quantum mechanics
    eigenfunctions, 55
    eigenvalues, 55
    Hamiltonian operator, 54
    postulates, 54
    state function, 54
quantum number, 57

radial distribution function, 130, 131
radii of curvature, 191
random alloy, 97
Raoult's law, 153
rate of entropy production, 347, 348
reaction affinity, 220
reaction complex, 250
reaction enthalpy, 222, 229
reaction isotherm, 232
reaction path, 250
real gas at moderate pressures, 127
redox couple, 267, 269
reduced mass, 79
reduced variables, 113
reduction potentials, 270
regular mixture, 169
relative partial molar enthalpy, 164
relaxation effect, 361
relaxation time, 278
replicas, 295
repulsion, 117
reversible, 27, 29, 343
reversible process, 11, 344
rigid rotator, 79

rotational motion, 79
rotational temperature, 82, 87

Sackur, Otto, 91
Schottky, Walter, 332
Schottky defects, 329
Second Law of Thermodynamics, 27, 46, 139, 344
Seebeck effect, 353
separable, 57, 79
separation of variables, 81
shear stress, 329
silver halides, 329
silver oxide, 240
simple harmonic, 81
solidus, 176
solubility in two immiscible liquids, 184
solubility product, 280
solutes, 179
solution, 179
solvent, 179
Sommerfeld's integral, 316
spectroscopic entropy, 77
spherical harmonics, 81
spherical polar coordinates, 79
square well, 118
standard affinity, 232
standard cell potential, 282
standard chemical potential, 126
standard concentration, 224
standard enthalpy, 224
standard enthalpy of formation, 237
standard entropy, 91, 126
standard equilibrium constant, 233
standard Gibbs energy of formation, 232
standard heats of formation, 245
standard hydrogen electrode, 270
standard molality, 224
standard pressure, 224
standard reaction enthalpy, 223
standard state, 223
standard states in biochemistry, 287
state function, 54
statistical mechanics, 295

statistical thermodynamics, 54, 239, 295
statistical thermodynamics of adsorption, 204
steady state, 3
steam distillation, 169
stiffness coefficients, 326
stoichiometric numbers, 219
strain components, 324
strong degeneracy, 314
strong electrolytes, 257
super-cooling, 214
superfluid, 308
surface area, 203
surface excess concentration, 196
surface phase, 194
surface potential, 266
surface tension, 189
surfactants, 197
symmetric state functions, 307
symmetry number, 83, 84, 248
system
    closed, 1
    heterogeneous, 1
    homogeneous, 1
    isolated, 1
    open, 1
    surroundings, 1
    thermodynamic, 1
    walls of, 1

tables of thermodynamic data, 225
Tait equation, 132
Temkin isotherm, 204
temperature
    Celsius, 5
temperature of zero, 49
temperature-composition diagram, 161
Tetrode, Hugo Martin, 91
theoretical calculation of equilibrium constants, 246
thermal conductivity, 354
thermal contact, 4
thermal de Broglie wavelength, 67
thermal diffusion, 352
thermal equilibrium, 4, 40

thermal expansion, 101
thermal properties, 296
thermocouple, 353
thermodynamic coordinates, 4
thermodynamic force, 347, 356, 357
thermodynamic functions, 311, 316
thermodynamic functions for a FD
    gas, 314
thermodynamic functions of ions, 285
thermodynamic potential, 30, 36, 40
thermodynamic probability, 63, 95
thermodynamic process, 343
thermodynamic temperature, 5
thermodynamics, 240
    non-equilibrium, 4
thermoelectric power, 354
thermometer, 5
thermostatics, 4
third and higher virial coefficients,
    130
Third Law of Thermodynamics, 98,
    99, 100, 104
three-component system, 177
translational energy levels, 60
translational energy states, 61
translational states, 68
transport number, 355, 357
transport properties, 277
triple point, 5, 139

unattainability of $T = 0\,\mathrm{K}$, 45, 105
undetermined multipliers, 63

van 't Hoff equation, 235
vapour phase chromatography, 185
variance, 172
vibrational motion, 79
vibrational partition function, 86
vibrational temperature, 86, 87
virial coefficient
    second, 119
virial coefficients, 10
virial equation, 7, 135
volatility ratio, 154
volume change on mixing, 150

Wagner, Carl, 332
wave number, 56
wave vector, 56
weak degeneracy, 309
well-behaved, 55
wetting coefficient, 200
work, 10

X-ray, 131

Young–Laplace equation, 192

Zeroth Law of Thermodynamics, 4, 5